Human Variation in the Americas

Visiting Scholar Conference Volumes

Processual and Postprocessual Archaeologies: Multiple Ways of Knowing the Past (Occasional Paper No. 10) *edited by Robert W. Preucel*

Paleonutrition: The Diet and Health of Prehistoric Americans (Occasional Paper No. 22) *edited by Kristin D. Sobolik*

New Methods, Old Problems: Geographic Information Systems in Modern Archaeological Research (Occasional Paper No. 23) *edited by Herbert D. G. Maschner*

Integrating Archaeological Demography: Multidisciplinary Approaches to Prehistoric Population (Occasional Paper No. 24) *edited by Richard R. Paine*

Studies in Culture Contact: Interaction, Culture Change, and Archaeology (Occasional Paper No. 25) *edited by James G. Cusick*

Material Symbols: Culture and Economy in Prehistory (Occasional Paper No. 26) *edited by John E. Robb*

Hierarchies in Action: *Cui Bono?* (Occasional Paper No. 27) *edited by Michael W. Diehl*

Fleeting Identities: Perishable Material Culture in Archaeological Research (Occasional Paper No. 28) *edited by Penelope Ballard Drooker*

The Dynamics of Power (Occasional Paper No. 30) *edited by Maria O'Donovan*

Hunters and Gatherers in Theory and Archaeology (Occasional Paper No. 31) *edited by George M. Crothers*

Biomolecular Archaeology: Genetic Approaches to the Past (Occasional Paper No. 32) *edited by David M. Reed*

Leadership and Polity in Mississippian Society (Occasional Paper No. 33) *edited by Brian M. Butler and Paul D. Welch*

The Archaeology of Food and Identity (Occasional Paper No. 34) *edited by Katheryn C. Twiss*

The Durable House: House Society Models in Archaeology (Occasional Paper No. 35) *edited by Robin A. Beck Jr.*

Religion, Archaeology, and the Material World (Occasional Paper No. 36) *edited by Lars Fogelin*

The Archaeology of Anthropogenic Environments (Occasional Paper No. 37) *edited by Rebecca M. Dean*

Human Variation in the Americas

The Integration of Archaeology and Biological Anthropology

Edited by
Benjamin M. Auerbach

Center for Archaeological Investigations
Southern Illinois University Carbondale
Occasional Paper No. 38

Printed in the United States of America
ISBN: 978-0-88104-095-1
Library of Congress Control Number: 2010929585

Copyedited by Mary Lou Wilshaw-Watts
Production supervised by Mary Lou Wilshaw-Watts
Designed and formatted by Linda Jorgensen Buhman, New Leaf Studio

Printed by the authority of the State of Illinois, October 2010, 500 copies, Purchase Order No. 96346

For information, write to the Center for Archaeological Investigations, Faner 3479, Mail Code 4527, Southern Illinois University, 1000 Faner Drive, Carbondale, IL 62901; phone (618-453-5031); or visit us online at www.cai.siuc.edu.

Contents

Figures

Tables

Acknowledgments

The 25th Annual Visiting Scholar Conference—"Human Variation in the Americas: A Meeting of Archaeology and Biological Anthropology"—and this volume, which developed from that meeting, emerged from numerous conversations I had with various researchers while visiting collections throughout North America during my dissertation data collection. Perhaps the most important occurred with Karen Weinstein over breakfast one morning in Anchorage, Alaska, during the 2006 American Association of Physical Anthropologists (AAPA) annual meeting. Starting with that discussion, we laid out the skeleton of what became an invited symposium we organized at the 2007 AAPA annual meeting: "Biological Variation and Evolutionary Dynamics in Ancient Populations of the Americas."

The visiting scholar conference was an expansion of this symposium, building on the core concept of the AAPA gathering by involving a broader range of researchers and perspectives. Both meetings, as well as this volume, promoted a dialogue about bringing together multiple sources of information to better understand diversity in the cultures and the biology of indigenous peoples from the Americas. Indeed, the spirit of this dialogue was strongly supported at the outset by the committee that oversees the Center for Archaeological Investigations Visiting Scholar program, who took a gamble hiring a biological anthropologist for a position traditionally filled by an archaeologist. I am grateful to them for providing the opportunity and for their active support for what I hope will be a signpost along the road of this continuing discussion among disciplines.

As with all edited volumes, this book could not have been possible without the combined efforts of numerous individuals. A great amount of thanks must be given to the participants in both the 2007 symposium and the 2008 conference, whose research and exchange of ideas made these meetings into engaging, productive, and thought-provoking events. While not all of these individuals have chapters in this book, much of the conversation to which they contributed shaped the research and interpretations presented herein. In this light, I would especially like to thank the many individuals, both invited speakers and audience members, who traveled to and took part in the Carbondale conference.

A recurring question (if not joke) that occurred in a few of the presentations during the conference was "When do we stop acknowledging our academic advisors?" Whenever that stage occurs, I have certainly not yet reached it. On that note, I would like to thank Chris Ruff, my doctoral advisor, for his help and continued guidance. Moreover, had it not been for Chris, I probably would never have noticed, let alone applied for, the visiting scholar position in 2007.

The conference itself would not have been possible without the help of a small army of people. Special thanks go to Andy Balkansky, Brian Butler, Donna Butler, Heather Lapham, Kathy Lundeen, and Paul Welch, whose input kept the design of the conference focused. Brian, Heather, and Kathy especially helped me learn not only the details of conference construction but also the quirks of organizing the event in Carbondale. Pat Eckert and the staff of the Division of Continuing Education did an excellent job advertising and running the logistics of the conference. Also, the graduate student volunteers—Ayla Amadio, Lauren Forsythe, Elliott Forsythe, Aimee Hosemann, Jessica Howe, Go Matsumoto, and Erica Muller—did yeomen's service in transporting participants, taking care of the book room, and running some of the conference's less visible (and often incidental) tasks. Thanks go to all of these people.

These and other members of the Department of Anthropology made my stay in Carbondale a great experience. I am grateful for the advice, great conversations, and warm welcome of the archaeologists noted above, as well as for that of Rob Corruccini, Susan Ford, Tracy Prowse, Ulrich Reichard, and Tony Webster throughout my brief tenure as visiting scholar. Both Susan and Paul were extremely generous with their time, conversations, and guidance, and I am especially appreciative of their help in my transition from graduate student to faculty member. Likewise, I am thankful for the many experiences, both within and outside the university, with the graduate students. It was especially a pleasure to teach the students enrolled in my graduate seminar on morphometrics—Lindsey Baker, Meadow Campbell, Ryan Campbell, Kyle Lubsen, Susannah Munson, and Matt Nowak—and I hope they gained new knowledge and perspectives from both the course and the conference. I look forward to interactions with many of these people in the future.

Just like the small army that helped with the production of the conference, a large number of people helped to make this volume possible. Again, I am grateful to Brian Butler for his support and for seeing this book through its final phases. I am highly appreciative of the work of Mary Lou Wilshaw-Watts, whose efforts as the copy editor went well beyond the basic duties of that office. Many thanks go to the dozens of anonymous reviewers who read through the individual chapters and provided essential feedback, strengthening the science and arguments of the authors herein. Given the diversity of subjects, regions, and methods covered in this book, the anonymous reviewers of the final volume deserve special acknowledgment, as they assessed the viability and cohesion of the work in its entirety. Finally, I am grateful to David Anderson, David Hunt, T. R. Kidder, and Ken Sassaman—the discussants for the conference and this volume—whose insights and candor have been irreplaceable, adding new dimensions to the conversation that participants may not have otherwise considered.

The visiting scholar position was an excellent opportunity, and I am happy to have been able to fill the role. However, it was the people in the Department of Anthropology and the Center for Archaeological Investigations that made the experience so enjoyable and memorable. Again, I give my deepest gratitude for your trust, friendship, and support.

Benjamin M. Auerbach
July 2010

1. | Finding Common Ground in the Archaeology and Biology of the Americas' Past

Benjamin M. Auerbach

In late December 1911, during the nascent days of North American professional archaeology and biological anthropology, the American Anthropological Association and the American Association for the Advancement of Science held a joint conference at the Smithsonian Institution. The proceedings of the session were subsequently published under the title "The problems of the unity or plurality and the probable place of origin of the American Aborigines" (Fewkes et al. 1912). As reflected by this publication, the gathering convened some prominent North American archaeologists, geologists, ethnologists, naturalists, and physical anthropologists of the time. The conference may represent the first professional meeting devoted to discussing evidence for the peopling of and, to a lesser extent, the subsequent settlement of the Americas.

Without question all the disciplines represented at the 1911 Smithsonian conference made significant progress in the following decades. For example, few, if any, anthropologists today would argue that the indigenous peoples of the Americas represent morphological or cultural homogeneity, a conclusion strongly proffered by William H. Holmes (Fewkes et al. 1912:30).[1] Yet some of the prevailing topics from the early twentieth century remain vigorously debated, including two central matters: From where (both geographically and ancestrally) did the original colonizers of the Americas come, and when? And how did the patterns of diversity, both cultural and morphologic, emerge among humans in the Americas in the millennia postcolonization?

Human variation is, admittedly, a broad subject, even within the confines of what may be gleaned from archaeological vestiges of biology and culture. As I

Human Variation in the Americas: The Integration of Archaeology and Biological Anthropology, edited by Benjamin M. Auerbach. Center for Archaeological Investigations, Occasional Paper No. 38. © 2010 by the Board of Trustees, Southern Illinois University. All rights reserved. ISBN 978-0-88104-095-1.

discuss in detail below, this volume targets specific themes and questions about the diversity of the past in the Americas, much as the 1911 conference used multiple fields and approaches to address a central question. *Variation* is difficult to define without context, and the chapters in this volume demonstrate various levels of focus in their investigations of diversity. What unites them, and therefore is the theme of this volume, is the attempt to explore population-level relationships through biology and through culture. No single edited book would be able to cover the breadth of human variation in the Americas—none likely ever will. Yet, the contributions herein present a cross section of current methods, perspectives, subjects, and data by which to explore human variation, whether in the Americas or elsewhere on the globe.

The following chapters primarily explore issues related to populations and their movements: colonizations of the Americas, postcolonization migration in the Americas, and factors influencing variation in the morphology (or physique) of those groups. In this endeavor, the authors of this volume bring multiple areas of evidence together to address research questions, with varying degrees of resolution. Some contributors take more "traditional" approaches, such as examining dental traits (Durand et al., Chapter 5) or stable isotopes (Cybulski, Chapter 4) to discern movements and relationships. Some utilize unique methods or observations to elucidate regional patterns that have otherwise been undocumented, such as Archaic mortuary practices along the Ohio River (Schmidt et al., Chapter 8).

Plural Perspectives on the Past

In reviewing the conclusions presented within each chapter, and in the two discussion chapters at the end of the volume, a compelling observation takes shape: the synthesis of approaches leaves a breadth of avenues for future research. It is clear that there are great gaps in knowledge; as Kemp and Schurr eloquently state in Chapter 2, "[W]e are more confident than ever in our uncertainty." A goal of this book, then, is to inspire investigators to look in new directions to pursue unanswered questions by gathering novel data or, as King (Chapter 10) and Jantz and colleagues (Chapter 11) demonstrate, by using archived data in novel ways.

In addition to utilizing more data, utilizing research tools from multiple disciplines is essential to both the studies herein and the investigations they might stimulate. Just as the 1911 conference highlighted the potential contributions of different disciplines in addressing these questions, modern investigators may employ tools and data from various specialties. Archaeologists are adding new sites (e.g., Jenkins 2007) and are revisiting previously excavated locations (e.g., Meltzer 2006). Biological anthropologists are using new forms of skeletal analysis and larger comparative samples to quantify patterns of morphological variation (e.g., Auerbach 2007; Jantz and Owsley 2005; Neves et al. 2007; Powell 2005). Moreover, genetic research has especially contributed to a renaissance in studies of human origins and diversity in the Americas (e.g., Goebel et al. 2008; Kemp and Schurr, Chapter 2). Often, however, these analyses have focused on one source of data to the exclusion of others or have incorporated findings from other disciplines in a limited or peripheral manner.

Furthermore, a problem emerging from the evidence provided by researchers in these disciplines is an unclear correspondence among their conclusions. As highlighted by papers like that of Goebel and colleagues (2008), the commensurability of data from archaeology, skeletal biology, and genetics (let alone linguistics or ethnography) is often poor and occasionally contradictory. Researchers are faced with the conundrum of relating genotypic diversity to phenotypic diversity and linking these to archaeological evidence of population movements, changes, and relationships. This book's authors certainly confront this problem. For example, as discussed in the majority of chapters (Chapters 2, 3, 4, 7, 10, 11, 12, and 13), there is ongoing disagreement about the ancestral origins, timing, and mechanics of the first colonizers of the Americas. In the face of fragmentary evidence, this conundrum is compounded by the different time periods represented by three sources of data: archaeological data dates to circa the Last Glacial Maximum, the earliest skeletal remains usable for osteometric analyses postdate the earliest archaeological evidence for colonization by millennia (an essential point often missed by researchers), and the genetic data may represent ancestral populations predating both the archaeological and the skeletal data. So, even with the gained knowledge and achievements of the last century, anthropologists still struggle to attain a comprehensive, holistic understanding of the past.

All of these facets of the past do interconnect, though their associations are often ambiguous. Although an oversimplification, one may think of the relationship of archaeology, skeletal biology, and genetics—not to exclude other indicators of culture or biology, but rather to focus on the three emphasized in this volume—as akin to an *n*-dimensional Russian nesting doll. The outer layers envelop and determine the shape of inner layers, but so, in turn, do inner layers influence the shape of the superficial; often the layers wholly switch places. In part this complexity is due to the different levels of inquiry inherent in each discipline (see Sassaman's commentary, Figure 13-1). However, this complexity is fundamentally a result of the constant interaction between elements of all three broad subjects, which thus determine each other. Plainly, biology influences culture and vice versa, on multiple levels. The biological characteristics of a people are varied and subject to stochastic and adaptive forces, but the perception of these characteristics by another people could influence population interactions and ultimately the exchange and survival of genetic lineages. Or, perhaps more commonly, interactions between peoples with different cultural relationships influence common or divergent biological characteristics, and eventually this leads to the persistence or extinction of lineages. Giving primacy to any one in shaping the others immediately demonstrates the plausibility of the reverse condition. The task that awaits for anthropological researchers, then, is disentangling these shifting relationships to foster an understanding of specific questions about the human experience and, within the context of this book, the human experience in the Americas.

Spurred by this continuing endeavor, the Center for Archaeological Investigations 2008 Visiting Scholar Conference—from which this volume emerged—

brought together archaeologists, skeletal biologists, and geneticists to have a dialogue about the development of a multidisciplinary perspective on the colonization and subsequent diversification of humans in the Americas. More specifically, the goal of both the conference and this volume was to address two issues:

> 1. What can current research and modern methods reveal about human biological variation in the archaeological record of the New World, especially prior to European colonization?
> 2. How well can the information elucidated by this research be interfaced with archaeological data and evidence? That is, are biological data and cultural data (as revealed by archaeology) consistent in the models of the past each suggests? How might biological data and cultural data be synergistically incorporated to yield a better conceptualization of the past?

These topics led to much discussion and debate in Carbondale, Illinois, in April 2008. As the papers in this volume demonstrate, that debate continued long after the conference attendees departed.

Checkpoints in Holistic Modeling

Of course, the Visiting Scholar Conference and this volume are not novel in their effort to bring modern archaeology and biological anthropology into a discourse. In 1976, Robert L. Blakely and others organized a session at the 11th Annual Meeting of the Southern Anthropological Society during which participants debated the role of biological anthropology in archaeological contexts (Blakely 1977).[2] A strong theme that emerged from that conference and its edited volume was the importance of biological evidence in archaeological interpretation of past cultures. As argued by the authors—through studies of affinity, mortuary practice, disease, and diet, among other subjects—"bioarchaeology" can add a richer, multifaceted biocultural component to a discourse that a reliance on artifacts, architecture, and their interpretation cannot achieve alone. That is, the coordination of these many sources better attains the holism to which anthropology classically aspires.

Other meetings followed the 1976 session (and its associated edited volume) as the dialogue between archaeology and biological anthropology continued. One of the most noteworthy was a session organized by Mary Lucas Powell, Patricia Bridges, and Ann Marie Wagner Mires at the 42nd Southeastern Archaeology Conference in 1985. This session likewise yielded an edited volume (Powell et al. 1991) in which some participants reflected on the progress of coordinating multiple sources of data in archaeological analyses and, therein, bioarchaeological research's emergence "out of the appendix" (Buikstra 1991). Powell and associates (1991) noted successful advancements to this end, though it was also emphasized that the merging of biological and archaeological evidence awaited further development. As argued by Clark Larsen more recently, "[A] broader concern is the need to increase even more the level of collaboration between archaeologists and bioarchaeologists" (Larsen 2006:372). Thus, the volume you now hold and the conference from which it issued may be thought of as, in part, "checkpoints" by which anthropological investigators of the Americas' past may assess the success of current attempts to achieve a more holistic picture using multidisciplinary combinations of evidence.

Consensus or Complexity?

In this context, what does the present volume suggest about the successful "marriage" of archaeology and biological anthropology in New World research? A survey of the contributions to this work demonstrates a lack of consensus. After reading the particular examples used by the researchers herein, the reader would most likely conclude that this marriage is a work in progress. However, the broader and more crucial message is that the authors in this volume are at least attempting to bring the different paradigms of these disciplines into concurrent consideration while exploring their respective subjects. As echoed by the discussants (Anderson, Chapter 12; Sassaman, Chapter 13), what remains most important is not the pursuit of an ultimate certitude reached by combining perspectives from the various disciplines. Indeed, by nature of their partial independence, the merging of archaeology and biological anthropology (in their many forms) to yield a single cohesive model of the past may ultimately prove unachievable. In the end, I leave it up to the reader to consider this after reflecting on the entire volume.

It is, therefore, a goal of this volume to provide multiple examples of how this apparent divide may (or may not) be bridged, despite the potential for creating an underlying editorial incoherence. The mere act of considering biological and archaeological evidence in parallel, among other sources, heralds the more fruitful, progressive steps all these disciplines must take to address the more fundamental questions about human origins and diversification in the Americas and, by extrapolation, in the world. Arguably, study of any single source of data without consideration of others often results in restricted, if not reductionist, arguments. The recently proposed Beringian Incubation Model (Tamm et al. 2007) is a good example, as genetic consensus cannot be corroborated yet by archaeology or skeletal biology (nor can they rule it out); see both discussants' archaeologically informed reactions, as well as Kemp and Schurr's discussion (Chapter 2). Anderson's (Chapter 12) compelling argument—for the inhabitation of now-submerged archipelagos south of Beringia—is noteworthy and may prove key for resolving this missing data problem.

Beyond an evaluation of the bioarchaeological program initiated over thirty years ago, the contributions following this introductory chapter demonstrate various attempts to reconcile evidence from different aspects of the disciplines under scrutiny. Setting the stage for considering the role of genetic anthropology in understanding the past, Kemp and Schurr (Chapter 2) make a strong case for the powerful tools genetic analyses provide in understanding human origins and patterns of diversification in the colonization of the Americas. Yet their chapter also highlights the need to better syncretize molecular studies with archaeological findings. As noted above, genetic and archaeological paradigms may simply lack congruence, despite their interrelatedness. Nevertheless, there are examples of the potential for combining disparate sources of knowledge to produce models of the past.

The subsequent two chapters focus on the subarctic northwestern regions of North America. Chatters's research in the Northwest (Chapter 3) makes the case that marked differences in cranial morphology may be linked to major shifts in traditions during the early Holocene, as revealed by the archaeological record. The complete biological replacement of one population with another is more difficult to conclude than the abrupt change in lithic artifacts suggests about culture though. In contrast, Cybulski's analysis of morphological and isotopic data from more recent northwestern populations (Chapter 4) reveals how some subsistence shifts, population movements, and even differential population survival may only be discernable via studies of skeletal biology. The important discoveries of human remains at Big Bar Lake and China Lake demonstrate significant movement among peoples and, potentially, populations that did not ultimately contribute to more recent variation in the region.

The detection of subregional population movements is certainly possible using both archaeological analysis and biological data, however, as demonstrated by the chapters on the Southwest and the Plains. Durand and colleagues' contribution (Chapter 5) on the Chaco Canyon and Middle San Juan regions provides one such situation in which discrete dental characteristics and archaeological models eventually corroborate each other and may yield a cohesive picture about migration. Watson's subsequent contribution (Chapter 6) tempers this, indicating that models of the diffusion of culture, language, and subsistence may be ambiguously supported by the available biological data. Part of this ambiguity, as I suggest in my contribution on the later occupation of the Plains (Chapter 7), may be largely due to the lack of instances of total population displacement or replacement in the Americas' past, coupled with cultural borrowing and multilingualism.[3]

In fact, as Schmidt and associates imply (Chapter 8), some cultural behaviors—in their case mortuary practices—may have transcended biological populations to become wider regional phenomena, as Watson argues may have been the case with the spread of agriculture in the Southwest (Chapter 6). In these instances, archaeology and biology occupy different planes of interaction, a problem highlighted by Sassaman in his commentary (Chapter 13). Widespread cultural practices occurring at a time before—most archaeologists would argue—highly organized societies existed in the modern Prairie region may require reevaluation of theories on overall population structure and models of interactions among Archaic peoples. This concept is also emphasized by Sassaman in Chapter 13 (as well as in his forthcoming book).

Studies of ancient DNA may help resolve the apparent contradictions that culture and biology present in understanding population interactions of the past, as Shinoda and colleagues explore in Chapter 9. Their analysis is unable to resolve cultural identities of archaeological groups in the Moquegua Valley. Yet, they are able to highlight the possible confounding that language may create in determining past relationships, while also successfully linking population movements to specific environmental (El Niño) changes.

Indeed, understanding and tracking past population movements and interactions may rest on the construction of detailed environmental models, linking peoples to the resources they sought and the need to maintain a certain ecological constraint (see Anderson's commentary, Chapter 12). Assuming sufficient time and selection pressures, this ecogeographic adaptation should be reflected in morphology, a trend argued by King (Chapter 10) and elsewhere by me (Auerbach 2007; Auerbach and Ruff 2010). That patterns of morphological variation reflect population and migration history as well—perhaps even supplanting climatic adaptation (Jantz et al., Chapter 11) as agents of change—cannot be discounted. Both migration history and environmental factors, therefore, should be taken into account when combined with broad and regional archaeological data.

Again, the common theme running through these papers is the need to bring archaeology and biology together when considering human colonization and diversification in the Americas. None of the authors of these chapters, though, would conclude that they have been able to reconcile their biological conclusions with the archaeology of their particular lines of inquiry. Furthermore, it is evident that, in some instances, the biological researchers have yet to reconcile contrasting conclusions among themselves. Much remains unresolved. Yet, as these studies show, there is great potential for mutual elucidation by juxtaposing the various sources of evidence. These papers additionally showcase the variety of methods available to researchers in analyzing human remains, from aDNA analysis to dental traits and from morphometrics—both osteological and anthropometric—to evidence of trauma and health.

Defining Diversity

Alongside the increasing number of analytical tools, however, a subtext with which many of the authors struggle, both in this volume and within biological anthropology as a whole, falls near the fulcrum of this book: How do we define *biological diversity*? As Kemp and Schurr reference in their overview (Chapter 2), the genetic diversity of the Americas—at least in mitochondria and the Y chromosome—is limited when contextualized within *Homo sapiens*. In this frame of reference, any discussion of biological diversity in the Americas may prove unsubstantial or minimal and requires broad sample comparisons or large local samples to detect any variation. Indeed, on the scales discussed in most of the studies in this volume, biological diversity is realistically limited to comparisons with larger regional or continental samples. Yet, this is misleading. Morphological diversity in the Americas may be nearly as extensive as the variation observed in Eurasia and Africa, based on postcranial studies (Auerbach 2007; Auerbach and Ruff 2010; King, Chapter 10) and on cranial analyses including early Holocene samples (Jantz and Owsley 2005; Neves et al. 2007). Like archaeological studies, as I note at the beginning of this chapter, patterns of similarity and variety are dependent on the scale one chooses when pursuing study. *Variation* must be clearly defined by the researcher, and the scalar level must be kept in mind by the reader. In short, the comparative framework is crucial.

A common complication in the analysis of biological diversity is the issue of biological versus statistical significance. Some tests may yield results that are statistically significant but are effectively without biological meaning. For example, the lengths of femora may be statistically different in two samples, but if the difference amounts to a few millimeters then the biological relevance is debatable. Conversely, because of statistical power issues, in small samples biologically relevant distinctions may be undetected or overlooked, a problem that plagues analyses of early or rare skeletal samples. Although this concern is not expressly considered by most of the authors herein, it is an ongoing issue in all biological studies in anthropology, and most researchers are cognizant of this problem and account for it in their interpretations. Just like the need to take heed of the scale of an analysis when considering variation, the reader should also note the biological relevance of any differences or similarities observed. Again, this is dependent, in part, on the constraints imposed by the study's author.

On the flip side, biological researchers should recognize the complexity of archaeological interpretations of past cultural diversity. Archaeological understanding of traditions, cultural horizons, and the relationships among them regularly undergo reevaluation and revision. The definition of past cultures is subject to new discoveries as much as paradigmatic shifts in the interpretation of social construction. A challenge for biological anthropologists, then, is to aid archaeologists in understanding past societies and in defining the interactions of peoples within them. In addition, biological anthropologists should be cautious about assumptions concerning group affiliations and cultures of humans whose remains are examined.

Finding Common Ground

Despite these limitations, and as demonstrated by the studies at hand, biological perspectives have much to offer archaeological research. The answer to the second question posed on page 4—regarding the commensurability of archaeological and biological data—is contingent on a variety of circumstances, including the factors under consideration, the overall scale of the study, and the available data. The studies of the Southwest in this volume (Durand et al., Chapter 5; Watson, Chapter 6) exemplify this. Both lack enough data to demonstrate an exceptionally complementary alignment of conclusions between biological and archaeological data, but it is clear that their higher-resolution data do lend greater support to some archaeological migration models rather than to others. Without a doubt, a strong message all the studies herein deliver is that more research involving larger samples, as well as more sources of and greater resolutions in data, is essential for future studies of the Americas. This, in turn, argues for more collaboration, better consistency in data collection, and utilization of more sophisticated analytical methods by biological anthropologists. More important, biological anthropologists and archaeologists *must* work together, regularly and with common research agendas. The separation between biological and archaeological studies is artificial; the door for collaboration is wide open.

The charge facing our disciplines is to find the common ground by which we may have not only a discourse but also a cooperative effort. This is hardly meant to imply that there are no such collaborations taking place. Indeed, many of the studies in this volume demonstrate that such combined undertakings have led to meaningful conclusions. Such collaborations should be common practice. Members of each discipline should regularly approach individuals from the others and discuss their common goals. As cited elsewhere (Buikstra 1991; Larsen 2006), bringing biologists in at the inception of archaeological projects—whether these are excavations or museum studies—is a key strategy by which this may occur. It is also essential for members of all of these disciplines to cultivate meaningful relationships with members of the indigenous nations of the Americas in order to bring them into the dialogue about their ancestors and the ancestors of groups no longer extant.

Following in the footsteps of the many researchers who attended that joint session a century ago, today's archaeological and biological investigators continue to traverse the road toward an understanding of human diversity and origins in the Americas. As this volume demonstrates, this journey is far from its end; research into the past of the Americas still grapples with the same questions debated throughout the last hundred years. By developing a regular, mutual discourse and collaboration between biological anthropologists and archaeologists, better resolutions to these questions are inevitable. It is my hope that the following papers add to that discussion and foster continued cooperation among the many disciplines striving to understand human variation in the Americas.

Notes

1. It is noteworthy that Aleš Hrdlička did not state so extreme a position in his interpretation of morphological variation in the Americas. Indeed, he argues that "since the peopling of [the Americas] was commenced, [humans have] developed numerous secondary, subracial, localized structural modifications. . . ." Although the terminology is obsolete, the sentiment is not.

2. This meeting was also where Jane Buikstra coined the modern interpretation of the term *bioarchaeology* in reference to employing biological data to interpret cultural phenomena in archeological contexts (Buikstra 1977, 2006).

3. Few of the authors in this volume discuss the implications of multilingualism as a complicating factor in understanding the complexities of group relationships and cultural variation, though it should not be ignored (Campbell 2000). Though it may be tempting to suggest that multilingualism is linked to complex patterns of gene flow—and undoubtedly shared communication certainly aided in this—the coincidence of these cannot be assumed (Goddard and Campbell 1994).

References

Auerbach, Benjamin M.
  2007   Human Skeletal Variation in the New World During the Holocene: Effects of Climate and Subsistence Across Geography and Time. Unpublished Ph.D. dissertation, Center for Functional Anatomy and Evolution, Johns Hopkins University, Baltimore.

Auerbach, Benjamin M., and Christopher B. Ruff
  2010   Stature Estimation Formulae for Indigenous North American Populations. *American Journal of Physical Anthropology* 141:190–207.

Blakely, Robert L. (editor)
  1977   *Biocultural Adaptation in Prehistoric America. Proceedings of the Southern Anthropological Society*, No. 11. University of Georgia Press, Athens.

Buikstra, Jane E.
  1977   Biocultural Dimensions of Archeological Study: A Regional Perspective. In *Biocultural Adaptation in Prehistoric America*, edited by Robert L. Blakely, pp. 67–84. University of Georgia Press, Athens.
  1991   Out of the Appendix and into the Dirt: Comments on Thirteen Years of Bioarchaeological Research. In *What Mean These Bones? Studies in Southeastern Bioarchaeology*, edited by Mary L. Powell, Patricia S. Bridges, and Ann Marie W. Mires, pp. 172–188. University of Alabama Press, Tuscaloosa.
  2006   Preface. In *Bioarchaeology: The Contextual Analysis of Human Remains*, edited by Jane E. Buikstra and Lane Beck, pp. xvii–xx. Academic Press, New York.

Campbell, Lyle
  2000   *American Indian Languages: The Historical Linguistics of Native America*. Oxford University Press, Oxford.

Fewkes, J. Walter, Aleš Hrdlička, William H. Dall, James W. Gidley, Austin Hobart Clark, William H. Holmes, Alice C. Fletcher, Walter Hough, Stansbury Hagar, Paul Bartsch, Alexander F. Chamberlain, and Roland B. Dixon
  1912   The Problems of the Unity or Plurality and the Probable Place of Origin of the American Aborigines. *American Anthropologist* 14:1–59.

Goddard, Ives, and Lyle Campbell
  1994   The History and Classification of American Indian Languages: What Are the Implications for the Peopling of the Americas? In *Method and Theory for Investigating the Peopling of the Americas*, edited by Robson Bonnichsen and D. Gentry Steele, pp. 189–207. Oregon State University, Corvallis.

Goebel, Ted, Michael R. Waters, and Dennis H. O'Rourke
  2008   The Late Pleistocene Dispersal of Modern Humans in the Americas. *Science* 319:1497–1502.

Jantz, Richard L., and Douglas W. Owsley
  2005   Circumpacific Populations and the Peopling of the New World: Evidence from Cranial Morphometrics. In *Paleoamerican Origins: Beyond Clovis*, edited by Robson Bonnichsen, Michael R. Waters, Dennis Stanford, and Bradley T. Lepper, pp. 267–276. Texas A&M University Press, College Station.

Jenkins, Dennis L.
  2007   Distribution and Dating of Cultural and Paleontological Remains at the Paisley Five Mile Point Caves in the Northern Great Basin: An Early Assessment. In *Paleoindian or Paleoarchaic? Great Basin Human Ecology at the Pleistocene-Holocene Transition*, edited by Kelly E. Graf and Dave N. Schmitt, pp. 57–81. University of Utah Press, Salt Lake City.

Larsen, Clark S.
  2006   The Changing Face of Bioarchaeology: An Interdisciplinary Science. In *Bioarchaeology: The Contextual Analysis of Human Remains*, edited by Jane E. Buikstra and Lane Beck, pp. 359–374. Academic Press, New York.
Meltzer, David J. (editor)
  2006   *Folsom: New Archaeological Investigations of a Classic Paleoindian Bison Kill*. University of California Press, Berkeley.
Neves, Walter A., Mark Hubbe, and Luís B. Piló
  2007   Early Holocene Human Skeletal Remains from Sumidouro Cave, Lagoa Santa, Brazil: History of Discoveries, Geological and Chronological Context, and Comparative Cranial Morphology. *Journal of Human Evolution* 52:16–30.
Powell, Joseph F.
  2005   *The First Americans: Race, Evolution, and the Origin of Native Americans*. Cambridge University Press, Cambridge.
Powell, Mary L., Patricia S. Bridges, and Ann Marie W. Mires (editors)
  1991   *What Mean These Bones? Studies in Southeastern Bioarchaeology*. University of Alabama Press, Tuscaloosa.
Tamm, Erika, Toomas Kivisild, Maere Reidla, Mait Metspalu, David Glenn Smith, Connie J. Mulligan, Claudio M. Bravi, Olga Rickards, Cristina Martinez-Labarga, Elsa K. Khusnutdinova, Sardana A. Fedorova, Maria V. Golubenko, Vadim A. Stepanov, Marina A. Gubina, Sergey I. Zhadanov, Ludmila P. Ossipova, Larisa Damba, Mikhail I. Voevoda, Jose E. Dipierri, Richard Villems, and Ripan S. Malhi
  2007   Beringian Standstill and Spread of Native American Founders. Electronic document, http://www.plosone.org/article/info:doi%2F10.1371%2Fjournal.pone.0000829, accessed September 6, 2007. doi:10.1371/journal.pone.0000829.

2. Ancient and Modern Genetic Variation in the Americas

Brian M. Kemp and Theodore G. Schurr

Summary Statement: A number of recent whole mitochondrial genome and nuclear DNA studies support the Beringian Incubation Model as the best description of the demographic history of the population that first entered the Americas approximately 20,000–15,000 B.P., having originated from a single source population located somewhere in Asia. Human presence in the Americas by at least 14,270–14,000 B.P. has been confirmed by archaeological evidence from the Monte Verde site in southern Chile and by the recovery of human coprolites in the Paisley Caves in southern Oregon. A secondary migration or expansion of humans, perhaps from the same source population, introduced additional mtDNA haplogroups into the northernmost areas of North America after the last glacial period. From the initial entrance of humans into the Americas, genetic drift has played a substantial role in shaping the Native American gene pool. On a continental scale, Native Americans exhibit simultaneously the highest measures of both homozygosity and interpopulational genetic distances (a classic example of the Wahlund effect). Genetic variation thus far detected in human remains and human byproducts (e.g., coprolites) that predate 5000 B.P. are consistent with this view, notwithstanding the small sample sizes.

Introduction

As an independent line of evidence to that provided by archaeological, paleontological, and/or morphological studies, molecular data[1] have proven instrumental to our understanding of the emergence and evolution of *Homo sapiens*. Information deciphered from the genomes of countless humans has also aided the reconstruction of the paths that our ancestors followed as they jour-

Human Variation in the Americas: The Integration of Archaeology and Biological Anthropology, edited by Benjamin M. Auerbach. Center for Archaeological Investigations, Occasional Paper No. 38. © 2010 by the Board of Trustees, Southern Illinois University. All rights reserved. ISBN 978-0-88104-095-1.

neyed across the world. Today, there is a general agreement between the genetic and archaeological evidence for the direction and shape of most of the major population movements[2] (Cavalli-Sforza et al. 1994; Forster 2004; Wells 2006).

The settlement of the Americas marks one of the most recent of these movements. The particular details of this process have been of long-standing anthropological interest (Fewkes et al. 1912) and the focus of a large amount of genetic research (Crawford 1998; Salzano and Callegari-Jacques 1988; Schurr 2004a). Discerning the location in Asia from where proto–Native Americans originated and estimating approximately when humans first colonized the Americas have been goals at the forefront of this research (Goebel et al. 2008; Schurr 2004a).

Intimately tied to the issue of the timing of this event is the route by which humans could have or *must* have entered the Americas from Beringia. The traditionally held view is that the Clovis archaeological complex—recently reassessed to span *maximally* from 13,250–12,800 B.P.[3] (Waters and Stafford 2007), though not all investigators agree with this redating (Haynes et al. 2007)—represents the first presence of humans in the Americas (Fiedel 2000). In this "Clovis-first" model, humans could have walked across the Bering Land Bridge, a landmass present during times of significantly lower sea levels, and then moved south through the ice-free corridor, an opening afforded by the retreating ice sheets that once covered most of the arctic and subarctic regions of North America.

However, evidence for a human presence in Americas south of the Cordilleran and Laurentide ice sheets prior to the opening of the ice-free corridor around 14,000–13,500 B.P. (Goebel et al. 2008) continues to mount. One of the most recent and undeniable pieces of evidence to support this conclusion was the discovery of human coprolites (desiccated human feces) from Paisley 5 Mile Point Cave in southern Oregon that are pre-Clovis in age, dating as early as 14,270–14,000 B.P.[4] (Gilbert, Jenkins, et al. 2008). Even with the associated uncertainty in the radiocarbon dates of these coprolites, the ice-free corridor has been argued to have been inhabitable prior to 13,000 B.P. (Mandryk et al. 2001). Thus the discovery of these coprolites has significantly widened the gap between first occupation of the landmass south of the glaciers and the time that it was possible for humans to enter via a land-based route. In addition, the Monte Verde site in southern Chile shows clear evidence for a pre-Clovis occupation in the Americas around 14,500 B.P. (Dillehay 1999). As it is quite unlikely that archaeologists have sampled from the first few millennia of Native America prehistory (Toth 1991), the date of first occupation of the Americas must be older. An entry prior to the opening of the ice-free corridor implies that humans *must* have initially used boats to enter the Americas along the Pacific coast.[5]

Attention has also been placed on estimating the amount and form of genetic and morphological variation that was brought to the Americas and the closely related topic of how many "waves" of migration contributed to the variation observed today (Eshleman et al. 2003; Schurr 2000, 2002, 2004a, 2004b). To understand the importance of resolving these issues, one can think about the pre-contact Americas as the ultimate "island" a number of humans migrated to but few returned from. In this case, the more precisely one can determine how much variation was carried *to* the Americas, the more precisely one can estimate how much variation has evolved *within* the Americas. It then becomes quickly apparent that resolving the time at which humans first entered the Americas and the timing of any subsequent "waves" is essential for estimating rates of evolution on the American "island." This would be true not only for genetic and biological evolution but also for rates of linguistic and cultural evolution. Moreover, by establishing the amount and nature of variation brought to the Americas, one can detect homoplasy in DNA sequences (i.e., recurrent mutations), as well as discern independent cultural innovations (as is clearly the case for agriculture in the New and Old Worlds) and biological adaptations (e.g., to high altitude; see Weinstein 2005).

A number of prior studies argued that the genetic variation found in the Americas is best accounted for by multiple human migrations from distinct source populations (see studies reviewed by Eshleman et al. 2003; Schurr 2000, 2002, 2004a, 2004b). Studies of variation in cranial morphology exhibited by Native Americans has also been explained by multiple migration events (see Neves and Hubbe 2005), although other researchers view the patterns of morphological change in the Americas as reflecting a process of adaptation and gene flow (see Gonzalez-Jose et al. 2008).

In line with the theme of the 2008 Twenty-Fifth Visiting Scholar Conference ("Archaeological and Biological Variation in the New World"), the purpose of this chapter is to review both the earliest and most recent studies[6] of genetic variation in Native American populations in order to uncover the trajectory the field has taken with respect to its attention to and appreciation of variation. In this discussion, there will be a particular focus on the variation reported from skeletal remains and other archaeological artifacts that predate 5000 B.P. The most recent ancient and modern DNA studies support an initial entry of humans into the Americas approximately 20,000–15,000 B.P. that emanated from a single source population located somewhere in Asia. This conclusion conflicts with the view held by some skeletal morphologists who maintain that multiple "stocks" of humans contributed biological variation to the Americas.

What Is Mitochondrial DNA?

The vast majority of Native American genetic studies have focused on variation found in the mitochondrial genome. Mitochondrial DNA (mtDNA) is an extranuclear genome found in the mitochondria of cells and has a number of characteristics that are useful in molecular anthropological and phylogenic studies. This genome is a small circular molecule and its entire sequence is known, including the positions of genes. The first whole human mitochondrial genome sequence, and consequently the reference to which all other mitochondrial sequences are compared, is merely 16,569 base pairs (bp) in length (Anderson et al. 1981; Andrews et al. 1999), which is trivial compared to the approximately 3 billion bp found in the nuclear genome. As a consequence of residing in the cytoplasm, human mtDNA is strictly maternally inherited (Giles et al. 1980), meaning that mutations occurring in it reflect only female history and movement. Moreover, mtDNA is particularly useful for discerning maternal ancestor-descendant relationships because it does not recombine during meiosis (Merriwether et al. 1991; Schurr et al. 1990).

The mitochondrial genome has a relatively high mutation rate compared to nuclear genes (Brown et al. 1979; Ingman et al. 2000; Kemp et al. 2007 and reference therein). This high rate generates a sufficient number of mutations to allow one to differentiate new haplotypes or haplogroups in populations that have recently diverged, such as Native Americans. The mitochondrial genome is also not under strong selection (Kivisild et al. 2006). As a result, the distribution of variation in this genome should reflect population history and not natural history (e.g., response to some selective force correlated with latitude such as climate; however, for a contrasting view see Mishmar et al. 2003).

Lastly, there are thousands of copies of the mitochondrial genome in each cell (Wallace 1994), compared with only two copies of each autosome and, in males, a single X and a single Y chromosome. For ancient DNA studies, the high copy number of the mtDNA and its ubiquitous presence in most cells partly compensate for the fact that DNA degrades with time and have made it the primary genetic system for the investigation of DNA from skeletal remains and other ancient human by-products (e.g., coprolites [Poinar et al. 2001]) and artifacts (e.g., quids [LeBlanc et al. 2007]).

Initial Studies of Native American mtDNAs

Native American tribes were some of the first human populations to be studied for mtDNA variation. The earliest of these molecular studies utilized restriction endonucleases (or "restriction enzymes") to screen humans for differences in their mitochondrial genomes. Restriction enzymes are used in nature by bacteria as a defense against invading viruses, which insert their DNAs into the bacterial host in an attempt to seize control of its genetic functions. Each restriction enzyme detects a specific order of nucleotides in a DNA sequence and cleaves the DNA at that particular location upon recognition.

As an example, the restriction enzyme *Hae*III is the third such enzyme purified from the bacterium *Haemophilus aegyptius* and recognizes the sequence GGCC. When this sequence is detected, the enzyme cleaves the DNA between the middle G and C, creating two pieces of DNA with GG at the end of one and CC at the end of the other. Using this example, if one person has the sequence GGCC at some place in her mtDNA and another has GGTC then the enzyme will cut the DNA of the former individual but not that of the latter. Here, two people have been differentiated genetically, and they can be scored as "+" (presence of site = cut) and "-" (absence of site = lack of cut), respectively, at that position in their genomes. This approach to detecting molecular variation is known as restriction fragment length polymorphism (RFLP) analysis.

Using six restriction enzymes[7] to detect mtDNA variation, Wallace and colleagues (1985) found that a certain mtDNA "morph"[8] (or combination of RFLPs) was exhibited by 40 percent of the Pimas (or Akimel O'odham), an indigenous population of the American Southwest. As this same morph had been previously detected in only one out of fifty-five "Orientals" (Blanc et al. 1983), Wallace and colleagues (1985) concluded that the proto–Native American population experienced a drastic founder effect after leaving Asia and entering the American continents.

Building on the foundation established by this report, Schurr and colleagues (1990) extended the previous sampling of Pimas[9] and added Native American samples from Central America (Maya) and South America (Ticuna). Using these six restriction enzymes[10] and others employed by Cann and colleagues (1987)[11] to survey the samples for known Asian polymorphisms and for the presence of a 9-bp deletion thought to be a marker specific to Asian and Asian-derived populations (Hertzberg et al. 1989), their study revealed that the mtDNAs of these populations could be placed into one of four haplogroups (i.e., groups of closely related haplotypes[12]). From these data, the authors concluded that Native American mtDNAs derive primarily from four Asian maternal lineages. Interestingly, some of the haplotypes were shared across these linguistically and geographically distant populations, suggesting a common origin for all "Amerinds."[13] However, the frequencies of the haplogroups differed substantially between these populations, with the Maya containing haplotypes belonging to all four of the haplogroups and Pimas and Ticuna containing haplotypes belonging to only three of them, albeit not the same three. Recognizing that the microevolutionary force of genetic drift best accounted for this observation, Schurr and colleagues (1990:620) noted that

> The shifting frequencies of founding haplotypes among tribes and the demonstration of the sequential accumulation of multiple new tribal mtDNA mutations indicate that reconstructing the migration patterns of Amerindians through their mtDNA phylogenies should be possible.

Two subsequent papers by Torroni and colleagues (1992, 1993) took mtDNA studies, and those specifically involving Native Americans, to the next level. These studies (1) increased the resolution of mtDNA analysis, (2) increased the number of populations and individuals sampled, and (3) devised a nomenclature for describing mtDNA variation. To the six restriction enzymes used in the studies discussed above, Torroni and colleagues (1992, 1993) added an additional eight,[14] substantially increasing the amount of information that could be obtained from an individual's mitochondrial genome. Analysis with this set of enzymes was estimated to screen approximately 15–20 percent of the genome, or around 2485-3313 bp[15] (Torroni et al. 1993). Additionally, in a subset of samples, they sequenced a 341 bp segment (nucleotide positions [nps] 16,030–16,370) of the control region (or D-loop) of the mtDNA genome, representing most of the first hypervariable region, or HVRI (Torroni et al. 1993).

By 1993, the mtDNA of 383 Native Americans from 17 populations had been examined by high-resolution RFLP analysis (Torroni et al. 1993), and over 97 percent could be placed into one of the four haplogroups originally described by Schurr and colleagues (1990). These haplogroups were given letter designations, beginning at the top of the alphabet. Accordingly, these first mitochondrial haplogroups received the names A, B, C, and D[16] (Torroni et al. 1992). The HVRI sequences obtained in 38 of these samples were partitioned into three haplogroups (A, B, and C), with members of haplogroup D clustering somewhat less distinctly, due to the lack of haplogroup D defining markers in HVRI.

As discovered by Schurr and colleagues (1990), most "Amerind" populations contained some frequency of at least three of these haplogroups.[17] In contrast, the Dogrib and Navajo (grouped linguistically into "Na-Dene"[18]) were almost fixed for haplogroup A,[19] and contained an appreciable frequency (approximately 27 percent) of a form of haplogroup A that exhibited a *Rsa*I site loss at np 16,329, caused by an A→G transition at np 16,331. Torroni and colleagues (1992, 1993) argued that this was evidence for the independent origins of Na-Dene Indians and Amerinds, who entered the Americas in two distinct migrations. Considering the intrahaplogroup variation reported in Amerinds, Torroni and colleagues (1993) argued that each haplogroup was introduced into the Americas by a single haplotype, the "founding haplotype" of the haplogroup. This deduction supported the observation of Wallace and colleagues (1985) that the proto–Native American population underwent an extreme founder effect.

With the genetic evidence available at the time, Torroni and colleagues (1993:581) concluded that "[t]he current study, together with previous studies . . . , confirms that *all* Native American mtDNAs fall into four distinct haplogroups (A–D)" [emphasis ours]. This statement would come to have a profound effect on the future of Native American mtDNA studies, locking the mind-set of many researchers into what might best be termed the "four-founding lineage paradigm."[20] As will be discussed below, in retrospect, the boldness of this statement was unwarranted given that between 2.6 percent (Torroni 1993:Table 2) and 6 percent (Torroni 1993:Table 4) of the samples they analyzed belonged to unknown haplogroups, which they attributed to non–Native American admixture. In retrospect, Torroni and colleagues (1993) should have emphasized, as did Schurr and colleagues (1990:613) a few years prior in their abstract, that "Amerind mtDNAs derived from *at least* four primary lineages" [emphasis ours]. Nonetheless, the four-founding lineage paradigm had been established, and what largely followed was a time when less, not more, mtDNA variation was surveyed in Native American populations.

The Four-Founding Lineage Paradigm

While researchers such as Wallace, Schurr, Torroni, and their colleagues were conducting high-resolution RFLP analyses of Native American mtDNAs, others were focused solely on sequence variation in the first (HVRI) or second (HVRII) hypervariable regions of the genome (e.g., Ginther et al. 1993; Ward et al. 1991, 1993). However, after it was "concluded" that only four haplogroups were present in the founding American population, high-resolution studies quickly fell out of favor. The cost of conducting such an intensive mutation detection process also limited its widespread use. Thus, in the 1990s, a researcher might have been left to ask, "Why look for variation beyond the markers that define haplogroups A, B, C, or D?"

More to the point, one could simply sequence the HVRI of a particular mtDNA to determine an individual's haplogroup, as demonstrated possible for at least haplogroups A, B, and C by Torroni and colleagues (1993). In sum, this meant that a study could benefit by being able to analyze more samples for known markers at the cost of conducting high-resolution RFLP analysis. Moreover, this approach meant that one could retrieve adequate comparative data from degraded DNA sources, such as frozen archived blood fractions (an approach called "Freezer Anthropology" [Merriwether 1999]) or even highly degraded DNA sources, such as ancient skeletal remains (e.g., Stone and Stoneking 1993, 1998).

The abandonment of the high-resolution RFLP approach for methods involving the genotyping of samples for haplogroup defining markers and/or sequencing the hypervariable regions of the D-loop resulted in the accumulation of numerous data sets from a variety of different Native American populations. These data revealed that the vast majority of mtDNAs in the Americas did belong to haplogroups A–D. However, the nonuniformity of haplotype definition in these studies generated interpretations of the resulting data that were sometimes inconsistent with each other.[21]

Terminology for mtDNA Variants

At this point, it might be useful to review the history of the terminology used to discuss mtDNA variation in Native American populations. Different descriptions of mtDNA variants over the past 15 years have sometimes created confusion about the units of genetic analysis in these studies and the relative comparability of mtDNA genotypes defined by RFLP analysis and HVRI sequencing and, hence, the number of founding haplotypes and lineages that have been brought to the Americas.

Because of the different approaches to detecting mtDNA haplogroups and characterizing haplotypes, parallel terminologies for describing these variants were employed. In the studies carried out in the Wallace lab, the terms *haplogroup* and *lineage* were used interchangeably to denote a distinct cluster of phylogenetically related mtDNAs. The smaller branches of a haplogroup were called *subhaplogroups* or *sublineages*. In addition, a unique mtDNA genotype defined by RFLP analysis was called a *haplotype*. A haplotype encompassed all of the mutations identified through this method, given that all of the nucleotides in the mtDNA genome are linked.

However, other researchers used different terms to describe Native American mtDNAs. While recognizing haplogroups on the basis of the defining RFLPs in these studies, the term *lineage*, not *haplotype*, was used to describe each unique HVRI sequence in a given population. Every set of related lineages was then called a *cluster* or sometimes a *clade*, with the nodal or root types of the clusters being considered their founding lineages.

In this regard, it should be reemphasized here that one can usually ascertain the mtDNA haplogroup to which a given HVRI sequence belongs by noting which combination of mutations is present in the mtDNA being analyzed. In other words, most of the founding haplogroups have a unique sequence "motif" that allows one to quickly delineate a particular maternal lineage from the rest. However, a RFLP haplotype may also have more than one HVRI sequence associated with it because of the more rapid mutation rate of the HVRI compared to the coding region in which the RFLPs occur. Thus, RFLP haplotypes and HVRI haplotypes/sequences may or may not identify the exact same mtDNA.

Therefore, while largely producing the same general picture of variation, RFLP analysis and HVRI sequencing generate slightly different assessments of haplotype and lineage diversity in human populations. On the other hand, as will be discussed below, over the past five years research employing whole mtDNA genome sequencing has enlarged our understanding of mtDNA variation in the Americas, both in terms of the number of haplogroups present in Native American populations and the number of distinct haplotypes that were brought by their founders.

In this chapter, we will use the following nomenclature to describe mtDNA variants and the maternal lineages to which they belong: (1) a *haplogroup* is a large cluster of phylogenetically related mtDNAs defined by mutations in the coding region; (2) a *subhaplogroup* is a smaller, derived clade within a haplogroup; and (3) a *haplotype/lineage* is a mtDNA that is distinct from another by one or more mutations.

Moving Beyond the Four-Founding Lineage Paradigm

A few early ancient DNA studies (e.g., Hauswirth et al. 1994; Pääbo et al. 1988) made claims of discovering new Native American founding lineages but suffered from failure to sufficiently demonstrate the authenticity of their data.[22] However, the assertion of the existence of new founding lineages was not limited to ancient DNA studies (see studies reviewed by Smith et al. 1999). For example, Easton and colleagues (1996) reported that additional haplogroups, which they called "X6" and "X7," were present among the Yanomama of Brazil and Venezuela. Putatively similar mtDNAs had also been identified in other Native American populations (Merriwether et al. 1994, 1995; Santos et al. 1996; Torroni et al. 1993). These X6/X7 haplotypes had the $+Dde$I 10,394 and $+Alu$I 10,397 sites that occur in Asian macrohaplogroup M but otherwise lacked the diagnostic RFLPs of haplogroups C and D, while differing between themselves by the presence or absence of the *Hae*III 16,517 site.

However, when the HVRI sequences from X6/X7 mtDNAs were analyzed with those from other Native American populations, they clustered within haplogroups C and D (Schurr and Wallace 1999; Stone and Stoneking 1998). These results suggested that X6/X7 mtDNAs were autochthonous haplogroup C and D mtDNAs that had lost the diagnostic markers of these maternal lineages (-*Hinc*II

13,259 and -*Alu*I 5,176, respectively) through back mutations, rather than belonging to an additional founding lineage deriving from haplogroup M. Furthermore, all subsequent studies of Yanomama mtDNA (Hunley et al. 2008; Merriwether et al. 2000) failed to corroborate this discovery, which may have originated from laboratory-based errors.[23]

Nevertheless, it was not clear that all of the nonhaplogroup A–D mtDNAs (or "others") in Native American populations resulted from back mutations, admixture, or lab errors. The bona fide existence of a fifth founding mtDNA haplogroup was demonstrated when Forster and colleagues (1996) discovered that 11 percent of the Nuu-chah-nulth and 5 percent of the Yakima belonged to haplogroup X[24] and, shortly after, similar haplotypes were reported at a frequency of 25 percent in the Ojibwa (Scozzari et al. 1997). Brown and colleagues (1998) further confirmed the presence of haplogroup X mtDNAs in native North Americans by comparing these haplotypes with putatively related mtDNAs from European and Central Asian populations. In addition, from a collection of 70 North American Indian samples previously determined to represent "others," Smith and colleagues (1999) discovered that 32 of them belonged to haplogroup X. These 32 individuals represented members of diverse linguistic groups residing in the Northeast (Algonquian), the Plains (Siouan), the Southwest (Kiowa-Tanoan), and California (Hokan).

While the wide distribution of haplogroup X in North America strongly suggested that it was an additional founding haplogroup, its presence in European populations left open the possibility that it was introduced through admixture after A.D. 1492. However, the form of haplogroup X in the Americas (X2a) was shown to be quite distinct from those found in Europe and Central Asia (X2e) (for nomenclature in haplogroup X, see Reidla et al. 2003), and the diversity within Native American populations suggested that the haplogroup was quite ancient in North America (Brown et al. 1998). Haplogroup X in the Americas prior to European contact was finally demonstrated by its presence in the 1340 ± 40 [14]C B.P. (3340 B.P.) human remains discovered near Vantage, Washington (Malhi and Smith 2002).[25] Thus, haplogroup X is now viewed as an additional founder lineage along with haplogroups A, B, C, and D, albeit a "minor" one that has an unusual geographic distribution compared to the other four (Brown et al. 1998; Dornelles et al. 2005; Perego et al. 2009; Smith et al. 1999).

Reevaluating the Five-Founding Lineage Paradigm

Even with the confirmation of haplogroup X in North American Indians, there were still some mtDNAs that did not belong to one of the five founding haplogroups and that required further explanation. In a study of mtDNA variation among the Cayapa of Ecuador, Rickards and colleagues (1999) claimed to have detected an additional Native American founding lineage. They found that approximately 22 percent of the population exhibited an HVRI sequence that did not appear to belong to haplogroups A, B, C, or D, and according to the authors, the lineage might have been specific to the Cayapa.[26] Their claim failed to find much support, in part because they did not screen the samples for the Native American haplogroup defining RFLPs, excepting the 9-bp deletion (Schurr 2004a). The "Cayapa haplotype" was later reported in one Brazilian (Alves-Silva et al. 2000), among northern Mexicans (Green et al. 2000) and Chilean Yaghan and Mapuche (Moraga et al. 2000) and, through RFLP analysis, was shown to belong to haplogroup D. However, as noted, its associated HVRI sequence was distinct from others typically seen for mtDNAs belonging to this haplogroup. Thus, it appeared to represent a subbranch of haplogroup D that had dramatically diverged in the Americas or possibly another founding haplotype for this haplogroup. However, the evidence was not considered conclusive, leaving this proposed founder haplotype in limbo (Bandelt et al. 2003).

The Cayapa haplotype made a very important appearance some years later in the 10,300-year-old On Your Knees Cave (OYKC) skeletal remains from Prince of Wales Island in southeastern Alaska (Kemp et al. 2007). The discovery of this sequence in remains of such antiquity alone strengthened the case for this being another founder haplotype. In addition, a nearly exhaustive search of published mtDNA data from extant and prehistoric Native Americans available at the time revealed that this haplotype and closely related derivative forms occurred in a mere 47 of 3,286 individuals (1.4 percent), including some from the Chumash, Nahua, and Quechua (see Kemp et al. 2007:Figure 1). Most of these 47 individuals resided near the coast or were located squarely in the West, a pattern that may have resulted from an ancient coastal migration. A number of haplogroup A haplotypes also have a coastal bias in North America, which may also be a signature of humans having migrated along this route (Eshleman et al. 2004).

Regarding the relationships between these individuals, the OYKC individual occupied the central location in a haplotype network, consistent with it being the ancestral form in the Americas (see Kemp et al. 2007:Figure 2). A similar haplotype was also detected in Asia, albeit in only 1 out of 3,824 Asian individuals screened, a Han Chinese from Qingdao (Yao et al. 2002). This evidence satisfied the criteria of Torroni and colleagues (1993:581–582) that founding haplotypes should (1) ". . . be widespread within the Amerinds and should be shared between tribes because they preceded the Amerind tribal differentiation," (2) ". . . be central to the branching of their haplogroup in the phylogenetic analysis, because all new haplotypes would have originated from them," and (3) ". . . still be possible to detect . . . in East Asian and Siberian populations, because they originated in Asia." The OYKC DNA study added an additional criterion in that "[t]he older a lineage, the higher its probability of representing, or closely resembling, the founding haplotype of the haplogroup of which it is a member" (Kemp et al. 2007:616).

While establishing the "OYKC/Cayapa" haplotype as the founder of a second American D subhaplogroup pushed the field to reexamine the nature of diversity within haplogroups A–D and X, Kemp and colleagues (2007) were unable to determine precisely where the haplotype fit into the ever-changing mtDNA nomenclature.[27] It has since been determined to belong to subhaplogroup "D4h3a" based on whole mtDNA genome sequencing (Perego et al. 2009).

Return to High Resolution mtDNA Studies: Whole Mitochondrial Genome Sequencing

Recently, Native American mtDNA studies have reembraced the high-resolution approach of the late 1980s and early 1990s.[28] However, the new approach represents an improvement over the high-resolution RFLP method for detecting variation in that whole mitochondrial genomes are now being routinely sequenced. To state the obvious, once the entire mtDNA sequence is known, there is no more information to be gained from this genome. This rekindled appreciation for variation likely stems from the discovery of additional founding haplogroups (X2a) and subhaplogroups (D4h3a). In addition, others noted mtDNA structure within the Americas that was suggestive, but not demonstrable proof, of additional Native American founding haplotypes (e.g., Malhi et al. 2002). Lastly, continued improvement in sequence chemistry and analysis meant that longer sequences could be read with more accuracy and processed faster and more cheaply than ever before.

The first of the recent whole mitochondrial DNA studies conducted by Tamm and colleagues (2007) was rather elegant. The authors began with the observations that humans were present in northeastern Siberia at the Yana Rhinoceros Horn Site around 30,000 B.P. (Pitulko et al. 2004) and in South America around 15,000 B.P. (Dillehay 1999). From these observations, they proposed two simple scenarios for the peopling of the Americas: (1) the proto–Native American population resided in Beringia, being isolated from central Asian populations, until conditions permitted entering the Americas (this is called the "Beringian Incubation Model," or BIM[29]) or (2) the proto–Native American population did not reach Beringia until just prior to 15,000 B.P. and subsequently entered the Americas (the "Direct Colonization Model," or DCM).

Each of these scenarios makes predictions as to what should be observed in the genetic record of Native Americans. Under BIM, humans would have carried lineages that accumulated unique mutational changes while in Beringia, that is, prior to entering the Americas. Alternatively, DCM predicts that Native American founding lineages will be essentially identical to haplotypes found in Asian populations. In addition, the structure of mtDNA variation in the Americas would be indicative of the speed at which humans initially spread throughout the continents. If humans spread rapidly, then one would expect to find founding haplotypes shared far and wide. However, if the movement were more protracted, then one would expect to find nested sets of variation in different geographic locations within the American continents.

Through the compilation of 599 earlier published complete mitochondrial genomes from Asia and the Americas and 27 novel sequences from Native Americans, Tamm and colleagues (2007) found evidence that previously unrecognized variation was exhibited by the first humans who entered the Americas. In addition, their analysis helped place the Native American mitochondrial haplotypes in the nomenclature being developed for worldwide populations. They noted that haplogroups A, B, and X were founded by single haplotypes that defined subhaplogroups A2, B2, and X2a, respectively. Haplogroup C was founded by three, or possibly four, independent haplotypes defining subhaplogroups C1b, C1c, C1d, and C4c. The last haplotype was detected in only two Ijka speakers from Colombia and, as Tamm and colleagues (2007) suggest, deserves further attention.[30] Haplogroup D was brought to the Americas on two haplotype backgrounds, those of subhaplogroups D1 and D4h3 (for haplogroup, subhaplogroup, and founding haplotype defining mutation, see Tamm et al. 2007:Table 1). More recently, Perego and colleagues (2009) have further clarified on the structure of D4h3 through the analysis of 44 whole genomes belonging to this subhaplogroup. Their study revealed a deep division between the sequences exhibited by Native Americans and that from a single Han Chinese individual (as discussed above, originally discovered in Asia by Yao et al. 2002), respectively referred to as clades D4h3a and D4h3b.

In line with the discussion above (Terminology for mtDNA Variants), Tamm and colleagues (2007) found evidence for five haplogroups present in indigenous American populations (A–D, X) that were introduced by nine founding lineages/ haplotypes, each representing one of nine subhaplogroups or clades of haplogroups A, B, C, D, and X. As eight of these subhaplogroups are found widespread in the Americas, these data are consistent with a very rapid spread of humans throughout the American continents. Interestingly, the American haplotypes were a few mutational steps away from their sister Asian haplotypes. That the Native American mutations have not been identified in Asian matrilines lends support to the idea that the proto–Native American population was isolated for some time from other Asian populations, during which it accumulated these new variants (i.e., during an "incubation" period). Tamm and colleagues (2007) also estimated an age of approximately 17,000–10,000 B.P. for A2, B2, C1b, C1c, C1d, and D1, given the amount of diversity observed in these subhaplogroups today. Two additional subhaplogroups, D2a and D3, which possibly arose in the same source population (as will be discussed below), were introduced much later into the Americas, and subhaplogroup A2a was carried back into western Siberia, where it is found today among the Evenks, Koryaks, and Sel'kups (Schurr et al. 1999; Tamm et al. 2007). Overall, their study provided compelling evidence that the initial founders of the Americas emerged from a single Beringian source population.

Within six months[31] of the publication of the Tamm and colleagues (2007) findings, three independent research teams reported results from the collection and/or analysis of whole mitochondrial genomes that generally supported the BIM and/or a single origin for the first Americans. Kitchen and colleagues (2008) compiled sets of previous published data[32] on which they conducted Bayesian skyline plot analyses (Drummond et al. 2005). Based on the coalescent, this analytical tool infers effective population size backward through time from a set of sequence data and has been demonstrated to be an effective means for reconstructing population dynamics. In an updated version of this analysis,[33] Mulligan and colleagues (2008) found that the proto–Native American population was isolated from central Asian populations for 7,000–15,000 years, during which it experienced little or no growth wherever it resided, which was suggested by the authors to have been in Beringia (Kitchen et al. 2008). This part of the demographic profile corresponds to the "incubation" phase proposed by Tamm and colleagues (2007), during which lineages would have accumulated novel mutations that would differentiate them from other Asian populations. Following this period, the population underwent a bottleneck upon entering the Americas approximately 17,000–16,000 B.P., after which a major reexpansion occurred (Mulligan et al. 2008), likely as a consequence of entering a large landmass that was filled with megafauna and the like but, importantly, not occupied by other humans. Note that while the degree of this bottleneck was likely exacerbated by the paleoclimatic conditions in Beringia, this founder effect, recognized more than twenty years prior by Wallace and colleagues (1985) represents just one of a series of founder effects that shaped the human gene pool since our species exited Africa (Ramachandran et al. 2005).

Similarly, Fagundes, Kanitz, and Bonatto (2008) produced 58 new whole mitochondrial genome sequences from populations in North and South America. Combined with 28 previously published sequences the authors found that subhaplogroups A2, B2, C1, D1, and X2a contained similar levels of variation and are, thus, of similar antiquity in the Americas. Moreover, the study identified Native American "exclusive mutations" in these subhaplogroups that differentiated them from similar haplotypes found in Asia. Again, these findings were consistent with an "incubation" period experienced by the proto–Native American population in Asia. The authors used a mutation rate from Mishmar and colleagues (2003) that estimated the incubation period to have lasted at least 5,000 years. Employing the same Bayesian skyline plot analysis as did Kitchen and colleagues (2008), Fagundes, Kanitz, and Bonatto (2008) further found that proto–Native Americans ended their initial differentiation from Asian populations by undergoing a population bottleneck around 23,000–19,000 B.P. and later rebounded in a population expansion in the Americas 18,000–15,000 B.P.[34]

A study conducted by Achilli and colleagues (2008) provided an additional 14 novel Native American whole mitochondrial sequences. Compared with 107 previously published whole genome sequences and 47 coding region[35] sequences, the authors found that the "common" subhaplogroups A2, B2, C1, and D1 were founded by six haplotypes (A2, B2, C1b, C1c, C1d, and D1).[36] Identical to that noted above, they also discovered that these founding haplotypes were a few mutational steps from similar types in Asian populations, again suggesting that an incubation period preceded the entrance of humans into the Americas. Today, these subhaplogroups exhibit a fair degree of "star-likeness" which resulted from a population expansion, presumably the one that followed shortly after humans first entered the Americas. With one exception (C1d),[37] Achilli and colleagues (2008) estimated the founder haplotypes to be about 18,000–21,000 years in age.

The results of these four recent studies are consistent with BIM.[38] In addition, all of the studies support the concept of a single source population having given rise to most, if not all, of the first inhabitants of the Americas. This consensus stands in opposition to the conclusion that the Americas were populated by multiple "waves of migrations" reached by some, but not all, of the earliest mtDNA studies (see Eshleman et al. 2003; Schurr 2000, 2002, 2004a, 2004b). The inferences drawn from these whole mitochondrial genome studies differ most dramatically in population event dates and age estimates of the founding haplotypes/haplogroups in the Americas. Nevertheless, the age of current estimates (approximately 20,000–15,000 B.P.) is more in line with the available archaeological data than estimates of 40,000–20,000 B.P. for the entrance of humans into the Americas provided by earlier mtDNA studies (see Eshleman et al. 2003; Schurr 2004a). This result is likely due, in part, to their better approximations of the number of haplotypes present in the founding American population.

Using sequence data to date events in human prehistory continues to be a questionable practice. However, the ability to do so should improve with consideration of the fact that the observed rate of molecular evolution may not be constant over time (Ho et al. 2007; Kemp et al. 2007) and with the use of proper points of calibration (Endicott and Ho 2008). For instance, Ho and Endicott's (2008) reanalysis of the Fagundes and colleagues (2008) and Achilli and colleagues (2008) data using calibration points internal to the human mitochondrial tree produced an average age of 14,000 B.P. for the founding lineages, in this case removing approximately 6,000 years from the original estimates.

The "Other" Genome: Thinking Outside the Mitochondrial DNA Genome

Investigations of Y chromosomal and autosomal variation among Native Americans are also consistent with an origin from a single source population and the Beringian Incubation Model.[39] The Y chromosome is passed intergenerationally through only males and can, therefore, be considered an analogue to mtDNA for tracing male movement/history. By contrast, autosomal DNA (i.e., nonsex chromosomal DNA found in the nucleus) is biparentally inherited and, therefore, reflects both male and female history.

There is a limited amount of Y chromosomal variation exhibited by Native American males compared to males in Asian populations (Bortolini et al. 2003; Karafet et al. 1999, 2001; Lell et al. 2002; Zegura et al. 2004), which is consistent with the notion that the peopling of the Americas was accompanied by a population bottleneck. Similar to mtDNA, Y chromosomal variation is affected more strongly by genetic drift (randomness) than is autosomal variation, because of the smaller effective population size of the Y chromosome. Currently, only Y chromosome haplogroups C and Q are believed to be indigenous to the Americas, with other haplogroups most likely being the product of postcontact admixture (Bolnick et al. 2006; Malhi et al. 2008; Zegura et al. 2004). However, Lell and colleagues (2002) and Bortolini and colleagues (2003) have suggested that certain R1-M173 haplotypes may also have been brought to the Americas by the proto–Native American migrants. R1-M173 haplotypes appear at low frequencies in only indigenous populations of North and Central America, including the Athapaskan-speaking Chipewyans (Bortolini et al. 2003; Lell et al. 2002). Although quite diverse in eastern Siberia, there is some debate as to whether those seen in Native Americans are indigenous in origin, owing to their also being commonly observed in European populations (Lell et al. 2002; Tarazona-Santos and Santos 2002).

One form of haplogroup Q in the Americas, Q3-M242*, is shared with Asian populations. Equivalent to the Native American derived mtDNA lineages, a sub-branch of haplogroup Q3-M242*, called Q3-M3*, is widespread in the Americas (Bortolini et al. 2003; Karafet et al. 1999, 2001; Lell et al. 2002; Underhill et al. 1996; Zegura et al. 2004). The defining mutation of this haplogroup likely arose during the peopling of the Americas or slightly beforehand, perhaps during the incubation period, and is known to be at least 10,300 years old in the Americas (Kemp et al. 2007). Q3-M3* is also the most widespread Y-chromosome type in the American continents (Bolnick et al. 2006; Malhi et al. 2008; Underhill et al. 1996; Zegura et al. 2004), and its distribution is consistent with the notion that all Native American males originated from a single source population. Haplogroup Q3-M3* has also been observed among the Chukchi, Evenks, and Siberian Eskimos and was probably introduced into northeastern Siberia through gene flow back across the Bering Strait (Karafet et al. 1997; Lell et al. 1997).

A recently described autosomal marker is also consistent with an origin from a single source population for all Native Americans. Through an analysis of 377 microsatellite markers[40] from the HGPD-CEPH[41] human genome diversity cell line panel, Zhivotovsky and colleagues (2003) discovered that a 275-bp allele at D9S1120[42] was present in all of the Native American populations sampled (Colombian, Karitiana, Maya, Pima, and Surui) yet absent from the other 47 worldwide populations represented in the panel. This allele has nine repeats of the sequence TAGA and, hence, has been nicknamed the "9 repeat allele" (9RA).

In order to better characterize the distribution of the 9RA, Schroeder and colleagues (2007) screened an additional 13 North American populations[43] representing members of "Na-Dene," Aleut-Eskimo, and North American "Amerinds,"[44] in addition to seven Altaic populations.[45] Their results indicated that the 9RA was ubiquitous among Native American populations, being present in approximately 31.7 percent of their samples. They also detected the 9RA in approximately 24 percent of the Chukchi and 18 percent of Koryaks, a finding that again suggested possible gene flow back across the Bering Strait, as discussed above for Y-chromosome haplogroup Q3-M3* and mtDNA subhaplogroup A2a. These results suggested that the 9RA arose during the incubation phase of a single source population, perhaps the "Beringian" population suggested by the mtDNA researchers. More recently, Schroeder and colleagues (2009) found that 91 percent of chromosomes that exhibit the 9RA are flanked by an identical genetic background that spans some 76.26 kilobases, a haplotype they coined the "American Model Haplotype" (AMH). As the remaining 9RA chromosomes share a portion of the AMH, it appears that all chromosomes carrying the marker are identical by descent. Importantly, this indicates that all Native American populations share a large portion of their ancestry, again supporting the notion of them having arisen from a single source population.

While mtDNA subhaplogroups D2a and D3 were later introduced into the Americas and are largely associated with Athapaskans and Aleut-Eskimos (Rubicz et al. 2003; Starikovskaya et al. 1998; Tamm et al. 2007; Zlojutro et al. 2006), 22–29 percent of the individuals in these populations exhibit the 9RA (Schroeder et al. 2007). This evidence lends itself to two possible scenarios: (1) very few women introduced D2a and D3 into the northernmost parts of the Americas such that they did not disrupt the frequency of the already established 9RA, or (2) the women who carried D2a and D3 to the Americas also carried the 9RA, probably having originated from the same source population that gave rise to the first Americans and the mtDNA and Y-chromosome founding haplotypes.

In explaining the distribution of the 9RA, only models that emphasized substantial isolation of Native American populations from the Old World resulted in similar theoretical expectations for the unusual distribution of such a marker (Schroeder et al. 2009). This view stands in interesting opposition to the recent models presented by Ray and colleagues (2010) that emphasize the importance of gene flow with Asia in shaping the Native American gene pool. Presently, it is unclear how much these observations are affected by having studied Native American populations with substantial non–Native American admixture or, more important, if it can explain the difference between the findings of the two research teams as they studied different sets of populations (Ray et al. 2010; Schroeder et al. 2009). Nevertheless, large-scale modeling such as that presented in Ray and colleagues (2010) will continue to add novel insights about the entrance of humans into the Americas.

Recently, Halverson and Bolnick (2008) have found patterns of ABO blood group[46] variation in pre-Columbian Native American populations to be consistent with a population bottleneck at the time that humans entered the Americas. In a study of the distribution of ABO alleles in contemporary Native American populations from Mesoamerica and South America, Estrada-Mena and colleagues (2010) drew a similar conclusion. As it has long been recognized that most Native American populations are nearly fixed for the O allele (Mourant 1976), the most interesting observation drawn by Estrada-Mena and colleagues (2010) is that a derived O allele, called $O^{1v(G542A)}$, is found in every Native American population screened to date at the sequence level (Nahua, Mazahua, Maya, Mexican mestizos, Cayapa, and Bolivian Aymara). Thus far, this allele has not been observed in Asian populations screened at the same level (Chinese, Japanese, and Korean). Thus, despite the current limited sampling of both Asian and Native American populations, the $O^{1v(G542A)}$ allele may end up being confirmed as yet another Native American specific marker that evolved during the "Beringian Incubation."

Neither of these recent studies on the ABO system in Native Americans has sufficiently addressed the observation that Eskimos and some Athapaskan populations exhibit appreciable frequencies of the A and B alleles (Szathmary 1979). These observations may very well be the product of a secondary expansion into the Americas as described above. Nevertheless, advances in our knowledge of variation in the ABO system at the sequence level (Estrada-Mena et al. 2010) among Native Americans mirrors the attention now placed on whole mtDNA sequencing.

Based on genotyping 36 markers in the vicinity of the 44 exon of the dystrophin gene, Bourgeois and colleagues (2009) described six common X-chromosome haplotypes, called B001–B006 found in worldwide populations. Comparatively, Native American populations have much higher frequencies of haplotypes B004 and B006 and lack haplotype B005. Thus it appears that X-chromosome variation among Native Americans has been similarly shaped by the founder effect associated with the peopling of the Americas (Bourgeois et al. 2009).

Limited but Highly Structured Variation in the Americas

Having examined the evidence concerning the extent of genetic variation that was initially brought to the Americas, it is also worthwhile to explore what has happened to that variation over the past 15,000 years or so. In North America, for example, one finds remarkable mtDNA structure according to geographic regions (Figure 2-1). As argued above, if the proto–Native American population originated from a single source population, then the frequencies depicted on this map suggest that genetic drift was a dominant force shaping mtDNA variation in Native Americans, most likely when populations were very small prior to the Formative period in North America. This structured pattern is also consistent with the idea that "tribalization" began early in the prehistory of the Americas (Malhi et al. 2002; Torroni et al. 1993).

Genetic structure in the nuclear genomes of Native Americans supports just such a scenario. Through a comparison of variation at 678 autosomal microsatellite markers in 422 indigenous individuals from North, Central, and South American populations to thousands of others from global populations, Wang and colleagues (2007:Table 1) found that Native Americans exhibit the highest levels of homozygosity and the largest interpopulation genetic distances (evaluated by F_{ST}). This pattern indicates that, compared with other continental "populations," the Native American gene pool contains the least amount of genetic variation while simultaneously being the most highly structured. This outcome is consistent with the scenario that the proto–Native American population underwent a genetic bottleneck, probably upon entry into the Americas and, after colonizing these continental regions, has been highly subject to genetic drift. The patterning of the continent-wide genomic diversity among Native Americans is consistent with a single major colonization that first entered the Americas via the coast (Wang et al. 2007).

Ancient DNA Studies: Genetic Variation in the Americas Prior to 5000 B.P.

If one could somehow study directly the remains of a population of the first Americans, one would be provided with an unparalleled opportunity to measure the amount of genetic variation carried to the Americas and to test directly for the hypothetical founder lineages. However, no such population has been identified. The finite number of remains that have been discovered in the Americas that are greater than 5,000 years old *and* available for destructive analysis, as is necessary for ancient DNA (aDNA) studies, have traditionally been the limiting factors in this line of scientific inquiry. Moreover, many of the oldest Native American remains no longer contain DNA (Smith et al. 2005).

Even though the physical remains of the very first Americans have not been discovered, and most likely will never be, genetic data recovered from human remains and artifacts that represent ancient populations of great antiquity allow one to get closer and closer to measuring directly the amount of genetic variation that

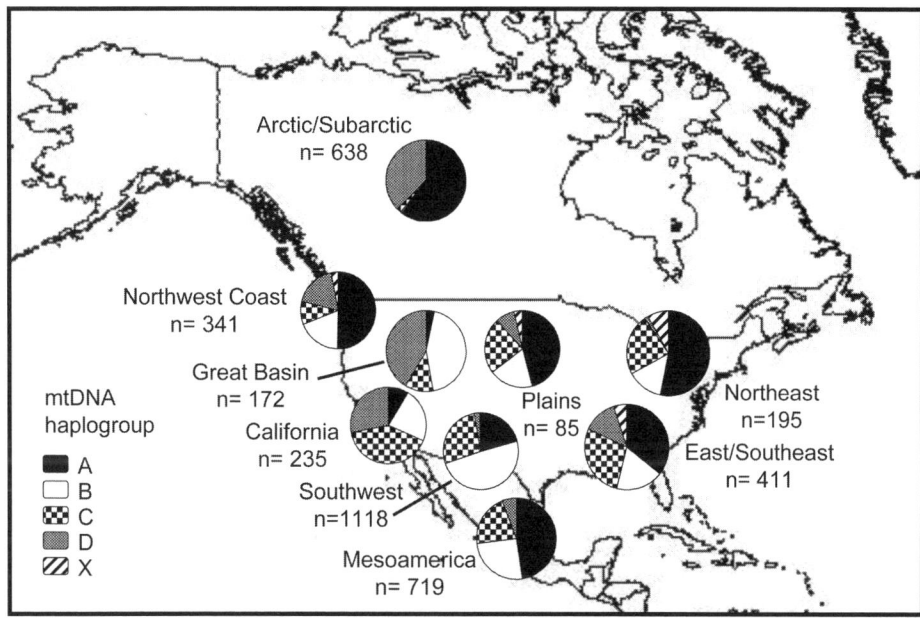

Figure 2-1. *Distribution of mtDNA haplogroup frequencies across North America from various previous studies (Bolnick and Smith 2003; Budowle et al. 2002; Carlyle et al. 2000; De la Cruz et al. 2008; Eshleman 2002; González-Oliver et al. 2001; Green et al. 2000; Huoponen et al. 1997; Johnson and Lorenz 2006; Kaestle and Smith 2001; Kemp et al. 2005, 2010; Lorenz and Smith 1996; Malhi 2001; Malhi et al. 2003, 2004; Merriwether, et al. 1995; Merriwether and Ferrell 1996; Monroe et al. 2006; Parr et al. 1996; Rubicz et al. 2003; Schurr et al. 1990; Scozzari et al. 1997; Shields et al. 1993; Smith et al. 1999; Stone and Stoneking 1998; Torroni et al. 1992, 1993, 1994; Ward et al. 1991, 1993).*

was present in the founding population. This is precisely the reason that Kemp and colleagues (2007) suggested an additional criterion be added to those proposed by Torroni and colleagues (1993) for establishing a lineage as a founder (discussed above under Reevaluating the Five-Founding Lineage Paradigm). In this case, it is important to discuss the available data from remains that predate 5000 B.P., which will be done in chronological order starting with the oldest evidence.[47]

Pleistocene DNA

The oldest human DNA from the Americas has come not from bones or teeth but rather from ancient feces (coprolites). Recently, Gilbert, Jenkins, and colleagues (2008) were able to determine the mitochondrial haplogroups of a number of humans that defecated in Paisley 5 Mile Point Cave in south-central Oregon, on the edge of the Great Basin. Three samples dating around 12,300 [14]C B.P. (approximately 14,270–14,000 cal yr B.P.) have revealed the presence of sub-haplogroups A2 and B2. These are extremely important data points as they place humans in the Americas prior to the Clovis period and prior to the opening of the ice-free corridor. An additional coprolite from the caves, dated to approximately 11,000–10,000 ^{14}C B.P., also revealed the presence of subhaplogroup B2.

Early Holocene DNA (10,300–8000 B.P.)

As previously noted, the earliest presence of D4h3 in the Americas was found in the 10,300–year-old OYKC remains from the northern tip of Prince of Wales Island in southeastern Alaska (Kemp et al. 2007). In addition, molecular sex determination indicated that the individual was male and his Y chromosome belonged to haplogroup Q3-M3*, establishing a minimum date for the emergence of the defining marker of this haplogroup (Kemp et al. 2007). Kaestle and Smith (2001) determined that the 9200 ± 60 ^{14}C B.P. (11,020 B.P.) individual[48] discovered at the Wizards Beach site in western Nevada belonged to haplogroup C. The HVRI sequence of the Wizards Beach individual represents any one of the three proposed founder haplotypes of subhaplogroup C1 (Kaestle 1998; Tamm et al. 2007). The individual found in Hourglass Cave in the Rocky Mountains in Colorado who died approximately 8000 B.P. was determined to be a member of haplogroup B (Stone and Stoneking 1996). Molecular sex typing confirmed the morphological assessment that this individual was male. As in the case of the Wizards Beach individual, the HVRI sequence of Hourglass Cave man also represented the proposed founder haplotype for his haplogroup (Stone and Stoneking 1996; Tamm et al. 2007).

Mid-Holocene DNA (8000–5000 B.P.)

An additional coprolite from Paisley Caves, dated to approximately 6600 ^{14}C B.P. was determined to have been made by a member of subhaplogroup B2 (Gilbert, Jenkins, et al. 2008). At 4950 ± 170 ^{14}C B.P. (6950 B.P.), two individuals from the China Lake site in the British Columbia interior were determined to belong to a form of haplogroup M that has yet to be identified in any extant Native American population (Malhi et al. 2007). As these individuals both exhibited a very rare supratrochlear spur on their left humeri, it was believed by the physical anthropologists who examined the remains that these males were full siblings and possibly twins (see Cybulski et al. 2004, this volume). This haplotype possibly represents a sixth founding haplogroup, one that may have gone extinct in the last 5,000 years or so. However, further analysis of this mtDNA is required to rule out prehistoric gene flow from the Beringian region, where haplogroup M derived lineages not seen in the Americas are common. Given the profound effect that genetic drift has had on the Native American gene pool (Wang et al. 2007), this haplotype may still exist in unstudied populations.

In total, eleven samples that predate 5000 B.P. have yielded well-preserved mtDNA and/or sex chromosomal DNA. Notwithstanding the small sample size and the fact that none of these data represent that of a "population," some trends are worth noting. One of the eleven samples (9.1 percent) belongs to subhaplogroup D4h3, a sublineage or clade that today is exhibited by only 1.4 percent of Native Americans. More noteworthy are the two samples that may represent a haplogroup that has possibly become extinct in the Americas. These observations are consistent with the notion that genetic drift has long been a powerful evolutionary force in the Americas and further indicate that additional genetic variation may be found in the ancient Americas. In fact, this process could account for the observation of "other" haplogroups in many aDNA studies in the Americas conducted in the past 20 years.

However, to date, most of the putative founder lineages and haplotypes suggested in the past have been shown to be either derivatives of haplogroups A–D and X or haplotypes introduced by admixture or the by-products of contamination. Nevertheless, continued aDNA research on very ancient Native American remains and/or coprolites will be crucial for addressing the possibility of additional founding lineages being present in the Americas. It is imperative that when aDNA researchers discover a previously unrecognized form of DNA in the Americas they authenticate the finding and do not simply place it in an "other" category.

An Asian Origin for Certain, but Where? [49]

While all of the genetic studies discussed in this chapter point to an Asian origin for Native Americans, one could ask whether the precision of this conclusion is an improvement over those that were held almost 100 years ago. For example, based on physical characteristics, Hrdlička (1912:11) argued for an East Asian origin, but noted that

> . . . difficulties arise when we endeavor to assign the origin of the Indian to some particular branch of the yellowish-brown population. We find that he stands quite as closely related to some of the Malaysian peoples as to Tibetans, the Upper Yenisei natives, and some of the northeastern Asiatics.

Prior to the discovery of mitochondrial haplogroup X in the Americas, mtDNA studies suggested either a south-central Siberian (Torroni et al. 1993) or a Mongolian (Kolman et al. 1996; Merriwether et al. 1996) origin for Native Americans, as these populations exhibited haplogroups A, B, C, and D and, importantly, the founder Native American haplotypes of these haplogroups proposed by Torroni and colleagues (1993). The reasoning here was simple in that researchers were looking for the least geographically distant population from the Americas that contained all of the haplogroups exhibited by Native Americans.

After haplogroup X was determined to be an additional Native American founding haplogroup (as discussed above), researchers began looking for a population in Asia that contained all five haplogroups. Derenko and colleagues (2001) discovered haplogroup X in populations of the Altai mountain region in southern Siberia, where haplogroups A, B, C, and D were also present. However, the forms of haplogroup X found in the Altaians (X2e) were intermediate between the sublineages found in North America (X2a) and Europe (also see Brown et al. 1998), being at least two mutational steps away from those in the former and at least one mutational step away from the latter (see Derenko et al. 2001:Figure 1). Nevertheless, the Lake Baikal–Altai mountain region stood as the most probable origin for Native Americans (see, for example, Schurr 2004a:Figures 6 and 7).

The recent whole mitochondrial, nuclear, and Y-chromosomal DNA studies, in concert with the ancient DNA research discussed above, have made it more difficult to determine a precise Asian origin for Native Americans than when less data were available. Ironically, the field has come full circle back to the statement made by Hrdlička in 1912, as the specific place of origin in Asia remains uncertain. The difference is that, today, we are more confident than ever in our uncertainty. No Asian populations sampled to date contain all of the founder Native American mtDNA and Y-chromosome haplogroups, the 9RA at D9S1120, and the $O^{1v(G542A)}$ allele. To complicate matters further, the closest relatives to the 9RA "AMH" described by Schroeder and colleagues (2009) are the Baloch, Brahui, Burusho, Pathan, and Sindhi populations of southern Asia (Afghanistan, Iran, and Pakistan) that have AMHs with 12 repeats (or 12RA). In total, these observations are hardly surprising, given the strong support for the Beringian Incubation Model discussed above. If this model is correct, then the proto–Native American population accumulated unique mutations when in isolation from other Asian groups prior to entering the Americas, leaving behind populations lacking these genetic variants.

In light of this evidence, the logical question arises as to the fate of this "Beringian" population, the answer to which is far from certain. Did all the members of this population enter the Americas? Is that population now extinct? Either of these scenarios seems unlikely. The best chance of identifying a population in Asia from which Native Americans originated will come from aDNA studies of remains that date around 20,000–15,000 B.P. The view that some population in Asia today will resemble the proto–Native American population is not plausible, as it ignores the fact that populations on the other side of the Bering Strait have undergone an equal degree of genetic evolution since their separation.

Conclusions

The flurry of recent whole mitochondrial genome studies support the Beringian Incubation Model as the best description of the demographic history of the population that would first enter the Americas approximately 20,000–15,000 B.P., having originated from a single source population located somewhere in Asia. Variation exhibited by Native Americans in their nuclear genomes likewise supports this scenario. A human presence in the Americas by at least 14,270–14,000 B.P. has been confirmed by the recovery of human coprolites in the Paisley Caves in southern Oregon. These data indicate that humans must have initially entered the Americas along the Pacific Coast, as the ice-free corridor was not open at this time.

A secondary migration or expansion of humans introduced additional mtDNA haplogroups into the northernmost areas of North America after the last glacial period. However, these migrants either originated from the same source population as did the initial migrants or were so few in number that they did not substantially disrupt the existing pattern of nuclear genetic variation (i.e., at D9S1120).

Since the time that humans entered the Americas, genetic drift has played a substantial role in shaping the Native American gene pool. On a continental scale, Native Americans exhibit the highest measure of homozygosity, and this is likely the by-product of having undergone a major population bottleneck around the time of settlement of the Americas. Native Americans also exhibit the highest interpopulation genetic distances, probably as a result of maintaining relatively small populations sizes during the Paleo-Indian period in conjunction with the early process of tribalization. Genetic variation thus far detected in human remains and human by-products (e.g., coprolites) that predate 5000 B.P. are consistent with this view, notwithstanding the small sample sizes.

As for how these conclusions can be reconciled with studies of cranial morphological variation found in the Americas is not certain. It may be that the only way to directly test for the presence of an earlier migration that did not contribute genes to extant Native Americans is through aDNA studies.

Acknowledgments

The ideas in this chapter were strengthened through many conversations over the past few years with David Glenn Smith, Ripan S. Malhi, Kari B. Schroeder, Cara Monroe, David Meltzer, and Tom Dillehay. Also thanks to Cara Monroe for helping to create Figure 2-1. Washington State University and the University of Pennsylvania provided support for the authors' travel to "Archaeological and Biological Variation in the New World," the Twenty-Fifth Annual Visiting Scholar Conference held at Southern Illinois University Carbondale. Lastly, we thank David Glenn Smith and two anonymous reviewers for their critical review of the manuscript.

Notes

1. Most of the genetic markers targeted for reconstructing population history are effectively neutral, having little or no effect on the phenotype.

2. However, major debate continues over the proper use of the molecular clock to date these events (see recent debate between Bandelt [2008] and Howell and colleagues [2008]).

3. The minimal span of the Clovis archaeological complex is now dated from 13,125–12,925 B.P., a mere 200 years (Waters and Stafford 2007). Note that all dates in this chapter are given in calendar years, unless otherwise noted.

4. While pre-Clovis human remains have yet to be discovered, the feces they left in dry rock shelters are apparently quite abundant. The oldest human remains thus far discovered in the Americas, the Arlington Springs "Man" from Santa Rosa Island off the coast of Santa Barbara, California, are Clovis in age (Johnson et al. 2002).

5. Ironically, the location of Arlington Springs "Man," the oldest skeletal remains discovered in the Americas, suggests that Native Americans must have used boats in Clovis times to be able to reach the Channel Islands.

6. Studies that have been conducted since the time of the last major review papers on the peopling of the Americas from a genetic perspective (Eshleman et al. 2003; Mulligan et al. 2004; Schurr 2000, 2002, 2004a).

7. *Ava*II, *Bam*HI, *Hae*II, *Hinc*II, *Hpa*I, and *Hpa*II

8. *Hinc*II morph 6

9. Schurr and colleagues (1990) noted that the Pima sample also included some Papago (or Tohono O'odham), a genetically and linguistically closely related population (Kemp et al. 2010; Malhi et al. 2003). This fact was not noted by Wallace and colleagues (1985), despite their using the same collection of samples.

10. These were the same six enzymes used by Wallace and colleagues (1985), with the exception that *Msp*I was used in place of *Hpa*II. As these two enzymes recognize and cut the same sequence CCGG, there is no difference to the approach of Schurr and colleagues (1990).

11. These enzymes included *Alu*I, *Dde*I, *Hha*I, *Hinf*I, *Mbo*I, *Rsa*I, and *Taq*I.

12. In this study, *haplotype* was defined as a unique combination of restriction enzyme recognition sites and length variants at the COII/tRNALys intergenic region. Haplotypes that shared phylogenetically important mutations (RFLPs) were assigned to a specific "haplotype group," or haplogroup for short. Interestingly, Schurr and colleagues (1990) grouped haplotype "AM4" with others that exhibited the 9-bp deletion, which today are known to belong to haplogroup B. Subsequent analysis indicated that this haplotype should actually be placed in the haplogroup containing haplotypes "AM6–9" (confirmed in the hypervariable region sequence data reported by Torroni and colleagues [1993]), namely, haplogroup A. Therefore, haplotype AM4 is a member of haplogroup A that independently gained the 9-bp deletion. The last twenty years of research has demonstrated that the 9 bp deletion is prone to homoplasy, having arisen multiple times in unrelated mtDNA lineages around the world (see studies discussed by Schurr and Wallace [2002] and Kemp and colleagues [2005]).

13. *Amerind* is a proposed linguistic macrogroup comprised of all Native American languages that do not belonging to Na-Dene or Aleut-Eskimo (Greenberg et al. 1986). This classification is not supported by all linguists (Campbell 1997; Nichols 1990), and molecular studies have failed to find genetic unity of populations classified as Amerind, to the exclusion of those classified as Na-Dene or Aleut-Eskimo (Bolnick et al. 2004; Hunley and Long 2005). While we use the term *Amerind* here to contextualize the earliest Native American mtDNA studies, the category of Amerind is not a meaningful genetic unit for Native American populations and we recommend that its use should be otherwise abandoned.

14. For a total of 14 restriction enzymes: *Alu*I, *Ava*II, *Bam*HI, *Dde*I, *Hae*II, *Hae*III, *Hha*I, *Hinf*I, *Hinc*II, *Hpa*I, *Hpa*II/*Msp*I, *Mbo*I, *Rsa*I, and *Taq*I.

15. A rough calculation, given the 16,569 bp length of the reference sequence (Anderson et al. 1981; Andrews et al. 1999).

16. Torroni and colleagues (1992) used the letters A, B, C, and D, but called them "clusters," whereas Torroni and colleagues (1993) later adopted the term *haplogroup*, which is the preferred term today. Naming mitochondrial DNA lineages has come a long way in the past 20 years, but still suffers from inconsisten-

cies in nomenclature. The high-resolution data obtained through the continued collection of whole genome sequences from worldwide populations will likely alleviate this problem by facilitating adoption of a standardized nomenclature system, such as that used to name Y-chromosome lineages (YCC 2002).

17. The Kuna of Panama were the only Amerind population to contain fewer than three haplogroups, being fixed for haplogroup A (Torroni et al. 1993).

18. *Na-Dene* is another questionable linguistic grouping, particularly with the inclusion of Haida (Campbell 1997), with equivocal evidence of being genetically distinct (Malhi et al. 2004 and references therein. See note 19 about admixture in the Haida.). However, Na-Dene as Athapaskan-Tlingit-Eyak has found support among linguists as having possible links to the Yeniseic language family in Siberia (Vajda 2010) and, thus, is far less controversial than *Amerind*.

19. It is interesting to note that in their abstract, Torroni and colleagues (1993:563) state "[t]his analysis revealed the presence of four haplotype groups (haplogroups A, B, C, and D) in the Amerind but only *one* haplogroup (A) in the Na-Dene, and confirmed the independent origins of the Amerinds and the Na-Dene" [emphasis ours]. However, they observed (1993:Table 2, Table 4) that, in addition to exhibiting haplogroup A, (1) Navajo contain haplogroup B, (2) Apache contain haplogroups B, C, and D, and (3) Haida contain haplogroup D. Torroni and colleagues (1993) suggested that these haplotypes were acquired by the Haida, Navajo, and Apache through admixture with Amerind populations. Indeed, subsequent work confirmed that many of the nonhaplogroup A mtDNAs were introduced into the southern Athapaskans (Navajo and Apache) through gene flow with indigenous Southwest populations (Malhi et al. 2003, 2008; Smith et al. 2000). However, one of the Apache haplogroup D samples sequenced by Torroni and colleagues (1993) belongs to subhaplogroup D2, a mtDNA lineage originating from the north. This type is found among Aleuts and Eskimos (Derbeneva et al. 2002; Rubicz et al. 2003; Starikovskaya et al. 1998) and has most recently been discovered in the 3,500–4,500-year-old frozen hair of a Palaeo-Eskimo from Greenland (Gilbert, Kivisild, et al. 2008). Therefore, the Na-Dene populations are not genetically monotypic with regard to mitochondrial haplogroups, as originally argued by Torroni and colleagues (1993).

20. Here a lineage is referred to as a *haplotype*. In the example of the study conducted by Torroni and colleagues (1993), four founding mtDNA haplotypes (maternal lineages) that just happened to represent four different haplogroups were detected. Today there are still four "major" haplogroups in the Americas, plus a minor one, X. However, multiple founding haplotypes representing subhaplogroups of the five founding lineages appear to have been carried to the Americas (Tamm et al. 2007). The distinction is subtle but extremely important, as the accuracy in reconstructing the evolutionary history of Native Americans hinges partly on the estimated amount of variation introduced into the Americas. In this case, it is important to know that haplogroup D was originally introduced in the form of more than one haplotype, as the mutational steps that separate them would have occurred in Asia, not in the Americas (Kemp et al. 2007).

21. See, for example, the necessary measures taken by Kemp and colleagues (2007) to compare D-loop sequences collected over a 14-year period.

22. From the results of the Little Salt Springs study (Pääbo et al. 1988), it is clear which haplogroup the young woman did not belong to rather than which haplogroup she did. None of the Windover sequences produced by Hauswirth and colleagues (1994) resemble those of Native Americans or those derived from Asia; instead they most closely resemble European mtDNA types, probably as a result of modern DNA contamination.

23. The possibility that X6 and X7 were the result of laboratory error is suggested here because of the overall poor quality of sequence data presented in the research report, which was scientifically scrutinized by Bandelt and colleagues (2002).

24. Not to be confused with X6 and X7 reported by Easton and colleagues (1996).

25. Prior to the Malhi and Smith (2002) study, Stone and Stoneking (1998) noted that one individual from the ancient Norris Farms Oneota had a HVRI sequence that probably belonged to haplogroup X. However, they did not confirm this by screening the individual's mtDNA genome for the coding region mutations that define haplogroup X.

26. The basic HVRI motif of the Cayapa haplotype is 16223(T), 16241(G), 16301(T), 16342(C), and 16362(C), relative to the reference sequence (Anderson et al. 1981; Andrews et al. 1999).

27. For the reasons discussed in note 16.

28. There were only a few studies between 2007 and the late 1980s and early 1990s that went beyond simply screening the haplogroup defining markers and/or sequencing variable portions and lengths of the D-loop (e.g., Bandelt et al. 2003; Derbeneva et al. 2002; Silva Jr. et al. 2002).

29. The Beringian Incubation Model was inspired by the works of Emöke J. E. Szathmary, as cited by Bonatto and Salzano (1997).

30. The status of C4c as an additional founder lineage is strengthened by its uniqueness to those C4 haplotypes found in Asia (Tamm et al. 2007). It has recently been detected among the Shuswap of British Columbia (Malhi et al. 2010).

31. Tamm and colleagues (2007) published on September 5, Kitchen and colleagues (2008) on February 13, Fagundes and colleagues (2008) on February 28, and Achilli and colleagues (2008) on March 12. It was a truly amazing occurrence that all four of these research teams published virtually simultaneously on a single scenario for the peopling of the Americas.

32. These data consisted of 77 coding region sequences.

33. Fagundes, Kanitz, and Bonatto (2008) found that the use of admixed individuals by Kitchen and colleagues (2008) had skewed their analysis. In this case, Mulligan and colleagues (2008) removed these data and reconducted the Bayesian skyline analysis.

34. While the scenario proposed by Fagundes and colleagues (2008) is similar to that of Kitchen and colleagues (2008), the dates do conflict. The likely sources of these discrepancies are the different data sets and priors used in their respective analyses.

35. Representing the mitochondrial genome sans the D-loop sequence.

36. At first glance, it may appear that Achilli and colleagues (2008) dismiss C4c, D4h3, and X2a as founder lineages because they are not mentioned in the abstract as "successful" founders. However, upon closer inspection, their study focused on exploring diversity in the more common haplogroups. The corresponding author of the study, Antonio Torroni, confirmed that their research team consider C4c, D4h3, and X2a "successful" founders, pointing out that those haplotypes highlighted in red in their Figure 1 are all considered founders (personal communication, August 30, 2008).

37. Achilli and colleagues (2008) estimated an age of 10,900 ± 2900 B.P. for C1d, which is similar to the 9500 ± 3400 B.P. age estimated by Tamm and colleagues (2007).

38. At the 2008 Society for American Archaeology conference in Vancouver, British Columbia, many archaeologists were not swayed by these recent molecular studies and cautioned that there is no evidence for this occupation of Beringia. Notwithstanding much of Beringia today is under water, "Beringia" as part of the "Beringian Incubation Model" need not mean Beringia proper. The molecular data suggest only that the proto–Native American population that resided somewhere in Asia *must* have been isolated. To geneticists, Beringia seemed too perfect a homeland.

39. Evidence from the nuclear genome will not be discussed in the same detail as was the mtDNA evidence.

40. Microsatellites, also known as short tandem repeats (STRs), are genetic markers consisting of repeated short sequences, such as GATA/GATA/GATA. Alternate alleles, or forms, of these markers are recorded by their lengths and/or repeat number, which ultimately dictate their length. Microsatellites exhibit numerous alleles and are, therefore, useful for identification purposes, as it is unlikely that two humans will have the same repeat profiles over a number of these markers. STRs are the basis for DNA profiling.

41. Human Genome Diversity Project-Centre d'Etude du Polymorphisme Humain

42. D9S1120 stands for DNA 9th-Chromosome Segment 1120.

43. Aleut, Inuit, Apache, Dogrib, Cherokee, Chippewa, Huichol, Mixtec, Northern Paiute, Sioux, Seri, Jemez, and Creek.

44. Schroeder and colleagues (2007) sought to test whether the allele crosscut the linguistic divisions proposed by Greenberg and colleagues (1986). See note 13 for a discussion about *Amerind*.

45. Chukchi, Koryaks, Evenk, "Southern Altai," "Northern Altai," Altai Kazakh, and Mongolians.

46. The ABO blood groups are coded for by alternate alleles at a genetic locus on the ninth chromosome.

47. In this section of the paper, we report dates as they are found in the original publications. That is, we have not done our own conversions from radiocarbon years to calendar years. In this case, for example, we discuss the On Your Knees Cave remains that date to 10,300 B.P. (Kemp et al. 2007) prior to those of the Wizards Beach remains that date to 9200 ± 60 [14]C B.P. (Kaestle and Smith 2001), even though in calendar years the latter may be of equivalent or greater antiquity.

48. The Wizards Beach individual was identified as Museum ID Ahur 2023 (Kaestle and Smith 2001:Table 3).

49. For a more detailed discussion of genetic evidence available by 2001 for the Asian origin(s) of Native Americans, see Schurr (2004b).

References

Achilli, Alessandro, Ugo A. Perego, Claudio M. Bravi, Michael D. Coble, Qing-Peng Kong, Scott R. Woodward, Antonio Salas, Antonio Torroni, and Hans-Jürgen Bandelt
    2008   The Phylogeny of the Four Pan-American mtDNA Haplogroups: Implications for Evolutionary and Disease Studies. PLoS ONE 3(3):e1764. doi:10.1371/journal.pone.0001764.

Alves-Silva, Juliana, Magda da Silva Santos, Pedro E. M. Guimarães, Alessandro C. S. Ferreira, Hans-Jürgen Bandelt, Sérgio D. J. Pena, and Vania Ferreira Prado
    2000   The Ancestry of Brazilian mtDNA Lineages. *American Journal of Human Genetics* 67:444–461.

Anderson, S., A. T. Bankier, B. G. Barrel, M. H. L. De Bruijn, A. R. Coulson, J. Drouin, I. C. Eperon, D. P. Nierlich, B. A. Roe, F. Sanger, P. H. Schreier, A. J. H. Smith, R. Staden, and I. G. Young
    1981   Sequence and Organization of the Human Mitochondrial Genome. *Nature* 290: 457–465.

Andrews, Richard M., Iwona Kubacka, Patrick F. Chinnery, Robert N. Lightowlers, Douglass M. Turnbull, and Neil Howell
    1999   Reanalysis and Revision of the Cambridge Reference Sequence for Human Mitochondrial DNA. *Nature Genetics* 23:147.

Bandelt, Hans-Jürgen
    2008   Clock Debate: When Times Are A-changin': Time Dependency of Molecular Rate Estimates: Tempest in a Teacup. *Heredity* 100:1–2.

Bandelt, Hans-Jürgen, Corinna Herrnstadt, Yong-Gang Yao, Qing-Pong Kong, Toomas Kivisild, Chiara Rengo, Rosaria Scozzari, M. Richards, Richard Villems, Vincent Macaulay, Neil Howell, Antonio Torroni and Ya-Ping Zhang
    2003   Identification of Native American Founder mtDNAs Through the Analysis of Complete mtDNA Sequences: Some Caveats. *Annals of Human Genetics* 67:512–524.

Bandelt, Hans-Jürgen, Lluís Quintana-Murci, Antonio Salas, and Vincent Macaulay
    2002   The Fingerprint of Phantom Mutations in Mitochondrial DNA Data. *American Journal of Human Genetics* 71:1150–60.

Blanc, Hugues, Kuang-Ho Chen, Melanie A. D'Amore, and Douglas C. Wallace
    1983   Amino Acid Change Associated With the Major Polymorphic Hinc II Site of Oriental and Caucasian Mitochondrial DNAs. *American Journal of Human Genetics* 35:167–76.

Bolnick, Deborah A., Daniel I. Bolnick, and David Glenn Smith
    2006   Asymmetric Male and Female Genetic Histories among Native Americans from Eastern North America. *Molecular Biology and Evolution* 23:2161–2174.

Bolnick, Deborah A., Beth A. Shook, Lyle Campbell, and Ives Goddard
    2004   Problematic Use of Greenberg's Linguistic Classification of the Americas in Studies of Native American Genetic Variation. *American Journal of Human Genetics* 75:519–522.

Bolnick, Deborah. A., and David. G. Smith
    2003   Unexpected Patterns of Mitochondrial DNA Variation among Native Ameri-
            cans from the Southeastern United States. *American Journal of Physical Anthropology*
            122:336-354.
Bonatto, Sandro L., and Francisco M. Salzano
    1997   A Single and Early Migration for the Peopling of the Americas Supported by
            Mitochondrial DNA Sequence Data. *Proceedings of the National Academy of Sciences of
            the United States of America* 94:1866–1871.
Bortolini, Maria C., Francisco M. Salzano, Mark G. Thomas, S. Stuart, S. P. Nasanen,
C. H. Bau, M. H. Hutz, Z. Layrisse, Maria Luiza Petzl-Erler, Luiza T. Tsuneto,
Kim Hill, Ana M. Hurtado, D. Castro-de-Guerra, M. M. Torres, H. Groot, R. Michalski,
P. Nymadawa, Gabriel Bedoya, N. Bradman, Damian Labuda, and Andres Ruiz-Linares
    2003   Y-chromosome Evidence for Differing Ancient Demographic Histories in the
            Americas. *American Journal of Human Genetics* 73:524–539.
Bourgeois, S., V. Yotova, Sijia Wang, S. Bourtoumieu, C. Moreau, R. Michalski,
J. P. Moisan, Kim Hill, Ana M. Hurtado, Andres Ruiz-Linares, and Damian Labuda
    2009   X-Chromosome Lineages and the Settlement of the Americas. *American Journal
            of Physical Anthropology* 140:417–428.
Brown, Michael D., Seyed H. Hosseini, Antonio Torroni, Hans-Jürgen Bandelt, Jon C.
Allen, Theodore G. Schurr, Rosaria Scozzari, Fulvio Cruciani, and Douglas C. Wallace
    1998   mtDNA Haplogroup X: An Ancient Link Between Europe/Western Asia and
            North America? *American Journal of Human Genetics* 63:1852–1861.
Brown, Wesley M., Matthew George Jr., and Allan C. Wilson
    1979   Rapid Evolution of Animal Mitochondrial DNA. *Proceedings of the National
            Academy of Sciences of the United States of America* 74:1967–1971.
Budowle, Bruce, Marc W. Allard, Constance L. Fisher, Alice R. Isenberg, Keith L.
Monson, John E. B. Stewart, Mark R. Wilson, and Kevin W. P. Miller
    2002   HVI and HVII Mitochondrial DNA Data in Apaches and Navajos. *International
            Journal of Legal Medicine* 116:212–215.
Campbell, Lyle
    1997   *American Indian Languages*. Oxford University Press, New York.
Cann, Rebecca L., Mark Stoneking, and Allan C. Wilson
    1987   Mitochondrial DNA and Human Evolution. *Nature* 325:31–36.
Carlyle, Shawn W., Ryan L. Parr, M. Geoffrey Hayes, and Dennis H. O'Rourke
    2000   Context of Maternal Lineages in the Greater Southwest. *American Journal of
            Physical Anthropology* 113:85–101.
Cavalli-Sforza, L. Luca, Paolo Menozzi, and Alberto Piazza
    1994   *The History and Geography of Human Genes*. Abridged paperback edition. Prince-
            ton University Press, Princeton, New Jersey.
Crawford, Michael H.
    1998   *The Origins of Native Americans: Evidence from Anthropological Genetics*. Cam-
            bridge University Press, Cambridge.
Cybulski, Jerome S., Harold Harry, Alan D. McMillan, and Scott Cousins
    2004   The China Lake and Big Bar Projects: Community Based Research in Physical
            Anthropology and Mortuary Archaeology. Ms. 4503, Library and Archives, Cana-
            dian Museum of Civilization, Gatineau.
De la Cruz, Isabel, Angelica González-Oliver, Brian M. Kemp, Juan A. Román,
David G. Smith, and Alfonso Torre-Blanco
    2008   Sex Identification of Infants Sacrificed to the Ancient Aztec Rain Gods in Tlate-
            lolco. *Current Anthropology* 49:519–526.

Derbeneva, Olga A., Rem I. Sukernik, Natalia V. Volodko, Seyed H. Hosseini,
Marie T. Lott, and Douglas C. Wallace
  2002   Analysis of Mitochondrial DNA Diversity in the Aleuts of the Commander
          Islands and Its Implictions for the Genetic History of Beringia. *American Journal of
          Human Genetics* 71:415–421.
Derenko, Miroslava V., Tomasz Grzybowski, Boris A. Malyarchuk, Jakub Czarny,
Danuta Miscicka-Sliwka, and Ilia A. Zakharov
  2001   The Presence of Mitochondrial Haplogroup X in Altaians from South Siberia.
          American Journal of Human Genetics 69:237–241.
Dillehay, Tom D.
  1999   Late Pleistocence Cultures of South America. *Evolutionary Anthropology* 7:206–
          216.
Dornelles, Claudia L, Sandro L. Bonatto, Loreta B. de Freitas, and Francisco M. Salzano
  2005   Is Haplogroup X Present in Extant South American Indians? *American Journal
          of Physical Anthropology* 127:439–448.
Drummond, Alexei J., Andrew Rambaut, Beth Shapiro, and Oliver G. Pybus
  2005   Bayesian Coalescent Inference of Past Population Dynamics from Molecular
          Sequences. *Molecular Biology and Evolution* 22:1185–1192.
Easton, Ruth D., D. Andrew Merriwether, Douglas E. Crews, and Robert E. Ferrell
  1996   mtDNA Variation in the Yanomami: Evidence for Additional New World Found-
          ing Lineages. *American Journal of Human Genetics* 59:213–225.
Endicott, Phillip, and Simon Y. Ho
  2008   A Bayesian Evaluation of Human Mitochondrial Substitution Rates. *American
          Journal of Human Genetics* 82:895–902.
Eshleman, Jason A.
  2002   Mitochondrial DNA and Prehistoric Population Movements in Western North
          America. Ph.D. dissertation, Department of Anthropology, University of California,
          Davis.
Eshleman, Jason A., Ripan S. Malhi, John R. Johnson, Frederika A. Kaestle, Joseph
Lorenz, and David Glenn Smith
  2004   Mitochondrial DNA and Prehistoric Settlements: Native Migrations on the
          Western Edge of North America. *Human Biology* 76:55–75.
Eshleman, Jason A., Ripan S. Malhi, and David Glenn Smith
  2003   Mitochondrial DNA Studies of Native Americans: Conceptions and Miscon-
          ceptions of the Population Prehistory of the Americas. *Evolutionary Anthropology*
          12:7–18.
Estrada-Mena, Benito, Javier Estrada, Raúl Ulloa-Arvizu, Miriam Guido, Rocío Méndez,
Ramón Coral, Thelma Canto, Julio Granados, Rodrigo Rubí-Castellanos, Héctor Rangel-
Villalobos, and Alejandro García-Carrancá
  2010   Blood Group O Alleles in Native Americans: Implications in the Peopling of
          the Americas. *American Journal of Physical Anthropology* 142:85–94.
Fagundes, Nelson J., Ricardo Kanitz, and Sandro L. Bonatto
  2008   A Reevaluation of the Native American mtDNA Genome Diversity and Its
          Bearing on the Models of Early Colonization of Beringia. PLoS ONE 3(9):e3157,
          doi:10.1371/journal.pone.0003157.
Fagundes, Nelson J. R., Ricardo Kanitz, Roberta Eckert, Ana C. S. Valls,
Mauricio R. Bogo, Francisco M. Salzano, David Glenn Smith, Wilson A. Silva Jr.,
Marco A. Zago, Andrea K. Ribeiro-dos-Santos, Sidney E. B. Santos, Maria Luiza
Petzl-Erler, and Sandro L. Bonatto
  2008   Mitochondrial Population Genomics Supports a Single Pre-Clovis Origin with

a Coastal Route for the Peopling of the Americas. *American Journal of Human Genetics* 82:583–592.

Fewkes, J. Walter, Aleš Hrdlička, William H. Dall, James W. Gidley, Austin Hobart Clark, William H. Holmes, Alice C. Fletcher, Walter Hough, Stansbury Hagar, Paul Bartsch, Alexander F. Chamberlain, and Roland B. Dixon
    1912   The Problems of the Unity or Plurality and the Probable Place of Origin of the American Aborigines. *American Anthropologist* 14:1–59.

Fiedel, Stuart J.
    2000   The Peopling of the New World: Present Evidence, New Theories, and Future Directions. *Journal of Archaeological Research* 8:39–103.

Forster, Peter
    2004   Ice Ages and the Mitochondrial DNA Chronology of Human Dispersals: A Review. *Philosophical Transactions of the Royal Society of London B: Biological Sciences* 359:255–264.

Forster, Peter, Rosalind Harding, Antonio Torroni, and Hans-Jürgen Bandelt
    1996   Origin and Evolution of Native American mDNA Variation: A Reappraisal. *American Journal of Human Genetics* 59:935–945.

Gilbert, M. Thomas P., Dennis L. Jenkins, Anders Götherström, Nuria Naveran, Juan J. Sanchez, Michael Hofreiter, Philip Francis Thomsen, Jonas Binladen, Thomas F. G. Higham, Robert M. Yohe II, Robert Parr, Linda Scott Cummings, and Eske Willerslev
    2008   DNA from Pre-Clovis Human Coprolites in Oregon, North America. *Science* 320:786–789.

Gilbert, M. Thomas P., Toomas Kivisild, Bjarne Grønnow, Pernille K. Andersen, Ene Metspalu, Maere Reidla, Erika Tamm, Erik Axelsson, Anders Götherström, Paula F. Campos, Morten Rasmussen, Mait Metspalu, Thomas F. G. Higham, Jean-Luc Schwenninger, Roger Nathan, Cees-Jan De Hoog, Anders Koch, Lone Nukaaraq Møller, Claus Andreasen, Morten Meldgaard, Richard Villems, Christian Bendixen, and Eske Willerslev
    2008   Paleo-Eskimo mtDNA Genome Reveals Matrilineal Discontinuity in Greenland. *Science* 320:1787–1789.

Giles, Richard E., Hugues Blanc, Howard M. Cann, and Douglas C. Wallace
    1980   Maternal Inheritance of Human Mitochondrial DNA. *Proceedings of the National Academy of Sciences of the United States of America* 77:6715–6719.

Ginther, C., D. Corach, G. A. Penacino, J. A. Rey, F. R. Carnese, M. H. Hutz, A. Anderson, J. Just, Francisco M. Salzano, and M. C. King
    1993   Genetic Variation among the Mapuche Indians from the Patagonian Region of Argentina: Mitochondrial DNA Sequence Variation and Allele Frequencies of Several Nuclear Genes. *Exs* 67:211–219.

Goebel, Ted, Michael R. Waters, and Dennis H. O'Rourke
    2008   The Late Pleistocene Dispersal of Modern Humans in the Americas. *Science* 319:1497–1502.

Gonzalez-Jose, Rolando, Maria C. Bortolini, F. R. Santos, and Sandro L. Bonatto
    2008   The Peopling of America: Craniofacial Shape Variation on a Continental Scale and Its Interpretation from an Interdisciplinary View. *American Journal of Physical Anthropology* 137:175–187.

González-Oliver, Angelica, Lourdes Marquez-Morfin, Jose C. Jimenez, and Alfonso Torre-Blanco
    2001   Founding Amerindian Mitochondrial DNA Lineages in Ancient Maya from Xcaret, Quintana Roo. *American Journal of Physical Anthropology* 116:230–235.

Green, Lance D., James N. Derr, and Alec Knight
2000   mtDNA Affinities of the Peoples of North-Central Mexico. *American Journal of Human Genetics* 66:989–998.
Greenberg, Joseph H., Christy G. Turner II, and Stephan L. Zegura
1986   The Settlement of the Americas: A Comparison of the Linguistic, Dental, and Genetic Evidence. *Current Anthropology* 27:477–497.
Halverson, Melissa S., and Deborah A. Bolnick
2008   An Ancient DNA Test of a Founder Effect in Native American ABO Blood Group Frequencies. *American Journal of Physical Anthropology* 137:342–347.
Hauswirth, William W., Cynthia D. Dickel, D. J. Rowold, and A. Hauswirth
1994   Inter- and Intrapopulation Studies of Ancient Humans. *Experientia* 50:585–591.
Haynes, Gary, David G. Anderson, Reid Ferring, Stuart J. Fiedel, Donald K. Grayson, C. Vance Haynes Jr., Vance T. Holliday, Bruce B. Huckell, Marcel Kornfeld, David J. Meltzer, Julie Morrow, Todd Surovell, Nicole M. Waguespack, Peter Wigand, and Robert M. Yohe II
2007   Comment on "Redefining the Age of Clovis: Implications for the Peopling of the Americas." *Science* 317:320; author reply 320.
Hertzberg, M., K. N. Mickleson, S. W. Serjeantson, J. F. Prior, and R. J. Trent
1989   An Asian-Specific 9-bp Deletion of Mitochondrial DNA Is Frequently Found in Polynesians. *American Journal of Human Genetics* 44:504–510.
Ho, Simon Y., and Phillip Endicott
2008   The Crucial Role of Calibration in Molecular Date Estimates for the Peopling of the Americas. *American Journal of Human Genetics* 83:142–146; author reply 146–147.
Ho, Simon Y. W., Beth Shapiro, Matthew J. Phillips, Alan Cooper, and Alexei J. Drummond
2007   Evidence for Time Dependency of Molecular Rate Estimates. *Systematic Biology* 56:515–522.
Howell, Neil, Corrina Howell, and Joanna L. Elson
2008   Molecular Clock Debate: Time Dependency of Molecular Rate Estimates for mtDNA: This Is Not the Time for Wishful Thinking. *Heredity* 101:107–108.
Hrdlička, Aleš
1912   The Bearing of Physical Anthropology on the Problems Under Consideration. *American Anthropologist* 14:8–12.
Hunley, Keith, and Jeffrey C. Long
2005   Gene Flow Across Linguistic Boundaries in Native North American Populations. *Proceedings of the National Academy of Sciences of the United States of America* 102:1312–1317.
Hunley, Keith L., J. E. Spence, and D. Andrew Merriwether
2008   The Impact of Group Fissions on Genetic Structure in Native South America and Implications for Human Evolution. *American Journal of Physical Anthropology* 135:195–205.
Huoponen, Kirsi, Antonio Torroni, P. R. Wickman, Daniele Sellitto, D. S. Gurley, Rosaria Scozzari, and Douglas. C. Wallace
1997   Mitochondrial DNA and Y Chromosome Specific Polymorphisms in the Seminole Tribe of Florida. *European Journal of Human Genetics* 5:25–34.
Ingman, Max, Henrik Kaessmann, Svante Pääbo, and Ulf Gyllensten
2000   Mitochondrial Genome Variation and the Origin of Modern Humans. *Nature* 408:708–713.
Johnson, John R., and Joseph G. Lorenz
2006   Genetics, Linguistics, and Prehistoric Migrations: An Analysis of California In-

dian Mitochondrial DNA Lineages. *Journal of California and Great Basin Anthropology* 26(1):31–62.

Johnson, John R., Thomas W. Stafford, Henry O. Aije, and Don P. Morrise
    2002   Arlington Springs Revisited. *Proceedings of the Fifth California Islands Symposium*, edited by David R. Browne, Kathryn L. Mitchell, and Henry W. Chaney, pp. 541–545. Santa Barbara Museum of Natural History.

Kaestle, Frederika Ann
    1998   Molecular Evidence for Prehistoric Native American Population Movement: The Numic Expansion. Ph.D. dissertation, Department of Anthropology, University of California, Davis.

Kaestle, Frederika A., and David Glenn Smith
    2001   Ancient Mitochondrial DNA Evidence for Prehistoric Population Movement: The Numic Expansion. *American Journal of Physical Anthropology* 115:1–12.

Karafet, Tatiana, L. Xu, R. Du, W. Wang, S. Feng, R. S. Wells, Alan J. Redd, Stephen L. Zegura, and Michael F. Hammer
    2001   Paternal Population History of East Asia: Sources, Patterns, and Microevolutionary Processes. *American Journal of Human Genetics* 69:615–628.

Karafet, Tatiana M., Stephen L. Zegura, Olga Posukh, Ludmila Osipova, A. Bergen, Jeffery Long, D. Goldman, William Klitz, Shinji Harihara, Peter de Knijff, Victor Wiebe, R. C. Griffiths, Alan R. Templeton, and Michael F. Hammer
    1999   Ancestral Asian Source(s) of New World Y-Chromosome Founder Haplotypes. *American Journal of Human Genetics* 64:817–831.

Karafet, Tatiana, Stephen L. Zegura, Jennifer Vuturo-Brady, Olga Posukh, Ludmila Osipova, Victor Wiebe, Francine Romero, Jeffery C. Long, Shinji Harihara, Feng Jin, Bumbein Dashyam, Tudevdagva Gerelsaikhan, Keiichi Omoto, and Michael F. Hammer
    1997   Y Chromosome Markers and Trans-Bering Strait Dispersals. *American Journal of Physical Anthropology* 102:301–314.

Kemp, Brian M., Angélica González-Oliver, Ripan S. Malhi, Cara Monroe, Kari Britt Schroeder, John McDonough, Gillian Rhett, Andres Resendéz, Rosenda I. Peñaloza-Espinosa, Leonor Buentello-Malo, Clara Gorodesky, and David Glenn Smith
    2010   Evaluating the Farming/Language Dispersal Hypothesis with Mitochondrial and Y-chromosomal DNA Variation in the Southwest and Mesoamerica. *Proceedings of the National Academy of Sciences of the United States of America* 107:6759–6764.

Kemp, Brian M., Ripan S. Malhi, John McDonough, Deborah A. Bolnick, Jason A. Eshleman, Olga Rickards, Cristina Martinez-Labarga, John R. Johnson, Joesph G. Lorenz, E. James Dixon, Terence E. Fifield, Timothy H. Heaton, Rosita Worl, and David Glenn Smith
    2007   Genetic Analysis of Early Holocene Skeletal Remains from Alaska and Its Implication for the Timing of the Peopling of the Americas. *American Journal of Physical Anthropology* 132:605–621.

Kemp, Brian M., Andrés Reséndez, Juan Alberto Román Berrelleza, Ripan S. Malhi, and David Glenn Smith
    2005   An Analysis of Ancient Aztec mtDNA from Tlatelolco: Pre-Columbian Relations and the Spread of Uto-Aztecan. In *Biomolecular Archaeology: Genetic Approaches to the Past*, edited by David M. Reed, pp. 22–46. Occasional Paper No. 32. Center for Archaeological Investigations, Southern Illinois University Carbondale.

Kitchen, Andrew, Michael M. Miyamoto, and Connie J. Mulligan
    2008   A Three-Stage Colonization Model for the Peopling of the Americas. PLoS ONE 3(2):e1596. doi:10.1371/journal.pone.0001596.

Kivisild, Toomas, Peidong Shen, Dennis P. Wall, Bao Do, Raphael Sung, Karen Davis, Giuseppe Passarino, Peter A. Underhill, Curt Scharfe, Antonio Torroni, Rosaria Scozzari, David Modiano, Alfredo Coppa, Peter de Knijff, Marcus Feldman, Luca L. Cavalli-Sforza, and Peter J. Oefner
>   2006   The Role of Selection in the Evolution of Human Mitochondrial Genomes. *Genetics* 172:373–387.

Kolman, Connie J., Nyamkhishig Sambuughin, and Eldredge Bermingham
>   1996   Mitochondrial DNA Analysis of Mongolian Populations and Implications for the Origin of New World Founders. *Genetics* 142:1321–1334.

LeBlanc, Steven A., Lori Kreisman, Brian M. Kemp, Shawn W. Carlyle, Anna Dhody, Francis Smiley, and Thomas Benjamin
>   2007   Quids and Aprons: Ancient DNA from Artifacts from the American Southwest. *Journal of Field Archaeology* 32:161–175.

Lell, Jeffrey T., Michael D. Brown, Theodore G. Schurr, Rem I. Sukernik, Yelena B. Starikovskaya, Antonio Torroni, Lorna G. Moore, Gary M. Troup, and Douglas C. Wallace
>   1997   Y Chromosome Polymorphisms in Native American and Siberian Populations: Identification of Native American Y Chromosome Haplotypes. *Human Genetics* 100:536–543.

Lell, Jeffrey T., Rem I. Sukernik, Yelena B. Starikovskaya, Bing Su, Li Jin, Theodore G. Schurr, Peter A. Underhill, and Douglas C. Wallace
>   2002   The Dual Origin and Siberian Affinities of Native American Y Chromosomes. *American Journal of Human Genetics* 70:192–206.

Lorenz, Joseph G., and David Glenn Smith
>   1996   Distribution of Four Founding mtDNA Haplogroups among Native North Americans. *American Journal of Physical Anthropology* 101:307–323.

Malhi, Ripan S.
>   2001   Investigating Prehistoric Population Movements in North America with Ancient and Modern DNA. Ph.D. dissertation, Department of Anthropology, University of California, Davis.

Malhi, Ripan S., Katherine E. Breece, Beth A. Schultz Shook, Frederika A. Kaestle, James C. Chatters, Steven Hackenberger, and David Glenn Smith
>   2004   Patterns of mtDNA Diversity in Northwestern North America. *Human Biology* 76:33–54.

Malhi, Ripan S., Jerome S. Cybulski, Raul Y. Tito, Jesse Johnson, Harold Harry, and Carrie Dan
>   2010   Brief Communication: Mitochondrial Haplotype C4c Confirmed as a Founding Genome in the Americas. *American Journal of Physical Anthropology* 141:494–497.

Malhi, Ripan S., Jason A. Eshleman, Jonathan A. Greenberg, Deborah A. Weiss, Beth A. Schultz Shook, Frederika A. Kaestle, Joseph G. Lorenz, Brian M. Kemp, John R. Johnson, and David G. Smith
>   2002   The Structure of Diversity Within the New World Mitochondrial DNA Haplogroups: Implications for the Prehistory of North America. *American Journal of Human Genetics* 70:905–919.

Malhi, Ripan Singh, Angelica Gonzalez-Oliver, Kari Britt Schroeder, Brian M. Kemp, Jonathan A. Greenberg, Solomon Z. Dobrowski, David Glenn Smith, Andres Resendéz, Tatiana Karafet, Michael Hammer, Stephen Zegura, and Tatiana Brovko
>   2008   Distribution of Y Chromosomes among Native North Americans: A Study of Athapaskan Population History. *American Journal of Physical Anthropology* 137:412–424.

Malhi, Ripan S., Brian M. Kemp, Jason A. Eshleman, Jerome Cybulski, David Glenn
Smith, Scott Cousins, and Harold Harry
  2007   Haplogroup M Discovered in Prehistoric North America. *Journal of Archaeological Science* 34:642–648.
Malhi, Ripan S., Holly M. Mortenson, Jason A. Eshleman, Brian M. Kemp, Joseph G.
Lorenz, Frederika A. Kaestle, John R. Johnson, Clara Gorodezky, and David Glenn Smith
  2003   Native American mtDNA Prehistory in the American Southwest. *American Journal of Physical Anthropology* 120:108–124.
Malhi, Ripan S., and David Glenn Smith
  2002   Brief Communication: Haplogroup X Confirmed in Prehistoric North America. *American Journal of Physical Anthropology* 119:84–86.
Mandryk, Carole A. S., Heiner Josenhans, Daryl W. Fedje, and Rolf W. Mathewes
  2001   Late Quaternary Paleoenvironments of Northwestern North America: Implications for Inland Versus Coastal Migration Routes. *Quaternary Science Review* 20:301–314.
Merriwether, D. Andrew
  1999   Freezer Anthropology: New Uses for Old Blood. *Philosophical Transactions of the Royal Society of London B: Biological Sciences* 354:121–129.
Merriwether, D. Andrew, Andrew G. Clark, Scott W. Ballinger, Theodore G. Schurr,
Himla Soodyall, Trefor Jenkins, Stephen T. Sherry, and Douglas C. Wallace
  1991   The Structure of Human Mitochondrial DNA Variation. *Journal of Molecular Evolution* 33:543–555.
Merriwether, D. Andrew, and Robert E. Ferrell
  1996   The Four Founding Lineage Hypothesis for the New World: A Critical Reevaluation. *Molecular Phylogenetics and Evolution* 5:241–246.
Merriwether, D. Andrew, William W. Hall, Anders Vahlne, and Robert E. Ferrell
  1996   mtDNA Variation Indicates Mongolia May Have Been the Source for the Founding Population for the New World. *American Journal of Human Genetics* 59:204–212.
Merriwether, D. Andrew, Brian M. Kemp, Douglas E. Crews, and James V. Neel
  2000   Gene Flow and Genetic Variation in the Yanomama as Revealed by Mitochondrial DNA. In *America Past, America Present: Genes and Languages in the Americas and Beyond*, edited by Colin Renfrew, pp. 89–124. McDonald Institute for Archaeological Research, Cambridge, England.
Merriwether, D. Andrew, Francisco Rothhammer, and Robert E. Ferrell
  1994   Genetic Variation in the New World: Ancient Teeth, Bone, and Tissue as Sources of DNA. *Cellular and Molecular Life Sciences* 50:592–601.
  1995   Distribution of the Four Founding Lineage Haplotypes in Native Americans Suggests a Single Wave of Migration for the New World. *American Journal of Physical Anthropology* 98:411–430.
Mishmar, Dan, Eduardo Ruiz-Pesini, Pawel Golik, Vincent Macaulay, Andrew G. Clarke,
Seyed Hosseini, Martin Brandon, Kirk Easley, Estella Chen, Michael D. Brown,
Rem I. Sukernik, Antonel Olckers, and Douglas C. Wallace
  2003   Natural Selection Shaped Regional mtDNA Variation in Humans. *Proceedings of the National Academy of Sciences of the United States of America* 100:171–176.
Monroe, Cara, Brian M. Kemp, and David Glenn Smith
  2006   Mitochondrial DNA Variation of Yuman Speaking Populations. Poster presented at the Languages and Genes conference, University of California, Santa Barbara.
Moraga, Mauricio L., P. Rocco, J. F. Miquel, F. Nervi, Elena Llop, R. Chakraborty,
Francisco Rothhammer, and P. Carvallo
  2000   Mitochondrial DNA Polymorphisms in Chilean Aboriginal Populations: Im-

plications for the Peopling of the Southern Cone of the Continent. *American Journal of Physical Anthropology* 113:19–29.

Mourant, Arthur E.
   1976   *The Distribution of the Human Blood Groups and Other Polymorphism.* Oxford University Press, London.

Mulligan, Connie J., Keith Hunley, Suzanne Cole, and Jeffrey C. Long
   2004   Population Genetics, History, and Health Patterns in Native Americans. *Annual Review of Genomics and Human Genetics* 5:295–315.

Mulligan, Connie J., Andrew Kitchen, and Michael M. Miyamoto
   2008   Updated Three-Stage Model for the Peopling of the Americas. PLoS ONE 3(9):e3199. doi:10.1371/journal.pone.0003199.

Neves, Walter A., and Mark Hubbe
   2005   Cranial Morphology of Early Americans from Lagoa Santa, Brazil: Implications for the Settlement of the New World. *Proceedings of the National Academy of Sciences of the United States of America* 102:18309–18314.

Nichols, Johanna
   1990   Linguistic Diversity and the First Settlement of the New World. *Language* 66:475–521.

Pääbo, Svante, John A. Gifford, and Allan C. Wilson
   1988   Mitochondrial DNA Sequences from a 7000-Year Old Brain. *Nucleic Acids Research* 16:9775–9788.

Parr, Ryan L., Shawn W. Carlyle, and Dennis H. O'Rourke
   1996   Ancient DNA Analysis of Fremont Amerindians of the Great Salt Lake Wetlands. *American Journal of Physical Anthropology* 99:507–518.

Perego, Ugo A., Alessandro Achilli, Norman Angerhofer, Matteo Accetturo, Maria Pala, Anna Olivieri, Baharak Hooshiar Kashani, Kathleen H. Ritchie, Rosaria Scozzari, Qing-Peng Kong, Natalie M. Myres, Antonio Salas, Ornella Semino, Hans-Jürgen Bandelt, Scott R. Woodward, and Antonio Torroni
   2009   Distinctive Paleo-Indian Migration Routes from Beringia Marked by Two Rare mtDNA Haplogroups. *Current Biology* 19:1–8.

Pitulko, Volodya V., P. A. Nikolsky, E. Y. Girya, A. E. Basilyan, V. E. Tumskoy, S. A. Koulakov, S. N. Astakhov, E. Y. Pavlova, and M. A. Anisimov
   2004   The Yana RHS Site: Humans in the Arctic Before the Last Glacial Maximum. *Science* 303:52–56.

Poinar, Hendrik N., Melanie Kuch, Kristin D. Sobolik, Ian Barnes, Arthur B. Stankiewicz, Tomasz Kuder, W. Geofferey Spaulding, Vaughn M. Bryant, Alan Cooper, and Svante Pääbo
   2001   A Molecular Analysis of Dietary Diversity for Three Archaic Native Americans. *Proceedings of the National Academy of Sciences of the United States of America* 98:4317–4322.

Ramachandran, Sohini, Omkar Deshpande, Charles C. Roseman, Noah A. Rosenberg, Marcus W. Feldman, and L. Luca Cavalli-Sforza
   2005   Support from the Relationship of Genetic and Geographic Distance in Human Populations for a Serial Founder Effect Originating in Africa. *Proceedings of the National Academy of Sciences of the United States of America* 102:15942–15947.

Ray, Nicolas, D. Wegmann, Nelson J. Fagundes, Sijia Wang, Andres Ruiz-Linares, and Laurent Excoffier
   2010   A Statistical Evaluation of Models for the Initial Settlement of the American Continent Emphasizes the Importance of Gene Flow with Asia. *Molecular Biology and Evolution* 27:337–345.

Reidla, Maere, Toomas Kivisild, Ene Metspalu, Katrin Kaldma, Kristiina Tambets, Helle-Viivi Tolk, Jüri Parik, Eva-Liis Loogvali, Miroslava V. Derenko, Boris Malyarchuk, Marina Bermisheva, Sergey Zhadanov, Erwan Pennarun, Marina Gubina, Maria Golubenko, Larisa Damba, Sardana Fedorova, Vladislava Gusar, Elena Grechanina, Ilia Mikerezi, Jean-Paul Moisan, André Chaventre, Elsa Khusnutdinova, Ludmila Osipova, Vadim Stepanov, Mikhail Voevoda, Alessandro Achilli, Chiara Rengo, Olga Rickards, Gian Franco De Stefano, Surinder Papiha, Lars Beckman, Branka Janicijevic, Pavao Rudan, Nicholas Anagnou, Emmanuel Michalodimitrakis, Slawomir Koziel, Esien Usanga, Tarekegn Geberhiwot, Corinna Herrnstadt, Neil Howell, Antonio Torroni, and Richard Villems
    2003   Origin and Diffusion of mtDNA Haplogroup X. *American Journal of Human Genetics* 73:1178–1190.

Rickards, Olga, Cristina Martínez-Labarga, J. Koji Lum, Gian Franco De Stefano, and Rebecca L. Cann
    1999   mtDNA History of the Cayapa Amerinds of Ecuador: Detection of Additional Founding Lineages for the Native American Populations. *American Journal of Human Genetics* 65:519–530.

Rubicz, Rohina, Theodore G. Schurr, Paul L. Babb, and Michael H. Crawford
    2003   Mitochondrial DNA Variation and the Origins of the Aleuts. *Human Biology* 75:809–835.

Salzano, Francisco M., and Sidia M. Callegari-Jacques
    1988   *South American Indians: A Case Study in Evolution*. Clarendon, Oxford.

Santos, Sidney E., Andrea K. Ribeiro-Dos-Santos, D. Meyer, and Marco A. Zago
    1996   Multiple Founder Haplotypes of Mitochondrial DNA in Amerindians Revealed by RFLP and Sequencing. *Annals of Human Genetics* 60:305–319.

Schroeder, Kari B., Mattias Jakobsson, Michael H. Crawford, Theodore G. Schurr, Simina M. Boca, Donald F. Conrad, Raul Y. Tito, Ludmilla P. Osipova, Larissa A. Tarskaia, Sergey I. Zhadanov, Jeffrey D. Wall, Jonathan K. Pritchard, Ripan S. Malhi, David G. Smith, and Noah A. Rosenberg
    2009   Haplotypic Background of a Private Allele at High Frequency in the Americas. *Molecular Biology and Evolution* 26:995–1016.

Schroeder, Kari B., Theodore G. Schurr, Jeffrey C. Long, Noah A. Rosenberg, Michael H. Crawford, Larissa A. Tarskaia, Ludmila P. Osipova, Sergey I. Zhadanov, and David G. Smith
    2007   A Private Allele Ubiquitous in the Americas. *Biology Letters* 3:218–223.

Schurr, Theodore G.
    2000   Mitochondrial DNA Variation in Native Americans and Siberians, and Its Implications for the Peopling of the New World. *American Scientist* 88:246–253.
    2002   A Molecular Anthropological View of the Peopling of the Americas. *Athena Review* 3(2):59–77.
    2004a The Peopling of the New World: Perspectives from Molecular Anthropology. *Annual Review of Anthropology* 33:551–583.
    2004b Molecular Genetic Diversity in Siberians and Native Americans Suggests an Early Colonization of the New World. In *Entering America: Northeast Asia and Beringia Before the Last Glacial Maximum*, edited by David B. Madsen, pp. 187–238. University of Utah Press, Salt Lake City.

Schurr, Theodore G., Scott W. Ballinger, Yik-Yuen Gan, Judith A. Hodge, D. Andrew Merriwether, Dale N. Lawrence, William C. Knowler, Kenneth M. Weiss, and Douglas C. Wallace
    1990   Amerindian Mitochondrial DNAs Have Rare Asian Mutations at High Fre-

quencies, Suggesting They Derived from Four Primary Maternal Lineages. *American Journal of Human Genetics* 46:613–623.

Schurr, Theodore G., Rem I. Sukernik, Yelena B. Starikovskaya, and Douglas C. Wallace
  1999   Mitochondrial DNA Variation in Koryaks and Itel'men: Population Replacement in the Okhotsk Sea-Bering Sea Region During the Neolithic. *American Journal of Physical Anthropology* 108:1–39.

Schurr, Theodore G., and Douglas C. Wallace
  1999   mtDNA Variation in Native Americans and Siberians, and Its Implications for the Peopling of the New World. In *Who Were the First Americans?: Proceedings of the 58th Annual Biology Colloquium*, edited by Robson Bonnichsen, pp. 41–77. Corvallis: Center for the Study of the First Americans, Corvallis, Oregon.

  2002   Mitochondrial DNA Diversity in Southeast Asian Populations. *Human Biology* 74:431–452.

Scozzari, Rosaria, Fulvio Cruciani, Piero Santolamazza, Daniele Sellitto, David E. C.
Cole, Laurence A. Rubin, Damian Labuda, Elisabetta Marini, Valeria Succa,
Guiseppe Vona, and Antonio Torroni
  1997   mtDNA and Y Chromosome-Specific Polymorphisms in Modern Ojibwa: Implications about the Origin of Their Gene Pool. *American Journal of Human Genetics* 60:241–244.

Shields, Gerald F., Andrea M. Schmeichen, Barbara L. Frazier, Alan Redd,
Mikhail I. Voevoda, Judy K. Reed, and Ryk H. Ward
  1993   mtDNA Sequences Suggest a Recent Evolutionary Divergence for Beringian and Northern North American Populations. *American Journal of Human Genetics* 53:549–562.

Silva, Wilson A., Jr., Sandro L. Bonatto, Adriano J. Holanda, Andrea K. Ribeiro-dos-Santos,
Beatriz M. Paixao, Gustavo H. Goldman, Kiyoko Abe-Sandes, Luis Rodriguez-Delfin,
Marcela Barbosa, Maria Luiza Paco-Larson, Maria Luiza Petzl-Erler, Valeria Valente,
Sidney E. B. Santos, and Marco A. Zago
  2002   Mitochondrial Genome Diversity of Native Americans Supports a Single Early Entry of Founder Populations into America. *American Journal of Human Genetics* 71:187–192.

Smith, David Glenn, Joseph Lorenz, Becky K. Rolfs, Robert L. Bettinger, Brian Green,
Jason Eshleman, Beth Schultz, and Ripan Malhi
  2000   Implications of the distribution of Albumin Naskapi and Albumin Mexico for New World Prehistory. *American Journal of Physical Anthropology* 111:557–572.

Smith, David Glenn, Ripan S. Malhi, Jason A. Eshleman, Frederika A. Kaestle, and
Brian M. Kemp
  2005   Mitochondrial DNA Haplogroups of Paleoamericans in North America. In *Paleoamerican Origins: Beyond Clovis*, edited by Robson Bonnichsen, Bradley T. Lepper, Dennis Stanford, and Michael R. Waters, pp. 243–254. Texas A&M University Press, College Station.

Smith, David Glenn, Ripan S. Malhi, Jason Eshleman, Joseph G. Lorenz, and
Frederika A. Kaestle
  1999   Distribution of mtDNA Haplogroup X Among Native North Americans. *American Journal of Physical Anthropology* 110:271–284.

Starikovskaya, Yelena B., Rem I. Sukernik, Theodore G. Schurr, Andres M. Kogelnik, and
Douglas C. Wallace
  1998   mtDNA Diversity in Chukchi and Siberian Eskimos: Implications for the Genetic History of Ancient Beringia and the Peopling of the New World. *American Journal of Human Genetics* 63:1473–1491.

Stone, Anne C., and Mark Stoneking
1993 Ancient DNA from a Pre-Columbian Amerindian Population. *American Journal of Physical Anthropology* 92:463–471.
1996 Genetic Analysis of an 8000 Year-Old Native American Skeleton. *Ancient Biomolecules* 1:83–87.
1998 mtDNA Analysis of a Prehistoric Oneota Population: Implications for the Peopling of the New World. *American Journal of Human Genetics* 62:1153–1170.

Szathmary, Emöke J. E.
1979 Blood Groups of Siberians, Eskimos, Subarctic, and Northwest Coast Indians: The Problem of Origins and Genetic Relationships. In *The First Americans: Origins, Affinities, and Adaptations*, edited by William S. Laughin and Albert B. Harper, pp. 185–209. Gustav Fischer, New York.

Tamm, Erika, Toomas Kivisild, Maere Reidla, Mait Metspalu, David Glenn Smith, Connie J. Mulligan, Claudio M. Bravi, Olga Rickards, Cristina Martinez-Labarga, Elas K. Khusnutdinova, Sardana A. Fedorova, Maria V. Golubenko, Vadim A. Stepanov, Marrina A. Gubina, Sergey I. Zhadanov, Ludmila P. Osipova, Larisa Damba, Mikhail I. Voevoda, Jose E. Dipierri, Richard Villems, and Ripan S. Malhi
2007 Beringian Standstill and Spread of Native American Founders. PloS ONE 2(9):e829. doi:10.1371/journal.pone.0000829.

Tarazona-Santos, Eduardo, and Fabrício R. Santos
2002 The Peopling of the Americas: A Second Major Migration? *American Journal of Human Genetics* 70:1377–1380.

Torroni, Antonio, Yu-Sheng Chen, Ornella Semino, Augusta Silvana, Beneceretti Santachiara, C. Ronald Scott, Marie T. Lott, Marcus Winter, and Douglas C. Wallace
1994 mtDNA and Y-Chromosome Polymorphisms in Four Native American Populations from Southern Mexico. *American Journal of Human Genetics* 54:303–318.

Torroni, Antonio, Theodore G. Schurr, Margaret F. Cabell, Michael D. Brown, James V. Neel, Merethe Larsen, David G. Smith, Carlos M. Vullo, and Douglas C. Wallace
1993 Asian Affinities and Continental Radiation of the Four Founding Native American mtDNAs. *American Journal of Human Genetics* 53:563–590.

Torroni, Antonio, Theodore G. Schurr, Chin-Chuan Yang, Emöke J. E. Szathmary, Robert C. Williams, Moses S. Schanfield, Gary A. Troup, William C. Knowler, Dale N. Lawrence, Kenneth M. Weiss, and Douglas C. Wallace
1992 Native American Mitochondrial DNA Analysis Indicates That the Amerind and the Nadene Populations Were Founded by Two Independent Migrations. *Genetics* 130:153–162.

Toth, Nicholas
1991 The Material Record. In *The First Americans: Search and Research*, edited by Thomas D. Dillehay and David J. Meltzer, pp. 53–76. CRC Press, Boca Raton, Florida.

Underhill, Peter A., Li Jin, Rachel Zemans, Peter J. Oefner, and L. Luca Cavalli-Sforza
1996 A Pre-Columbian Y Chromosome-Specific Transition and Its Implications for Human Evolutionary History. *Proceedings of the National Academy of Sciences of the United States of America* 93:196–200.

Vajda, Edward
2010 A Siberian Link with Na-Dene Languages. *Archaeological Papers of the University of Alaska* 5:75–156.

Wallace, Douglas C.
1994 Mitochondrial DNA Sequence Variation in Human Evolution and Disease. *Proceedings of the National Academy of Sciences of the United States of America* 91:8739–8746.

Wallace, Douglas C., Katherine Garrison, and William C. Knowler
  1985   Dramatic Founder Effects in Amerindian Mitochondrial DNAs. *American Journal of Physical Anthropology* 68:149–156.
Wang, Sijia, Cecil M. Lewis Jr., Mattias Jakobsson, Sohini Ramachandran, Nicolas Ray, Gabriel Bedoya, Winston Rojas, Maria V. Parra, Julio A. Molina, Carla Gallo, Guido Mazzotti, Giovanni Poletti, Kim Hill, Ana M. Hurtado, Damian Labuda, William Klitz, Ramiro Barrantes, Maria Catira Bortolini, Francisco M. Salzano, Maria Luiza Petzl-Erler, Luiza T. Tsuneto, Elena Llop, Francisco Rothhammer, Laurent Excoffier, Marcus W. Feldman, Noah A. Rosenberg, and Andres Ruiz-Linares
  2007   Genetic Variation and Population Structure in Native Americans. PLoS Genet 3(11):e185. doi:10.1371/journal.pgen.0030185.
Ward, Ryk H., Barbara L. Frazier, Kerry Dew-Jager, and Svante Pääbo
  1991   Extensive Mitochondrial Diversity Within a Single Amerindian Tribe. *Proceedings of the National Academy of Sciences of the United States of America* 88: 8720–8724.
Ward, Ryk H., Alan Redd, Diana Valencia, Barbara Frazier, and Svante Pääbo
  1993   Genetic and Linguistic Differentiation in the Americas. *Proceedings of the National Academy of Sciences of the United States of America* 90:10663–10667.
Waters, Michael R., and Thomas W. Stafford Jr.
  2007   Redefining the Age of Clovis: Implications for the Peopling of the Americas. *Science* 315:1122–1126.
Weinstein, Karen J.
  2005   Body Proportions in Ancient Andeans from High and Low Altitudes. *American Journal of Physical Anthropology* 128:569–585.
Wells, Spencer
  2006   *Deep Ancestry: Inside the Genographic Project*. National Geographic, Washington, D.C.
Yao, Yong-Gang, Qing-Peng Kong, Hans-Jürgen Bandelt, Toomas Kivisild, and Ya-Ping Zhang
  2002   Phylogeographic Differentiation of Mitochondrial DNA in Han Chinese. *American Journal of Human Genetics* 70:635–651.
YCC
  2002   A Nomenclature System for the Tree of Human Y-Chromosomal Binary Haplogroups. *Genome Research* 12:339–348.
Zegura, Stephen L., Tatiana M. Karafet, Lev A. Zhivotovsky, and Michael F. Hammer
  2004   High-Resolution SNPs and Microsatellite Haplotypes Point to a Single, Recent Entry of Native American Y Chromosomes into the Americas. *Molecular Biology and Evolution* 21:164–175.
Zhivotovsky, Lev A., Noah A. Rosenberg, and Marcus W. Feldman
  2003   Features of Evolution and Expansion of Modern Humans, Inferred from Genomewide Microsatellite Markers. *American Journal of Human Genetics* 72:1171–1186.
Zlojutro, Mark, Rohina Rubicz, Eric J. Devor, Victor A. Spitsyn, Sergei V. Makarov, Kristin Wilson and Michael H. Crawford
  2006   Genetic Structure of the Aleuts and Circumpolar Populations Based on Mitochondrial DNA Sequences: A Synthesis. *American Journal of Physical Anthropology* 129:446–464.

3. Peopling the Americas via Multiple Migrations from Beringia: Evidence from the Early Holocene of the Columbia Plateau

James C. Chatters

Summary Statement: Explanations for the marked craniofacial differences between Paleo-Americans and later Amerinds typically invoke microevolutionary processes *or* multiple migrations. This paper offers the hypothesis that both explanations may be correct: The difference might be explained by multiple episodes of immigration from post-Pleistocene Beringian populations, who bore the derived characteristics common to both Amerinds and north Asian peoples. Evidence is presented that one such event occurred in northwestern North America between 9200 and 8000 B.P. During that time, radiocarbon-dated tool assemblages show that the Old Cordilleran Tradition moved south from Alaska to dominate the Northwest coast by 8200 B.P. By 8000 B.P., the new tradition had replaced the Western Stemmed Tradition on the Columbia Plateau. Coincident stylistic changes in both lithics and fiber arts indicate that the expansion was not only of ideas but also of the people who bore them. Consistent with this evidence, the remains of Western Stemmed Tradition practitioners exhibit the craniofacial characteristics of Paleo-Americans, whereas Old Cordilleran Tradition practitioners are indistinguishable from local late Holocene Amerinds. This finding, along with the well-documented late Holocene immigration of Athabascans into temperate North America, should alert archaeologists to the possibility that ethnic expansions and contractions contributed to regional and diachronic variability in both culture and human skeletal morphology.

Human Variation in the Americas: The Integration of Archaeology and Biological Anthropology, edited by Benjamin M. Auerbach. Center for Archaeological Investigations, Occasional Paper No. 38. © 2010 by the Board of Trustees, Southern Illinois University. All rights reserved. ISBN 978-0-88104-095-1.

Introduction

Craniofacial morphometric analyses conducted over the past two decades have consistently shown that the earliest inhabitants of the Americas were outside the range of variation for modern Native American (Amerind) and Northeast Asian groups (Brace et al 2001; Gonzalez-Jose et al. 2001, 2003; Jantz and Owsley 2001; Mena et al. 2003; Neves and Pucciarelli 1991; Powell and Neves 1999; Steele and Powell 1992, 1993). The early skeletal population, referred to collectively as Paleo-American or Paleo-Indian, is highly morphologically diverse (Jantz and Owsley 2001; Powell and Neves 1999; Powell 2004), but most individuals share the characteristics of a long, narrow cranium; relatively narrow cheek bones; and forward positioned or prognathous face. This pattern contrasts sharply with the short, round crania and broad, coronally flattened faces common to both Northeast Asians and Amerinds. The Paleo-Americans are, in fact, often morphometrically farther from Amerind groups than any other modern population. They, like many of the other late Pleistocene *Homo sapiens* from Eurasia are usually most similar to modern peoples of the southern Pacific Rim.

Explanations for the extreme difference between Paleo-Americans and later Amerinds fall into two camps. Making the assumption that craniofacial form is under genetic control and can thus be used to trace population-historical relationships through time (Jantz and Owsley 2001; Perez et al. 2007), many scholars see the morphological change as evidence of a demographic replacement. Noting the frequently abrupt temporal discontinuity between the two morphologies toward the end of the early Holocene, they suggest that the generalized Paleo-American population arrived during the terminal Pleistocene in one or more initial peopling events and was replaced or genetically swamped by an early Holocene migration of Northeast Asians (Jantz and Owsley 2001; Gonzalez-Jose et al. 2005; Lahr 1996; Neves and Hubbe 2005; Neves and Pucciarelli 1991; Steele and Powell 1992, 1993).

Genetics seemingly provides no support for the multiple migration hypothesis. Nearly all pre-Columbian inhabitants of the Americas, both ancient and modern, belong to one of five mitochondrial (mtDNA) haplogroups and two Y chromosome haplogroups (Fagundes et al. 2008; Gilbert et al. 2008; Goebel et al. 2008; Kemp and Schurr, this volume; Schurr 2004); all modern groups thus far studied also share the private gene 9RA, which occurs only in the Americas and adjacent eastern Siberia (Schroeder et al. 2007).

Powell (2004) takes an alternative position, suggesting that microevolutionary processes alone can account for the extreme changes in craniofacial form (Powell and Neves 1999). The small, dispersed, mobile bands that initially peopled the hemisphere, he argues, would have been highly subject to founder effect and drift. These effects, plus the selective forces from differentiating early Holocene climatic conditions and cultures caused a highly variable founding population to diverge into regional patterns. As human populations grew, gene flow resulted in similar morphologies throughout the hemisphere. This position is seemingly supported by American mtDNA frequencies, which show the highest intergroup variability in the world. It is, however, contradicted by the narrow range of variability in the frequency of the 9RA in nuclear DNA. Craniofacial morphology is presumably polygenic, so the individual alleles that generate it are unlikely to be more subject to drift than 9RA, let alone to drift collectively.

The conclusion that can be drawn from the conference presented in these pages, along with recent findings of molecular anthropologists, is that both of these positions are probably correct. Tamm and colleagues (2007; see also Kitchen et al. 2008; Mulligan et al. 2008) have recently reviewed the accumulation of mutations within autochthonous Amerind mitochondrial subhaplogroups and their distribution throughout the Americas. The number of mutations within those autochthonous subgroups that are shared by groups now living in North and South America demonstrates not only that all New World populations must be derived from the same parent population but also that said parent population lived somewhere in greater Beringia, isolated from the rest of humanity, for at least 5,000 years before the first bands made the journey southward into temperate North America and beyond.

Assuming that the proposed isolation occurred, the Americas could have experienced multiple episodes of immigration from the Beringian source population without leaving a separate genetic signature, beyond founder effects, in the relative frequencies of the parental haplogroups and subhaplogroups. Immigration events could have been separated widely in both time and space, each daughter population being derived from a different subgroup of the Beringian parent population. Peoples living in Beringia for thousands of years between initial and later emigration events would have diverged both culturally and genetically as a result of microevolutionary processes. Indeed, as Bever (2001) has shown, at least three separate technological traditions—the Nenana, Denali, and Northern Paleo-Indian—existed in eastern Beringia during the terminal Pleistocene. Living as small mobile groups dispersed widely across a subcontinent-scale landscape, the Beringians who remained behind after the first people departed on their southward journey would have been highly subject to genetic drift. They would also have been subjected to the strong selective pressures of harsh, arctic climatic conditions for many more generations than had the first of those southward emigrants. Therefore, it is reasonable to expect later groups migrating out of Beringia to have differed morphologically from the people who preceded them, perhaps exhibiting phenotypes that reflect the selective pressures of a cold climate.

In a paper published several months after this conference, another group of researchers reached similar conclusions. Gonzalez-Jose and collaborators (2008) conducted a geometric morphometric analysis of craniofacial form intended to explore the reasons for the absence of Northeast Asian characteristics among Paleo-Americans and the subsequent Holocene appearance of such derived characteristics throughout the hemisphere. Comparing the craniofacial profiles of early Old World *Homo sapiens* and Paleo-Americans against those of modern populations from America and Asia, they concluded that the Paleo-Americans represented one or more late Pleistocene episodes of immigration by peoples with generalized craniofacial characteristics. Like their Old World ancestors, these peoples exhibited a high degree of morphological variability. After this original migration,

Gonzalez-Jose and colleagues (2008) suggest, peoples who remained in the Arctic evolved large, broad, coronally flattened faces in response to the extreme climate (Howells 1973; Roseman 2004). Gene flow from the Arctic then distributed these distinctive characteristics southward into both eastern Asia (e.g., Hanihara 1994) and the Americas, leading to the broad geographic distribution of these climatically derived characteristics.

This reconstruction of events finds support in the form of at least two recent episodes of southward migration by Arctic-adapted peoples that occurred after 1000 B.P. Linguistic, and to some extent archaeological, evidence shows that one Athabascan group moved southward along the east flank of the Cascade Range or Coast Mountains of the Pacific Northwest to take up residence on the coast of southern Oregon and northern California by around 700–600 B.P. (Moratto 1984). Two other groups, which gave rise to the Navajo, Apache, and Kiowa Apache moved shortly afterward down the east flank of the Rocky Mountains into the southern Great Plains and American Southwest (Foster 1996; Matson and Magne 2007).

It is not at all unreasonable to suggest that similar earlier expansions took place, more or less gradually replacing the generalized Paleo-American morphology with one more closely resembling that of northeastern Asians. Here I present archaeological and physical anthropological evidence that may represent one such earlier expansion, which occurred on the Columbia Plateau around 8000 B.P. I begin with an introduction to the geography of the Columbia Plateau, then present evidence for marked differences in the craniofacial morphology between early Holocene Paleo-Americans and middle and later Holocene peoples who more closely resemble modern Native Americans and, by extension, Northeast Asians and Alaskans. This is followed by archaeological evidence that may represent an episode of cultural replacement coinciding with the change in human morphology. I close with a consideration of possible selective forces that might have enabled newcomers moving down the Pacific Coast to displace the early Holocene incumbents of the Columbia Plateau.

Geography and Resources of the Columbia Plateau

The Columbia Plateau lies in the northwestern-most interior of the United States, framed between the Cascade Range and the Rocky Mountains in parts of Washington, Oregon, and Idaho. Ecologically, it ranges from arid steppes in its lowest southeastern corner to grasslands and conifer forests along the mountain fronts. The Columbia River is the master stream of the region and was the greatest single source of water and food for most of its human inhabitants. Salmon and other anadromous salmonids, which spawned in the river basin and reared in the Pacific, were the foundation of regional subsistence. Freshwater mollusks were also abundant. Various root-producing species were the most productive plant resources. The historical dynamics of these resources in relation to one another, as well as people's ability to process and store them, played a major part in the region's cultural history (Chatters 1995, 1998; Chatters et al. 2010; Prentiss and Chatters 2003), including the events described herein.

The Change in Craniofacial Morphology

The skeletal evidence used in this paper comes from three periods: one preceding the hypothesized ethnic replacement event, a second within two millennia after the event, and the third from the late Holocene. Human remains predating the proposed event, which occurred within two centuries of 8000 B.P., have been found in four localities on or immediately adjacent to the Columbia Plateau: Buhl (Green et al. 1998), Marmes Rockshelter (Krantz 1979), Gore Creek (Cybulski et al. 1981), and Kennewick (Chatters 2000). One additional, undocumented skull, known only as Stick Man, has been found in museum collections and sourced to the Columbia Basin, near the town of Quincy (Chatters et al. 2000). The oldest skeletal material from the Columbia Plateau that postdates 8000 B.P. comes from Prospect, Oregon (Cressman 1940); Marmes Rockshelter, Washington (Breschini 1979); and DeMoss (Green et al. 1986), Braden (Harten 1980), and Clark's Fork (Pennefeather-Obrien and Strezewski 2002), Idaho.

Of the five pre-8000 B.P. skeletal samples, only Buhl, Kennewick, and Stick Man include skulls intact enough for craniofacial measurement. The earliest of these is Buhl, the well-preserved partial skeleton of a young adult female dating to 10,675 ± 95 B.P. Green and colleagues (1998) assert that she bore characteristics typical of modern Amerinds, but subsequent morphometric analyses of the original team's measurements (Brace et al. 2001), and measurement of a three-dimensional digital reconstruction based on photographs (Herrmann et al. 2006), have shown that, like other Paleo-Americans, she falls outside the norm for all modern peoples but lies closest to the Ainu and Polynesians. Kennewick is the nearly complete skeleton of a middle-aged male dating 8410 ± 60 B.P. (Taylor et al. 1997). Stick Man is represented only by a near-complete neurocranium, a portion of which has produced three dates; the two from the most refined collagen average 8125 ± 50 B.P. (Chatters et al. 2000). Morphometrically, both Kennewick and Stick Man are closest to modern Polynesian populations; Kennewick would not be a typical member of any modern population (Powell and Rose 1999), whereas Stick Man could be a typical inhabitant of the Polynesian Moriori (Chatters et al. 2000). The groups that are morphometrically farthest from these individuals are modern Amerinds. All three skulls are characterized by a strong degree of facial forwardness and possess longer, narrower crania than are seen in historic populations of the Columbia Plateau.

Among the post-8000 B.P. collections, only Prospect, Braden, and DeMoss collections contain measurable crania. Although measurements have been reported for the Clark's Fork skull by Pennefeather-Obrien and Strezewski (2002) the authors of that study expressed no confidence in the results because the cranium had been badly reconstructed by the original investigator from incomplete parts. The Prospect specimen was found directly beneath tephra from Mt. Mazama (Cressman 1940), which would give it an age of approximately 6850 B.P. (Bacon 1983). It is the remains of what appears from its size to have been an adult male. Jantz and Owsley (2001) measured the skull and included it in their 2001 morphometric analysis of Paleo-American craniofacial variability. In that analysis, the Prospect specimen grouped comfortably among modern native North American groups. The Braden site is a small cemetery with a bone collagen date of 5790 ± 120 B.P. (Harten 1980). It contained two individual interments and the mass secondary grave of up to ten individuals, including four juveniles, four adult males, and two adult females. Four adult crania (three male and one female) are at least partially measurable; only one, an unnumbered specimen, is complete. DeMoss, which dates to 5965 ± 60 B.P., was the secondary interment of an as-yet-undetermined number of individuals, who are represented primarily by teeth (Green et al. 1986). Three measurable partial adult crania (two female and one male) are included in the collection. Morphometric analysis using the forensic program Fordisc 2.0 (Ousley and Jantz 1996) places the unnumbered Braden specimen firmly among modern Amerind and Northeast Asian populations. The only DeMoss skull complete enough for such a comparison, a female designated "A," has severe tumpline deformation, so a Fordisc analysis has not been attempted for her.

Superficially, and through separate morphometric analyses by various researchers, the skeletons predating and postdating 8000 B.P. have been found to fit the pattern observed throughout the Americas. The Paleo-Americans, although usually not fitting comfortably in any modern human group, more closely resemble Ainu and Polynesian peoples and are farthest from modern Amerinds. Middle Holocene skeletons appear to more closely resemble modern Amerinds and Northeast Asians.

To further illustrate the characteristics that distinguish these two groups and demonstrate the metric relationships between each group and late Holocene (i.e., modern) Amerinds from the Columbia Plateau, I conducted a simple metric exercise. In order to utilize the largest possible number of incomplete skulls in the available sample, I compared them using four indices: cranial, upper facial, cranial height, and facial forwardness. Cranial and upper facial indices are standard measures computed following Bass (1987). Cranial height and facial forwardness indices were developed specifically for this analysis in order to make best use of the measurements available on such incomplete crania. Each corrects for skull size by dividing the most frequently represented measure of the dimension in question by the geometric mean of cranial length and breadth. The cranial height index is thus computed

$$(100 \times \text{bregma radius}) / [(\text{ maximum cranial length + breadth}) / 2]$$

The facial forwardness index[1] is computed in the same manner:

$$(100 \times \text{nasion radius}) / [(\text{ maximum cranial length + breadth}) / 2]$$

Only male individuals were used in this analysis for three reasons. First, males and females are dimorphic to varying degrees, so the outcome of any comparison between populations will be affected to some extent by the proportion of males and females in each group as well as any differences in the degree of dimorphism. Second, the two females in the middle Holocene sample from De-

Moss and the one female from Braden have saddled parietals that indicate ha-bitual carrying of loads using a tumpline. The deformation resulting from that behavior appears to have made the skulls hyperdolicranic and may have affected other dimensions in DeMoss A, the most complete of these crania. Third, I am uncomfortable with the measurements of the Buhl skull reported by Herrmann and colleagues (2006). In 2000, I attempted to work with specialists to generate such a model from the same set of photographs Herrmann and colleagues used and was told that it could not be done without either having oblique photos of the skull (which were absent) or knowing the focal length of the camera and the distance between the lens and subject for each image. Limiting the comparison to adult males left Kennewick Man and Stick Man representing the early Holocene and three specimens from Braden and one each from DeMoss and Prospect rep-resenting the middle Holocene.

For a comparison with known Amerind people of the region, I used 13 male skulls from the Columbia Plateau that date between 3000 B.P. and the protohis-toric. They include a 3,000-year-old skull from site 45YK13 on the mid–Columbia River, an approximately 2,300-year-old skull from the Congdon site on the Co-lumbia River (Butler 1963), a 1,000-year-old skull from site 45OK66 at the mouth of the Columbia and Okanogan rivers (Chatters 2003), and 10 individuals collect-ed by amateurs from late prehistoric or protohistoric cemeteries in the Snake and Columbia river valleys and whose remains were turned over to Central Wash-ington University for repatriation.[2] These individuals were selected because they were the only male skulls in the collection lacking the intentional occipital and frontal flattening that marks most crania from the region. When compared to modern populations using Fordisc 2.0 (Ousley and Jantz 1996), all most closely compare with Chinese, Siberian, or Amerind populations. With the exception of measurements from the Prospect individual, which were provided by Richard Jantz of the University of Tennessee, Knoxville, all measurements were taken by me or by students under my supervision.

Results (Table 3-1) show that for cranial index and cranial height, the early Holocene individuals have longer and higher skulls on average than either of the later groups but that they are within the range and two standard deviations of modern Plateau peoples. There is a general trend toward narrower crania but no discrete break in cranial index or cranial height. The opposite is true for the facial forwardness and upper facial indices. Both of the early Holocene individuals fall outside the range of the late Holocene population and beyond two standard devi-ations of the late Holocene mean. The middle Holocene specimens, however, are not distinguishable from the late Holocene Columbia Plateau population; means for the two groups are nearly identical for all four measures. Figure 3-1 illustrates the marked difference in facial forwardness that distinguishes the early Holocene skulls from the two later groups. Sometime between 8000 B.P. and approximately 6000 B.P., people living on the Columbia Plateau lost the narrower, forward-positioned faces and acquired the broad, coronally flattened faces that are so char-acteristic of modern Amerinds. Taking the approximate age of the Prospect speci-men at face value, it appears that these characteristics had arrived at least by 6850 B.P. The change is rapid enough to be evidence of population replacement.

Table 3-1. *Craniofacial Measures Comparing a Chronological Series of Male Human Crania from the Columbia Plateau*

Measure	Early Holocene				Middle Holocene					Late Holocene[a]	
	Kennewick	Stick	Mean	Prospect[b]	DeMoss B	Braden 5	Braden 9	Braden no #	Mean	Range	Mean
Max length	189	196		183	195	194	189	188			
Max breadth	139	139		144	147	140	140	146			
Bregma radius	122	136		123			129	123			
Nasion radius	103	106		97				98			
Prosthion radius	110			98				101			
Bizygomatic br.	135			148				137			
Nasion-prosthion	75			72				72			
Cranial index	73.5	70.9	72.2 ± 1.9	78.7	75.4	72.2	74.1	77.7	75.6 ± 2.4	71.1–81.7	76.9 ± 3.1
Upper facial index	55.6		55.6	48.6				52.6	50.6 ± 2.8	46.7–51.1	49.5 ± 1.6
Height index[c]	74.4	81.2	77.8 ± 4.8	75.2			67	73.6	71.9 ± 4.3	70.8–75.8	73.5 ± 1.9
Forwardness index[d]	62.8	63.3	63.0 ± 0.3	59.3				58.7	59.0 ± 0.4	55.7–61.8	59.0 ± 1.6

[a] Data from 13 undeformed male skulls postdating 3000 B.P., housed at Central Washington University in fall 2008.
[b] Courtesy of Richard Jantz, University of Tennessee, Knoxville.
[c] Computed as (100 × bregma radius)/(mean of length + breadth).
[d] Computed as (100 × nasion radius)/(mean of length + breadth).

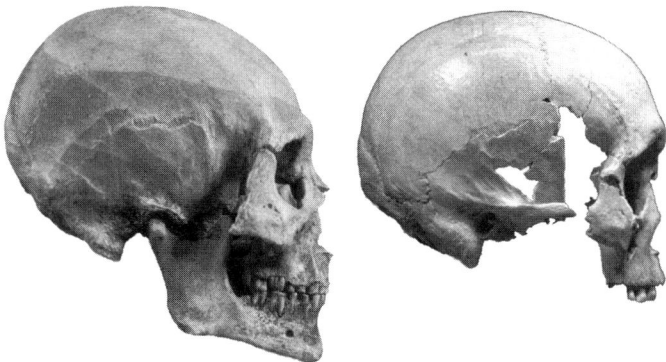

Figure 3-1. *Profile views of Kennewick Man (left, a cast) and the complete skull of the unnumbered adult male from the Braden site positioned in the Frankfort plane, showing marked differences in facial forwardness.*

Archaeological Evidence for an Ethnic Expansion

Change in craniofacial morphology does not a migration make, as Powell (2004) so emphatically points out. In the case of the Columbia Plateau, however, cultural behavior undergoes marked changes that approximately coincide with the change in facial morphology, greatly strengthening the case for an ethnic replacement at the end of the early Holocene.

The archaeological record of the Columbia Plateau is marked between 8200 and 7800 B.P. by a change from one long-standing cultural tradition to another. At that time, regional manifestations of the Western Stemmed Tradition, known locally as the Windust Phase (Leonhardy and Rice 1970), Phillipi Phase (Dumond and Minor 1983), and Goatfels Complex (Choquette 1996), were replaced by manifestations of the Old Cordilleran Tradition, which is represented along various segments of the Columbia River system by the Cascade (Leonhardy and Rice 1970), Okanogan (Grabert 1968), and Vantage phases (Nelson 1969). The two traditions differ in settlement and land-use patterns; mobility strategy; bone, stone, lithic, and food processing technologies; and subsistence emphases. The differences and the chronological relationship between the two traditions have been recently presented in detail by my colleagues and me (Chatters et al. 2010). The characteristics of the two traditions are only summarized here; supporting data can be found in the original text.

Western Stemmed Tradition

The Buhl, Kennewick, and Stick Man remains are associated with the Western Stemmed Tradition (WST). A large stemmed knife and tiny eyed needle, both common to the WST, were found with the Buhl woman; the projectile point found embedded in Kennewick Man's pelvis has a leaf shape (Fagan 1999), but its thin blade widens distally in a manner characteristic of both stemmed and foliate projectile points associated with the WST (Chatters and Prentiss 2010). Both Kennewick and Stick Man are coeval with the latest WST components in the region.

The WST is also known by Great Basin archaeologists and some students of Plateau prehistory as the Paleoarchaic (Beck and Jones 1997; Willig and Aikens 1988). The name is intended to emphasize the differences between the WST and the Paleo-Indian tradition, from which it differs in some aspects of technology and an inferred greater emphasis on small game, fish, and plant foods (Graf and Schmidt 2008).

In the Columbia Plateau, the WST was characterized by a wide-ranging pattern of land use that emphasized wetland habitats and included both lowland and upland portions of the Columbia Basin and the adjacent mountains. Mutually exclusive distributions of two unique tool types—bifacial crescents have been found only near the upland marshes, whereas plumb-bob shaped implements labeled "bola stones" occur, with but a single exception, along the floodplains of major rivers—indicate highly patterned use of specialized subsistence technologies. The settlement pattern took the form of a seasonal round in which different kinds of activities were performed each season. Campsites tended to be repeatedly occupied in a highly established pattern of behavior. Mobility was frequent on a subannual basis, but high frequencies of tool reuse are indicative of longer residence times than we see under the Old Cordilleran Tradition; mobility appears to have been semilogistical in Binford's (1980) terms. Use of lithics from distant sources combines with a mortuary pattern that included secondary cremation and isolated interments to indicate mobility over large geographic ranges.

The stone and bone assemblages contain a large complement of composite implements, indicating a high degree of planning depth (sensu Torrence 1983). Along with the crescents and bolas we find unilaterally barbed harpoons; composite bone shafts; and tiny eyed needles, as well as stemmed, broad-bladed projectile points; large ovate bifaces; large end and side scrapers; gravers; and a wide variety of retouched flake tools. Stone for lithics came sometimes from river cobbles but more often from vein quarries. Cryptocrystalline silicates were preferred over crystalline igneous rocks. Processing of plant foods appears to have been limited to surface roasting and a small amount of grinding, perhaps of small seeds. Milling stones occur occasionally, but except for a single possible case (Goldendale site; Warren et al. 1963), never more than a few are found in any large assemblage. Evidence for stone boiling, in the form of concentrations of small fire-affected rocks, is slight; earth ovens are absent. Projectile points were used as killing implements, but wear patterns and a high incidence of resharpening even on unbroken specimens indicate that these broad-bladed tools also served as knives. Lingual rounding on Kennewick Man's mandibular teeth is indicative of paramasticatory use of the mouth as an implement of technology.

During the WST, Columbia Plateau peoples focused their subsistence efforts on big game. Deer, elk, or bison rank first or second in number of identified specimens (NISP) in nearly all reported faunal assemblages, although birds, rabbits, and marmots often rank in the top four. Fish have not been found in significant numbers in any WST assemblage. There is also no evidence that plants held an important place in the diet, as shown by the rarity of milling stones and lack of technologies for reducing complex starches.

Old Cordilleran Tradition

The DeMoss, Braden, and Prospect skeletons are associated with later components of the Old Cordilleran Tradition (OCT), which persisted in the Columbia Plateau until around 5500 cal B.P. (Ames et al. 1998; Chatters 1995) and probably gave rise to at least some of the late Holocene cultural manifestations in the region. DeMoss and Braden skeletons were found associated with numerous large bifaces and foliate and side-notched projectile points that are common to the late Cascade Phase of the OCT. These are the earliest skeletal remains with measurable crania that are directly associated with artifacts of the OCT. Two partial skeletons were found in Marmes Rockshelter in strata containing OCT assemblages dating between 8900 and 7500 cal B.P., but neither skull included more than dispersed cranial fragments. The Prospect skeleton occurred in a geologic context coeval with the OCT and can be presumed to be the remains of an OCT practitioner.

The OCT as characterized by Chatters and colleagues (2010) is a combination of Carlson's (1996a) Pebble Tool and Microblade traditions, which have closely similar stone tool technologies that are consistent with the Old Cordilleran culture concept as developed by Butler (1961). Most scholars of Northwest Coast prehistory see Carlson's two traditions as a unified whole in which microlithic technology was utilized more or less intensively depending on the availability of high-quality lithic material. Microblades were employed in areas where high quality material was scarce and came primarily from pebble sources (Ames and Maschner 1999). All OCT assemblages include narrow foliate bifaces, cobble tools, and an assortment of flake tools, along with a maritime or riparian habitat choice. As I discuss later in this paper, this tradition is recognizable along the Northwest Coast from Glacier Bay to at least southern Oregon and inland from the confluence of the Chilcotin and Fraser rivers south into the northern Great Basin. Some anecdotal evidence suggests components of this tradition have been found as far east as the Rocky Mountain front and southward as far as Arizona (Joseph Randolph, personal communication 2009).

OCT practitioners focused their activities along major stream courses, although they sometimes ranged into the upper reaches of those streams and made use of alpine habitats. Unlike the WST folk, they made little use of upland environments in the Columbia Basin. There is also no evident patterning of tools or activities, by either season or geography. The reduction in activity patterning is matched by a much lower frequency of site reuse. Some sites do show multiple episodes of habitation, but far more represent what appear to be single occupation events. Mobility was frequent and entirely residential. Lower frequencies of tool reuse show residence time at each occupation site was shorter than it was during the WST. Smaller foraging ranges are indicated by an increase in the use of local cobble material for stone tools and, based on the date of the human skeletal remains discussed in this paper, the repeated use of cemeteries for interring the dead.

In the Columbia Plateau, OCT assemblages have simpler bone and stone tool technologies and more complex food-processing technologies than are found in the WST. Composite tools are less common and, when they do occur, are associated with fishing. Stone tools include small foliate dart points, larger foliate bifaces, small end and side scrapers, drills, cobble choppers, cobble-spall knives, and sometimes microblades. Bifaces typically dominate the assemblages. Cobble tools are also particularly abundant in OCT assemblages; cobbles were, in fact, the preferred source of lithic material in most regions occupied by this tradition. One interesting change in technology is visible in the projectile points. Unlike the WST points, which have broad blades with heavily worn and refreshed edges, the OCT projectiles have narrow blades and show no evidence of wear, with re-working only used to repair broken tips. Bone implements, which are rare, differ in kind from those found in the WST. They typically consist only of splinter awls and antler wedges, although large eyed needles, small barbs from composite fishhooks, and barbs from fish leisters have also been found. Finally, the absence of wear on the lingual surfaces of anterior teeth of the Braden skeletons indicates the mouth was no longer used as a tool.

Plants, small animals, and fish appear to have assumed a more important role in OCT subsistence. Fish, river mussels, and small mammals are more important than large game by NISP in assemblages from all sites except rock-shelters, which probably served primarily as hunters' bivouacs. Multiple food-processing technologies are also evident, showing a greater importance of plant foods. Grindings stones, including manos and edge-ground cobbles, are found in most sites and account for a much higher proportion of all tools. Hopper mortars and pestles appear in assemblages postdating 6850 B.P. (Bense 1972). Remains of earth ovens appear for the first time near the beginning of the OCT, along with an increase in small thermally affected rocks that were probably used for stone boiling. A low degree of tooth wear and high frequencies of caries and antemortem tooth loss in the DeMoss and Clark's Fork burials indicate that, by 5800 B.P., at least some OCT bands were depending very heavily on processed plant foods with high sugar contents. With the increase in plant foods, we might also assume that women of the OCT had taken on a more important role in provisioning the household.

The Old Cordilleran Tradition as an Ethnic Expansion

The OCT probably has its origins in one of the Dyuktai-descendant cultures that occupied eastern Beringia at the end of the Pleistocene (Prentiss and Clarke 2008). Its characteristic foliate projectile points and bifaces, cobble tools, and sporadic use of microblades are first clearly recognizable at the deepest levels of the Groundhog Bay 2 site, dating around 9200 B.P. (Ackerman 1968; Bever 2001) and in nearby On Your Knees Cave on Prince of Wales Island around 9150 B.P. (Dixon 2008). Assemblages of similar content can be seen appearing progressively southward over the next thousand years. Figure 3-2 illustrates this progression; data is represented in Table 3-2. By 9000 B.P., the first OCT assemblages appeared near the mouth of the Fraser River. They arrived shortly afterward in the interior, at the Fraser-Chilcotin confluence. By 8400 B.P. the OCT had reached southern interior British Columbia and had extended down the Pacific Coast to the California-Oregon border by at least 8250 cal B.P. It can then be seen mov-

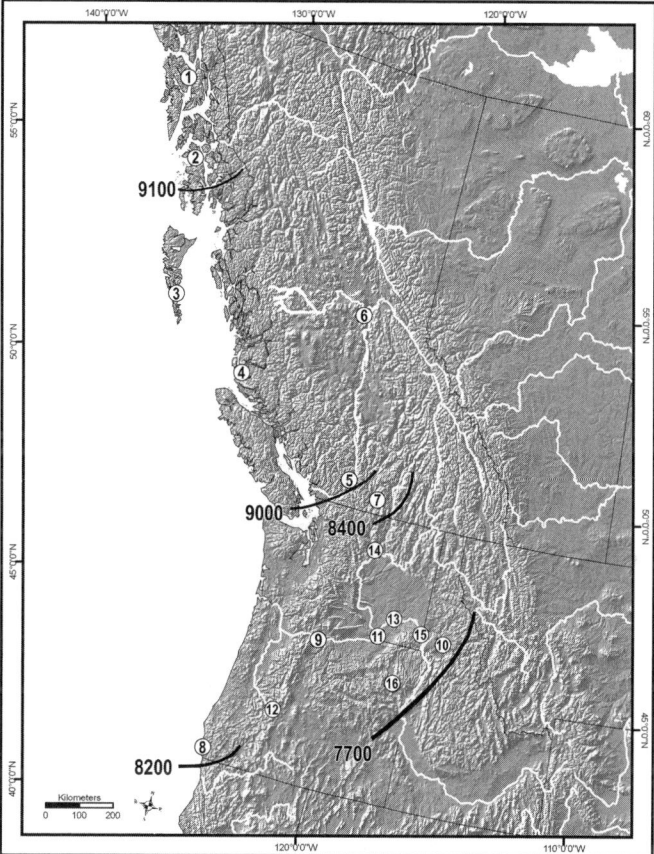

Figure 3-2. *Chronology of the Old Cordilleran Tradition expansion southward along the Northwest Coast and onto the Columbia Plateau. Dark lines mark the approximate frontier reached by the adjacent century B.P. Numbered points correspond to sites listed in Table 3-2.*

ing up the Columbia River, reaching The Dalles by around 8100 B.P. completely replacing the WST on the Columbia Plateau within the next hundred years. The pattern marks the unmistakable expansion of a way of life and may also represent an expansion of the people who practiced that lifeway.

While some students of regional prehistory (e.g., Ames et al. 1998; Davis 2001) see the change from one tradition to another as a smooth transition, an increasing number see the event as a sharp discontinuity; two lines of evidence strongly support the latter position. Both lines represent stylistic changes that coincide with the aforementioned changes in settlement, land use, technology, and subsistence. One is a change in basketry and sandal styles, the other a change in the most basic level of toolmaking behavior.

Dry caves along the northern fringe of the Great Basin, where the change from WST to OCT is also evident, contain basketry fragments and sandals dating

Table 3-2. *Earliest Dates on Old Cordilleran Tradition Components on the Northwest Coast and Plateau*

Map #	Site	^{14}C yr B.P.	Reference
1	Groundhog Bay 2	9202 ± 86	Bever 2001
2	On Your Knees Cave	9150 ± 50	Dixon 2008
3	Richardson Island[a]	8850 ± 60	Fedje et al. 2005
4	Namu	9000 ± 140[b]	Carlson 1996b
5	Milliken	9000 ± 150	Borden 1960
6	FlRQ-013	8770 ± 60	Bufford et al. 2008
7	Landels	8400 ± 90	Rousseau 1991
8	Indian Sands (Oregon)	8250 ± 80	Moss and Erlandson 1996
9	5-Mile Rapids	8090 ± 90	Butler and O'Connor 2004
10	10NP453	7980 ± 60	Ridenour 2006
11	Ash Cave	7940 ± 150	Butler 1961
12	Cascadia Cave	7910 ± 280	Newman 1966
13	Marmes Rockshelter	7840 ± 150[c]	Rice 1969; Sheppard et al. 1987
14	Plew Site	7730 ± 120[c]	Draper 1986
15	Thorn Thicket	7710 ± 180	Sprague and Combes 1966
16	Stockoff	7660 ± 780	McPherson et al. 1981

[a]Fedje and colleagues (2008) report large foliate bifaces from the Queen Charlotte Islands dating to as early as 10,600 B.P., but these appear to differ technologically from the Old Cordilleran Tradition. If included with it, however, they could push its presence on the Northwest Coast back more than a millennium before the earliest dates cited here.

[b]Carlson (2008) states that the 9000 B.P. date from Namu marks the earliest foliate projectile point, which places this component firmly in the OCT. Earlier dates of up to 9700 B.P. on deposits lacking projectile points or microblades could push the age of the OCT on the Northwest Coast considerably farther back in time.

[c]Date on mussel shell.

back more than 10,000 years B.P. Both underwent distinct changes at approximately the same time as the OCT arrived (Connolly and Barker 2004). Before 8200 B.P., all sandals were of the Fort Rock style, which has an enclosed toe, open heel, and complex method of attachment to the ankle; basketry was simple and undecorated. After 8200 B.P., and until the Numic expansion less than a thousand years ago, basketry was made with a complex array of decorations; sandals were of the open twined or spiral weft styles, which have a heel cup and open toe, and were cinched and tied to the ankle with a line that ran from the toe through a series of loops that extended from the edge of the sole to the top of the foot. Connolly and Barker (2004) see this as evidence for what they call a "population change."

The second line of evidence comes from stone tool selection and tool use at the Beech Creek site, located in Washington near the crest of the Cascade Range (Chatters and Prentiss 2010). WST and OCT assemblages are distinguishable in a series of occupations that extended from at least 8300 to as late as 6500 B.P. (Mack et al. 2010). Analysis showed that, in comparison to their OCT counterparts, WST toolmakers had selected larger flakes for utilization and retouch, selected flakes with steeper angles for both scraping and cutting functions, preferred tertiary percussion flakes over biface thinning flakes and crystalline volcanic rock over cryptocystalline silicates for utilization, made side scrapers with steeper angles, and produced consistently larger tools from their blanks. In addition, WST occupants used their tools more intensively and used what archaeologists identify as the same tool forms—particularly projectile points, bifaces, and end scrapers—in different ways. These contrasting preferences represent what Bettinger and colleagues (1996) have called "rote social learning" and Ames (2000) labels "passive style." They are learned, perhaps subconsciously, as a young person mimics the toolmaking behavior of an elder. If the WST and OCT were genetically related in a cultural sense, we would expect these elements of rote social learning to continue from one to the next. The fact that they do not corroborates the idea of an ethnic discontinuity on the Columbia Plateau.

Selective Forces and Old Cordilleran Tradition Expansion

The archaeological evidence may be construed as support for the idea that a population from the north—probably from an east Beringian source population—bearing facial characteristics resembling those of modern Northeast Asian peoples, began to expand southward along the Pacific Coast after 9200 B.P. and entered the Columbia Plateau within one or two centuries of 8000 B.P. This date was a critical juncture in the ecological history of the region. From around 9000 B.P. until as late as 7000 B.P., a Douglas fir and oak parkland covered the lowlands along the Pacific Coast of Washington, Oregon, and southern British Columbia, while an extensive grassland, dotted by small lakes and marshes, blanketed the Columbia Basin and lower mountain slopes between the Cascade Range and the Rocky Mountains (Barnosky et al. 1987). Both offered excellent habitat for big game (see Schalk et al. 2007). Along the northern Pacific Coast, sea levels rose above their modern levels, increasing the convolution, and therefore length, of the shoreline and thus the productivity of littoral resources (Moss and Erlandson 1996). Along the Washington and Oregon coasts, offshore winds blowing from the wasting Laurentide glaciers gave the region a continental climatic pattern and reduced upwelling off the coast. This meant lower near-shore productivity (Sancetta et al. 1992). This, along with the grassland cover of the interior Columbia Plateau, which produced more turbid runoff waters, likely meant low salmon productivity. These conditions changed between 8500 and 8000 B.P. The easterly winds ceased as the Laurentide ice sheet shrank, permitting a westerly flow of marine air and establishing the maritime climatic pattern the Pacific

Northwest experiences today. Westerly airflow meant the Cascade Range cast a rain shadow across the Columbia Plateau, converting grassland to desert at the same time the maritime climate allowed forests to move down slope. Forests along the coast remained open for another 1,500 years, however. At the same time, normal upwelling was established along the southern Northwest Coast. Salmon populations thus began to grow as the forests colonized the upper watersheds, improving water quality, and near-shore nutrient supplies increased. The upshot of this change was that the terrestrial big-game populations on the Columbia Plateau declined at the same time the coast was experiencing both continued terrestrial productivity and increasing aquatic food supplies. Throughout the region, a maritime-adapted population, with developed strategies for fish and invertebrate exploitation and new technologies for rendering plant foods digestible, would have had an advantage over a big-game supported population that lacked complex fishing and plant-processing technologies.

The new plant-processing technologies would have given the newly arriving culture selective advantage even without the change in resource distribution that came with climate change. As I have shown elsewhere (Chatters 2007), Paleo-American males and females show a great disparity in size and length of life.[3] The females are often very gracile and/or show signs of repeated nutritional stress. One possible explanation for this disparity may be that the importance of big game to the subsistence of most Paleo-American cultures, including the WST of the Columbia Plateau, conferred a greater societal value on strong adult males, to the extent that male children and male adults were fed better than females. Since it seems to have affected female longevity, this practice would have meant a low number of births per female (Chatters 2007). If females held a greater role in producing the food on which the society depended, as they would have after developing methods for breaking down complex starches into digestible sugars, they would have been better fed, in part through their own food-producing efforts. Better-fed females would live longer, potentially produce more offspring, and thus give a population with empowered females selective advantage over one in which females were undervalued. Such could have been one of the advantages that allowed the OCT to expand at the expense of the WST.

Discussion

Two issues warrant further consideration. The first takes the form of a caveat. It is important to note that the link between the Northeast Asian-like facial features and the apparent ethnic discontinuity at around 8000 B.P. is a cultural one. More than 1,200 years separate the Kennewick Man, Stick Man, and the earliest of the OCT-associated skeletons. That same span separates those more recent skeletons from the archaeologically indicated ethnic discontinuity itself. The wider, flatter face could have arrived via later gene flow or an immigration event that left no archaeological trace, as was the case with the Athabascans for several centuries after they immigrated to the Southwest (Foster 1996). The new facial features might also be the result of natural selection for a jaw structure more suited to consumption of plant foods. Plant foods are often tougher than meat or at least more chewing is required to extract an equivalent caloric return, either of which would require stronger masticatory muscles. In humans, individuals with a stronger bite have been shown to have shorter, wider faces and shorter, broader dental arcades (Ingervall and Helkimo 1978; Raadsheer et al. 1999). It is difficult to conceive of 1,500 years being enough time to effect as much difference as we see between the two samples, but the rate of change is dependent on the strength of selective advantage conveyed by the feature in question. The change could also be epigenetic. Research in animals has shown that a tougher diet, which requires greater force from the muscles that elevate the mandible, results in a change in craniofacial form. In humans, people who suffer from bruxism and those with severe dental wear exhibit the same broader face and palate as those with stronger musculature (Kiliaridis 2006). The differences between individuals seen in these studies are, however, very minor and do not reach the degree of difference seen among the Columbia Plateau skeletons. It is also important to note that dental wear differences may not be that great between Kennewick and the later individuals; Kennewick man is considered by some observers to have extreme dental wear, although it is not as extreme as I have seen in most of the later prehistoric skeletons from the Columbia Plateau.

A second important point is that the Athabascan example presented in the introduction shows that, assuming the Northeast Asian-like facial features did arrive on the Columbia Plateau with the immigration of the OCT, this might not be the only secondary expansion from the east Beringian parent population to occur at the end of the early Holocene. The Athabascan expansion began with the onset of the Little Climatic Anomaly, a Northern Hemisphere-wide warming that had many climatic and demographic impacts across North America (e.g., Prentiss and Chatters 2003; Stine 1994). Warming can improve primary productivity at northern latitudes at the same time populations in less mesic portions of the temperate zone are experiencing climatic deterioration. The vast landmass of subarctic and Arctic North America may have served at such times as a population incubator. When droughts or extreme temperature changes wrought local depopulations in the temperate zone, the door was opened to mobile bands from the north. Archaeologists and physical anthropologists should be alert to the possibility of such expansions, particularly around times of abrupt climatic change.

Conclusion

The archaeological record of the Columbia Plateau and adjacent Northwest Coast provide evidence that I hypothesize represents an ethnic expansion out of eastern Beringia between 9200 and 8000 B.P. Human skeletons associated culturally or temporally with the Western Stemmed Tradition, which existed on the Columbia Plateau from at least 10,700 until 8000 B.P., have the long heads and narrower, forward-positioned faces characteristic of Paleo-Americans found elsewhere in North America. Human skeletons temporally or archaeologically associated with the Old Cordilleran Tradition, which replaced the Western Stemmed, have the broad, coronally flattened faces that are hallmarks of Northeast Asians and modern Alaskan peoples and do not differ in those characteristics from late prehistoric inhabitants of the region. From this relationship, I offer the hypothesis that the expansion of the Old Cordilleran Tradition was one, perhaps the earliest, of the gene-flow events that ultimately gave most Amerinds their distinctively Northeast Asian appearance. I further posit that secondary migrations from the remnants of eastern Beringia into the Americas could have occurred sporadically until the Athabascan migrations of the last one thousand years. Multiple migrations may have significantly contributed to diachronic change in the morphology of America's inhabitants and their ultimate similarity to modern Asians. Those migrations need not, however, have entailed the movement of distinct populations from the Old World but instead may represent the later expansion of peoples from a Beringian parent population hundreds of generations after the Americas had first been populated by peoples we now call Paleo-Americans.

Acknowledgments

This paper arose from conversations and presentations that took place at the 25th Annual Visiting Scholar Conference at Southern Illinois University Carbondale in 2008; it bears almost no resemblance to the paper I presented at that venue. The final product owes much to comments by Ben Auerbach, Ken Ames, Anna Prentiss, and one anonymous reviewer. Special thanks to Dick Jantz for supplying the craniometric data on the Prospect skull and to Jason Cooper and Emily Gantz for their assistance in producing Figure 3-2.

Notes

1. Elsewhere (Chatters 2000) I have computed the forwardness index by dividing the sum of four radii (prosthion, nasion, subspinale, and zygomaxillary) by the geometric mean of length, breadth, basion radius, nasion radius, and prosthion radius, which is a more accurate reflection of facial position, but use of this measure would disqualify most specimens from the comparison.

2. Skulls used in this comparison were the Whisky Dick Canyon skeletons W, YY, and DA from the mid-Columbia River; ZZ1-478A; MCRP 58 and 76 from the Sanpoil River; MCRP 78 and 82 from Douglas County; MCRP 48 from the Wenatchee River; MCRP 187 from the Snake River; 45YK13 from Priest Rapids; 45OK66-6 from the mouth of the Okanogan River; and KL-41C from the Congdon site near the Dalles on the Columbia River.

3. Chatters (2007) found that among 9 males and 6 females from North America for which stature could be computed, males averaged 166.5 ± 3.9 cm tall; females, only 151.8 ± 5.9 cm. When male and female statures are considered by region, where they may arguably be considered to represent members of a population, similar disparities exist. The difference in stature on the Columbia Plateau, between Buhl (female) and the average stature of Gore Creek and Kennewick

(males), is at least 11.4 cm. The difference in central Texas, between Wilson Leonard (female) and the average of Putnam and Horn Shelter II, no. 1 (males) is 11.3 cm. In Colorado, the Gordon Creek female was 11.1 cm shorter than the Hourglass Cave male. Eleven females and 16 males for which age could be estimated indicate that males died at the average age of 32.8 ± 10.3 years, whereas females died at 22 ± 4.5 years.

References

Ackerman, Robert E.
   1968   *The Archaeology of the Glacier Bay Region, Southeastern Alaska.* Washington State University Laboratory of Anthropology, Report of Investigations 44, Pullman.
Ames, Kenneth M.
   2000   Review of the Archaeological Data. In *Cultural Affiliation Report,* complied by Francis P. McManamon. National Park Service, United States Department of the Interior. Electronic document, http://www.nps.gov/archeology/kennewick/ames.htm, accessed October 1, 2008.
Ames, Kenneth M., Don E. Dumond, Jerry R. Galm, and Rick Minor
   1998   Prehistory of the Southern Plateau. In *Plateau,* edited by Deward E. Walker Jr., pp. 103–119. Handbook of North American Indians, Vol. 12, William C. Sturtevant, general editor, Smithsonian Institution, Washington, D.C.
Ames, Kenneth M., and Herbert D. G. Maschner
   1999   *Peoples of the Northwest Coast: Their Archaeology and Prehistory.* Thames and Hudson, New York.
Bacon, Charles R.
   1983   Eruptive History of Mount Mazama and Crater Lake Caldera, Cascade Range, U.S.A. *Journal of Volcanology and Geothermal Research* 18:57–117.
Barnosky, Cathy W., Patricia M. Anderson, and Patrick J. Bartlein
   1987   The Northwestern U.S. During Deglaciation: Vegetational History and Paleoclimatic Implications. In *North American and Adjacent Oceans during the Last Deglaciation,* edited by William F. Ruddiman and H. E. Wright Jr., pp. 289–322. The Geology of North America DNAG Vol. K-3. Geological Society of America, Boulder, Colorado.
Bass, William M.
   1987   *Human Osteology: A Laboratory and Field Manual.* 3rd ed. Missouri Archaeological Society, Columbia.
Beck, Charlotte, and George T. Jones
   1997   The Terminal Pleistocene/Early Holocene Archaeology of the Great Basin. *Journal of World Prehistory* 11:161–236.
Bense, Judith A.
   1972   The Cascade Phase: A Study in the Effects of the Altithermal on a Cultural System. Ph.D. dissertation, Department of Anthropology,Washington State University, Pullman.
Bettinger, Robert L., Robert Boyd, and Peter J. Richerson
   1996   Style, Function, and Cultural Evolutionary Process. In *Darwinian Archaeologies,* edited by Herbert G. D. Maschner, pp. 113–164. Plenum Press, New York.
Bever, Michael R.
   2001   An Overview of Alaskan Late Pleistocene Archaeology: Historical Themes and Current Perspectives. *Journal of World Prehistory* 15:125–191.

Binford, Lewis R.

    1980   Willow Smoke and Dogs' Tails: Hunter-Gatherer Settlement Systems and Archaeological Site Formation. *American Antiquity* 45:4–20.

Borden, Charles E.

    1960   *DjRi3, an Early Site in the Fraser Canyon, British Columbia*. Bulletin 162. National Museum of Canada Contributions to Anthropology, Ottawa.

Brace, C. Loring, A. Russell Nelson, Noriko Seguchi, Hiroaki Oe, Leslie Serling, Pan Quifeng, Li Yongyi, and Dashtseveg Tumen

    2001   Old World Sources of the First New World Human Inhabitants: A Comparative Facial View. *Proceedings of the National Academy of Sciences of the United States of America* 98:10017–10022.

Breschini, Gary S.

    1979   The Marmes Burial Casts. *Northwest Anthropological Research Notes* 13(2):111–175.

Bufford, Aidan K. C., Frank Craig, Remi N. V. Farvecque, and Nicole Jackman

    2008   An Early Cordilleran Assemblage from the Nechako-Fraser Basin. In *Projectile Point Sequences in Northwestern North America*, edited by Roy L. Carlson and Martin P. R. Magne, pp. 293–302. Archeology Press, Simon Fraser University, Burnaby, British Columbia.

Butler, B. Robert

    1961   *The Old Cordilleran Culture in the Pacific Northwest*. Occasional Papers of the Idaho State College Museum No. 5, Pocatello.

    1963   Further Notes on the Burials and the Physical Stratigraphy at the Congdon Site, a Multi-Component Middle Period Site at The Dalles on the Lower Columbia River. *Tebiwa* 6(2):16–32.

Butler, Virginia L., and Jim E. O'Connor

    2004   9000 Years of Salmon Fishing on the Columbia River, North America. *Quaternary Research* 62:1–8.

Carlson, Roy L.

    1996a   Prologue: Introduction to Early Human Occupation in British Columbia. In *Early Human Occupation in British Columbia*, edited by Roy L. Carlson and Luke Dalla Bona, pp. 3–10. University of British Columbia Press, Vancouver.

    1996b   Early Namu. In *Early Human Occupation in British Columbia*, edited by Roy L. Carlson and Luke Dalla Bona, pp. 83–102. University of British Columbia Press, Vancouver.

    2008   Projectile Points from the Central and Northern Mainland Coast of British Columbia. In *Projectile Point Sequences in Northwestern North America,* edited by Roy L. Carlson and Martin P. R. Magne, pp. 61–78. Archaeology Press, Simon Fraser University, Burnaby, British Columbia.

Chatters, James C.

    1995   Population Growth, Climatic Cooling, and the Development of Collector Strategies on the Southern Plateau, Western North America. *Journal of World Prehistory* 9:341–400.

    1998   Environment. In *Plateau*, edited by Deward E. Walker, pp. 29–48. Handbook of North American Indians, Vol. 12, William C. Sturtevant, general editor, Smithsonian Institution, Washington, D.C.

    2000   The Discovery and First Analysis of an Early Holocene Human Skeleton from Kennewick, Washington. *American Antiquity* 65:261–316.

    2003   *Osteoarchaeology and Mortuary Practices of the Sinkaietck*. Applied Paleoscience, Bothell, Washington.

2007   Patterns of Death and the Peopling of the Americas. Manuscript submitted for publication in *Hombre Temprano en America III*, edited by Jose C. Jimenez-Lopez. Instituto National de Anthropologia y Historia, Mexico City.

Chatters, James C., Steven Hackenberger, Alan Busacca, Linda S. Cummings, Richard L. Jantz, Thomas W. Stafford Jr., and Royal E. Taylor

2000   A Possible Second Early Holocene Skull from Central Washington? *Current Research in the Pleistocene* 17:93–95.

Chatters, James C., Steven Hackenberger, Anna. M. Prentiss, and Jane-Leigh Thomas

2010   The Paleoindian to Archaic Transition in the Pacific Northwest: In Situ Development or Ethnic Replacement? In *On the Brink: Transformations in Human Organization and Adaptation at the Pleistocene-Holocene Boundary in North America,* edited by C. Britt Bousman and Bradley J. Vierra. Texas A&M Press, College Station, in press.

Chatters, James C., and Anna M. Prentiss

2010   Technological and Functional Analysis of Lithic Artifacts. In *Archaeological Data Recovery at the Beech Creek Site (45LE415), Gifford Pinchot National Forest, Washington,* edited by Cheryl A. Mack, James C. Chatters, and Anna M. Prentiss, pp. 49–123. Gifford Pinchot National Forest, Pacific Northwest Region, U.S. Forest Service, Trout Lake, Washington.

Choquette, Wayne

1996   Early Postglacial Habitation of the Upper Columbia Region. In *Early Human Occupation in British Columbia*, edited by Roy L. Carlson and Luke Dalla Bona, pp. 45–50. University of British Columbia Press, Vancouver.

Connolly, Thomas J., and Patrick Barker

2004   Basketry Chronology of the Early Holocene in the Northern Great Basin. In *Early and Middle Holocene Archaeology of the Northern Great Basin*, edited by Dennis L. Jenkins, Thomas J. Connolly, and C. Melvin Aikens, pp. 241–250. University of Oregon Anthropological Papers 62, Eugene.

Cressman, Luther S.

1940   Studies on Early Man in South Central Oregon. *Yearbook of the Carnegie Institution of Washington* 39:300–306.

Cybulski, Jerry S., Don E. Howes, James C. Haggarty, and Morley Eldridge

1981   An Early Human Skeleton from South Central British Columbia: Dating and Bioarchaeological Inference. *Canadian Journal of Archaeology* 5:49–59.

Davis, Loren G.

2001   Lower Salmon River Cultural Chronology: A Revised and Expanded Model. *Northwest Anthropological Research Notes* 35(2):229–247.

Dixon, E. James

2008   Bifaces from On-Your-Knees Cave, Southeast Alaska. In *Projectile Point Sequences in Northwestern North America*, edited by Roy L. Carlson and Martin P. R. Magne, pp. 11–18. Archeology Press, Simon Fraser University, Burnaby, British Columbia.

Draper, John A.

1986   45OK424 Site Report. In *The Wells Reservoir Archaeological Project. Vol 2: Site Reports*, edited by James C. Chatters. Central Washington Archaeological Survey, Archaeological Report 86-6. Central Washington University, Ellensburg.

Dumond, Don E., and Rick Minor

1983   *Archaeology in the John Day Reservoir: The Wildcat Canyon Site*. University of Oregon Anthropological Papers No. 30, Eugene.

Fagan, John L.

1999   *Analysis of the Lithic Artifact Embedded in the Columbia Park Remains.* Report submitted to the National Park Service, United States Department of the Interior, Washington, D.C.

Fagundes, Nelson J. R., Ricardo Knitz, Robert Eckert, Ana C. S. Valls, Mauricio R. Bogo, Francisco M. Salzano, David Glenn Smith, Wilson A. Silva Jr., Marco A. Zago, Andrea K. Ribiero dos Santos, Sidney E. B. Santos, Maria L. Petzl-Erler, and Sandro L. Bonnato
>2008   Mitochondrial Population Genomics Supports a Single Pre-Clovis Origin with a Coastal Route for the Peopling of the New World. *American Journal of Human Genetics* 82:583–592.

Fedje, Daryl, Quentin Mackie, D. McLaren, and Tina Christensen
>2008   A Projectile Point Sequence for Haida Gwaii. In *Projectile Point Sequences in Northwestern North America,* edited by Roy L. Carlson and Martin P. R. Magne, pp. 19–40. Archaeology Press, Simon Fraser University, Burnaby, British Columbia.

Fedje, Daryl, Martin P. R. Magne, and Tina Christensen
>2005   Test Excavations at Raised Beach Sites in Southern Haida Gwaii and Their Significance to Northwest Coast Archaeology. In *Haida Gwaii, Human History and Environment from the Time of the Loon to the Time of the Iron People*, edited by Daryl W. Fedje and Rolf W. Mathewes pp. 204–244. University of British Columbia Press, Vancouver.

Foster, Michael K.
>1996   Language and the Culture History of North America. In *Languages*, edited by Ives Goddard, pp. 64–110. Handbook of North American Indians, Vol. 17, William C. Sturtevant, general editor, Smithsonian Institution, Washington, D.C.

Gilbert, M. Thomas P., Dennis L. Jenkins, Anders Gotherstrom, Nuria Naveren, Juan J. Sanchez, Michael Hofreiter, Philip F. Thomsen, Jonas Binladen, Thomas F. G. Higham, Robert M. Yohe II, Robert Parr, Linda S. Cummings, and Eske Willerslev
>2008   DNA from Pre-Clovis Human Coprolites in Oregon, North America. *Science* 320:786–789.

Goebel, Theodore, Michael R. Waters, and Donald H. O'Rourke
>2008   The Late Pleistocene Dispersal of Modern Humans in the Americas. *Science* 319:1497–1502.

Gonzalez-Jose, Renaldo, Maria C. Bortolini, Fabricio R. Santos, and Sandro L. Bonatto
>2008   The Peopling of America: Craniofacial Shape Variation on a Continental Scale and Its Interpretation from an Interdisciplinary View. *American Journal of Physical Anthropology* 137:175–187.

Gonzalez-Jose, Renaldo, Silvia Dahinten, Maria Luis, Miquel Hernández, and Hector M. Pucciarelli
>2001   Craniometric Variation and the Settlement of the Americas: Testing Hypotheses by Means of R Matrix and Matrix Correlation Analyses. *American Journal of Physical Anthropology* 116:154–156.

Gonzalez-Jose, Renaldo, Antonio Gonzalez-Martin, Miquel Hernández, Hector M. Pucciarelli, Marina Sardi, Alfonso Rosales, and Silvina Van der Molen
>2003   Craniometric Evidence for Paleoamerican Survival in Baja California. *Nature* 425:62–65.

Gonzalez-Jose, Renaldo, Walter Neves, Marta M. Lahr, Sylvia Gonzalez, Hector Pucciarelli, Miquel H. Martinez, and Gonzalo Correal
>2005   Late Pleistocene/Holocene Craniofacial Morphology in Mesoamerican Paleoindians: Implications for the Peopling of the New World. *American Journal of Physical Anthropology* 128:772–780.

Grabert, Garland F.
  1968   *North-Central Washington Prehistory: A Final Report on Salvage Archaeology in the Wells Reservoir-Part 1.* Reports in Archaeology 1. Department of Anthropology, University of Washington, Seattle.

Graf, Kelly E., and Dave N. Schmidt (editors)
  2008   *Paleoindian or Paleoarchaic? Great Basin Human Ecology at the Pleistocene-Holocene Transition.* University of Utah Press, Salt Lake City.

Green, Thomas J., Bruce Cochran, Todd W. Fenton, James C. Woods, Gene L. Titmus, Larry Tieszen, Mary Ann Davis, and Suzanne J. Miller
  1998   The Buhl Burial: A Paleoindian Woman from Southern Idaho. *American Antiquity* 63:437–456.

Green, Thomas J., Max G. Pavesic, James C. Woods, and Gene L. Titmus
  1986   The DeMoss Burial Locality: Preliminary Observations. *Idaho Archaeologist* 9:437–456.

Hanihara, Tsunehiko
  1994   Craniofacial Continuity and Discontinuity of Far Easterners in the Late Pleistocene and Holocene. *Journal of Human Evolution* 27:417–441.

Harten, Louise B.
  1980   The Osteology of the Human Skeletal Material from the Braden Site, 10-WN-117, in Western Idaho. In *Anthropological Papers in Memory of Earl H. Swanson, Jr.,* edited by Louise B. Harten, Claude N. Warren, and Donald R. Tuohy, pp. 130–148. Idaho Museum of Natural History, Pocatello.

Herrmann, Nicholas P., Richard L. Jantz, and Douglas W. Owsley
  2006   Buhl Revisited: Three-Dimensional Photographic Reconstruction and Morphometric Re-evaluation. In *El Hombre Temprano en America y Sus Implicaciones de la Cuenca Mexico,* edited by Jose C. Jimenez-Lopez, Silvia Gonzales, Jose A. P. Padilla, and Frederico O. Pedraza, pp. 211–220. Instituto Nacional de Anthropologia e Historia, Mexico City.

Howells, William W.
  1973   *Cranial Variation in Man: A Multivariate Analysis of Patterns of Difference among Human Populations.* Papers of the Peabody Museum of Archaeology and Ethnology Vol. 79. Harvard University, Cambridge.

Ingervall, Bengt, and Eva Helkimo
  1978   Masticatory Muscle Force and Facial Morphology in Man. *Archives of Oral Biology* 23:203–206.

Jantz, Richard L., and Douglas W. Owsley
  2001   Variation among Early North American Crania. *American Journal of Physical Anthropology* 114:146–155.

Kiliaridis, Stavros
  2006   The Importance of Masticatory Muscle Function in Dentofacial Growth. *Seminars in Orthodontics* 12:110–119.

Kitchen, Andrew, Michael A. Miyamoto, and Connie J. Mulligan
  2008   A Three-Stage Colonization Model for the Peopling of the Americas. *PLoS One* 3:e1596. doi:10.1371/journal.pone.0001596.

Krantz, Grover S.
  1979   Oldest Human Remains from the Marmes Site. *Northwest Anthropological Research Notes* 13:159–173.

Lahr, Marta M.
  1996   *Evolution of Modern Human Diversity: A Study of Cranial Variation.* Cambridge University Press, Cambridge.

Leonhardy, Frank C., and David G. Rice
 1970   A Proposed Culture Typology for the Lower Snake River, Southeastern Washington. *Northwest Anthropological Research Notes* 19:161–168.

Mack, Cheryl A., James C. Chatters, and Anna M. Prentiss
 2010   *Archaeological Data Recovery at the Beech Creek Site (45LE415), Gifford Pinchot National Forest, Washington*. Gifford Pinchot National Forest, Pacific Northwest Region, U.S. Forest Service, Trout Lake, Washington.

McPherson, Penny J., David M. Hall, Vincent J. McGlone, and Nancy J. Nachtwey
 1981   *Archaeological Excavations in the Blue Mountains: Mitigation of Sites 35UN52, 35UN74, and 35UN95 in the Vicinity of Ladd Canyon, Union County, Oregon.* Western Cultural Resources Management, Inc., Boulder, Colorado.

Matson, Richard G., and Martin P. R. Magne
 2007   *Athapaskan Migrations: The Archaeology of Eagle Lake, British Columbia*. University of Arizona Press, Tucson.

Mena, Francisco, Omar Reyes, Thomas. W. Stafford Jr., and John Southon
 2003   Early Human Remains from Bano Nuevo-1 Cave, Central Patagonian Andes, Chile. *Quaternary International* 109:113–121.

Moratto, Michael J.
 1984   *California Archaeology.* Academic Press, New York.

Moss, Madonna L., and Jon M. Erlandson
 1996   The Pleistocene-Holocene Transition along the Pacific Coast of North America. In *Humans at the End of the Ice Age: the Archaeology of the Pleistocene-Holocene Transition*, edited by Lawrence Guy Strauss, Berit Valentin Eriksen, Jon M. Erlandson, and David R. Yesner, pp. 272–301. Plenum Press, New York.

Mulligan, Connie J., Andrew Kitchen, and Michael A. Miyamoto
 2008   Updated Three-Stage Model for the Peopling of the Americas. *PLoS One* 3:e3199 doi:10.1371/journal.pone.0003199.

Nelson, Charles M.
 1969   *The Sunset Creek Site (45-KT-28) and Its Place in Plateau Prehistory.* Report of Investigations 47. Washington State University Laboratory of Anthropology, Pullman.

Neves, Walter A., and Mark Hubbe
 2005   Cranial Morphology of Early Americans from Lagoa Santa, Brazil: Implications for the Settlement of the New World. *Proceedings of the National Academy of Sciences of the United States of America* 102:18309–18314.

Neves, Walter A., and Hector M. Pucciarelli
 1991   Morphological Affinities of the First Americans: An Exploratory Analysis Based on Early South American Human Remains. *Journal of Human Evolution* 21:261–273.

Newman, Thomas M.
 1966   *Cascadia Cave*. Occasional Papers of the Idaho State University Museum No. 18. Pocatello.

Ousley, Steven, and Richard L. Jantz
 1996   Fordisc 2.0: Personal Computer Forensic Discriminant Functions. Department of Anthropology, University of Tennessee, Knoxville.

Pennefeather-Obrien, Elaine E., and Michael Strezewski
 2002   An Initial Description of the Archaeology and Morphology of the Clarks Fork Skeletal Material, Bonner County, Idaho. *North American Archaeologist* 23:101–115.

Perez, S. Ivan, Valeria Bernal, and Paula N. Gomez
 2007   Evolutionary Relationships among Prehistoric Human Populations: An Evaluation of Relatedness Patterns Based on Facial Morphometric Data Using Molecular Data. *Human Biology* 79:25–50.

Powell, Joseph F.
2004   *The First Americans: Race, Evolution, and the Origin of Native Americans*. Cambridge University Press, Cambridge.

Powell, Joseph F., and Walter A. Neves
1999   Craniofacial Morphology of the First Americans: Pattern and Process in the Peopling of the Americas. *Yearbook of Physical Anthropology* 42:153–188.

Powell, Joseph F., and Jerome C. Rose
1999   *Report on the Osteological Assessment of the "Kennewick Man" Skeleton (CEN-WW.97. Kennewick)*. Report submitted to the National Park Service, United States Department of the Interior, Washington, D.C.

Prentiss, Anna M., and David S. Clarke
2008   Lithic Technological Organization in an Evolutionary Framework: Examples from North America's Pacific Northwest Region. In *Lithic Technology: Measures of Production, Use and Curation*, edited by William Andrefsky Jr. Cambridge University Press, Cambridge.

Prentiss, William C., and James C. Chatters
2003   Systems as Individuals: Toward a Taxic Macroevolutionary Theory of Culture Change. *Current Anthropology* 44:33–58.

Raadsheer, Maarten C., T. M. van Eijden, F. C. van Ginkel, and B. Prahl-Andersen
1999   Contribution of Jaw Muscle Size and Cranial Morphology in Human Bite Force Magnitude. *Journal of Dental Research* 78:31–42.

Rice, David G.
1969   *Preliminary Report: Marmes Rockshelter Archaeological Site, Southern Columbia Plateau*. Washington State University Laboratory of Anthropology, Pullman.

Ridenour, Dora I.
2006   Results of Investigations at We'epts Pa'axat (45NP453), Clearwater Region, North Central Idaho. Unpublished Master's thesis, Department of Sociology, Anthropology, and Criminal Justice, University of Idaho, Moscow.

Roseman, Charles C.
2004   Detecting Interregionally Diversifying Natural Selection on Modern Human Cranial Form by Using Matched Molecular and Morphometric Data. *Proceedings of the National Academy of Sciences of the United States of America* 101:12824–12829.

Rousseau, Michael K.
1991   Landels: An 8500 Year-Old Deer Hunting Camp. *Midden* 23(4):6–9.

Sancetta, Constance, Mitchell Lyle, Linda Heuser, Rainer Zahn, and John Platt Bradbury
1992   Late Glacial to Holocene Changes in Winds, Upwelling, and Seasonal Production of the Northern California Current System. *Quaternary Research* 38: 359–370.

Schalk, Randall F., Stephen M. Kenady, and Michael C. Wilson
2007   Early Post-Glacial Ungulates on the Northwest Coast: Implications for Hunter-Gatherer Ecological Niches. *Current Research in the Pleistocene* 24:182–184.

Schroeder, Kari B., Theodore G. Schurr, Jeffrey C. Long, Noah A. Rosenberg, Michael H. Crawford, Larissa A. Tarskaia, Ludmilla P. Osipova, Sergey I. Zhadanov, and David G. Smith
2007   A Private Allele Ubiquitous in the Americas. *Biology Letters* 3:218–223.

Schurr, Theodore
2004   The Peopling of the New World: Perspectives from Molecular Anthropology. *Annual Review of Anthropology* 33:551–583.

Sheppard, John C., Peter Wigand, Carl E. Gustafson, and Meyer Rubin
1987   A Re-evaluation of the Marmes Rockshelter Radiocarbon Chronology. *American Antiquity* 52:118–125.

Sprague, Roderick, and John D. Combes
    1966   *Excavations in the Little Goose and Lower Granite Dam Reservoirs, 1965.* Report of Investigations 37. Washington State University Laboratory of Anthropology, Pullman.
Steele, D. Gentry, and Joseph F. Powell
    1992   Peopling of the Americas: Paleobiological Evidence. *Human Biology* 64:303–336.
    1993   Paleobiology of the First Americans. *Evolutionary Anthropology* 2:138–146.
Stine, Scott
    1994   Extreme and Persistent Drought in California and Patagonia during Mediaeval Time. *Nature* 369:546–549.
Tamm, Erika, Toomas Kivisild, Maere Reidla, Mait Metspalu, David G. Smith,
Connie J. Mulligan, Claudio M. Bravi, Olga Richards, Cristina Martinez-Labarga,
Elsa K. Khusnutdinova, Sardana A. Federova, Maria V. Golubenko, Vadim A. Stepanov,
Marina A. Gubina, Sergey I. Zhadanov, Ludmila P. Ossipova, Larisa Damba,
Mikhail I. Voevoda, Jose E. Dipierri, Richard Villems, and Ripan S. Malhi.
    2007   Beringian Standstill and Spread of Native American Founders. *PLoS One* 2(9):e829. doi:10.1371/journal.pone.0000829.
Taylor, Royal E., Donna Kirner, Jon Southon, and James Chatters
    1997   Radiocarbon Dates of Kennewick Man. *Science* 280:1171–1172.
Torrence, Robin
    1983   Time Budgeting and Hunter-Gatherer Technology. In *Hunter-Gatherer Economy in Prehistory, a European Perspective*, edited by Geoffrey Bailey, pp. 11–22. Cambridge University Press, Cambridge.
Warren, Claude N., Alan L. Bryan, and Donald R. Tuohy
    1963   The Goldendale Site and Its Place in Plateau Prehistory. *Tebiwa* 32(2):68–185.
Willig, Judith A., and C. Melvin Aikens
    1988   The Clovis-Archaic Interface in Far Western North America. In *Early Occupation in Far Western North America: The Clovis-Archaic Interface*, edited by Judith A. Willig, C. Melvin Aikens, and John L. Fagan, pp. 1–40. Nevada State Museum Anthropological Papers No. 21, Carson City.

4. | Human Skeletal Variation and Environmental Diversity in Northwestern North America

Jerome S. Cybulski

Summary Statement: Body size, mass, limb proportions, and long-bone strength were compared in representative skeletal samples from the Northwest Coast and Plateau culture areas. Ecological differences and subsistence behavior were studied by stable isotope ratios (δ^{13}C alone or δ^{13}C and δ^{15}N depending on sample availability) and dental pathology (tooth caries and periapical abscesses). Intra-area and temporal variation were examined especially for the Plateau where some of the earliest North American remains are known, including individuals from Kennewick, Washington, Clark's Fork, Braden, and Buhl in Idaho, and Gore Creek, China Lake, and Big Bar Lake in British Columbia. Ancient DNA findings were considered as appropriate. All told, 245 sites of collection were investigated.

Isotope distributions and the frequencies of dental lesions differed between the adjacent culture areas, reflecting ecological differences. Plateau samples also showed variation in δ^{13}C and δ^{15}N values by time and space commensurate with expected food resources. People of the Plateau, especially the men, were taller and leaner than those of the Northwest Coast and had longer and stronger lower limb bones, better adapted overall for hunting and gathering. People of the Northwest Coast had short bodies with strongly developed upper limbs and relatively short lower limbs, well suited for a marine oriented subsistence base dependent on canoe travel. Kennewick Man was typically Plateau adapted. Somewhat greater variability in Plateau females suggests coastal immigrants, possibly accounting for the genetic characterization of a 5000-year-old woman from Big Bar Lake.

Human Variation in the Americas: The Integration of Archaeology and Biological Anthropology, edited by Benjamin M. Auerbach. Center for Archaeological Investigations, Occasional Paper No. 38. © 2010 by the Board of Trustees, Southern Illinois University. All rights reserved. ISBN 978-0-88104-095-1.

Introduction

Variation in human body size, body mass, and limb proportions has been related to climate or latitude and postcranial robusticity and bone strength to variation in habitual behavior, patterns of mobility, subsistence behavior, and physique (Katzmarzyk and Leonard 1998; Ruff 1994, 2002; Stock 2006; Westcott 2006). This paper compares and contrasts relevant metrical postcranial derivatives in two adjacent aboriginal culture areas of northwestern North America, the Northwest Coast and Plateau. Of long-standing interest to anthropologists for studies of social and political organization, economic complexity, technological innovation, and language diversity, the two bordering areas, sometimes identified together as the Pacific Northwest, present largely different ecologies defined by the natural habitats in which their human populations functioned and adapted. Those differences are mirrored by differences in body size, limb proportions, and robusticity and by differences in bone chemistry and dental pathology related to subsistence behavior. In addition to skeletal metrics, stable isotope ratios and frequencies of dental caries and periapical abscess lesions are also compared in this paper.

This study places its emphasis on the two cultural areas as unit samples. However, intra-area variation, especially on the Plateau, is also considered and attention is given to variation among individuals as represented by some of the earliest archaeologically known human remains from North America. They include finds dated between 10,000 and 8000 B.P. (radiocarbon years before present) at Gore Creek, British Columbia, and Kennewick, Washington (Chatters 2000; Cybulski et al. 1981) and later remains from China Lake and Big Bar Lake in British Columbia, dated to about 5000 B.P. (Cybulski et al. 2007; Malhi et al. 2007). The place of these individuals is discussed within the larger Northwest Coast versus Plateau picture. Other early as well as later remains are considered in comparative context and, for the China Lake and Big Bar Lake skeletons, ancient DNA findings are also considered.

Materials and Methods

Geographical Extents

The Northwest Coast and Plateau are two of ten culture areas which have been defined for aboriginal North America. Although each of the two has occasionally been contracted or expanded geographically and ethnographically, depending on author, such alterations have been minor since the culture area concept was introduced in the late 1800s. The definitions used here closely follow those authored in volume 7 and volume 12 of the Handbook of North American Indians (Suttles, ed. 1990; Walker 1998; see also Richards and Rousseau 1987 for the extent of the Canadian Plateau as considered here), tempered in part by the available skeletal remains that have been reported and studied for each area.

The Northwest Coast is an elongate strip that extends along the north Pacific Coast from Yakutat Bay in the Gulf of Alaska to the Oregon-California border, a distance of about 2,500 km. The lower half (approximate) of its length is separated from the adjacent Plateau by the Coast Mountains of British Columbia and the Cascade Range of Washington and Oregon. The Plateau is a larger cultural landmass (about 60 percent larger in geographical area) that extends eastward to the Rocky Mountains and includes south-central and southeastern British Columbia, much of Washington State and Idaho, northern Oregon and the western interior of that state, and a significant portion of Montana that abuts the Plains culture area to the east. To the south is the Great Basin culture area and to the north is the Subarctic.

Population

At the time of initial contact with Europeans (ca. 175 B.P.), the Northwest Coast was the second most densely populated culture area of North America after California. According to kilometric figures provided by Kroeber for different tribal groups (1939:134–142) and more recent population estimates for those same groups proposed by Boyd (1990), the Northwest Coast was inhabited by almost 190,000 people or about 47 per 100 km^2. This high density appears to have been largely the result of the coastal peoples' successful maritime adaptation to rich offshore, littoral, sublittoral, and related riverine food resources. Numerous coastal villages lined the beaches and bluffs of the multiple fjords, inlets, bays, sounds, and rivers found especially along the northern two-thirds of the coast including the Alaska panhandle, the coast of British Columbia, and the northern coast of Washington State. The people of this segment, particularly British Columbia from where most of the human skeletal remains have been studied, epitomized Northwest Coast culture with its highly settled nature, social stratification, complex technologies for exploiting marine and riverine resources and for woodworking, and a sophisticated monumental art style (Drucker 1965).

The population of the Plateau was much smaller at 87,000 people (Boyd 1998) and a calculated density of 13 per 100 km^2 (utilizing Kroeber's 1939 area figures). The economy was more land oriented, as might be expected, although at the time of contact with Europeans (ca. 150 B.P.) dense permanent settlements were linearly established along major rivers such as the Fraser and Thompson in British Columbia and the middle and upper Columbia south and north of the Canada-U.S. border, and their tributaries. Those rivers emptied into the sea, providing seasonal salmon runs. While this would seemingly define a dependence on fishing by at least late in the prehistoric period and perhaps earlier, permanent and semipermanent camps were established in upland areas for hunting and root gathering. Indeed, the resource base appears to have been much more varied on the Plateau than on the coast (Cybulski 2006a:542).

Human Skeletal Remains and Sample Representation

Just as there was a disparity in population size between the Northwest Coast and Plateau, there is a corresponding disparity in the overall sample sizes of human skeletal remains that could be considered for study. In addition to my own data, the collection of which began in 1967 (e.g., Cybulski 1975a) and continues to the present day (e.g., Cybulski et al. 2007), I've depended heavily on the work of others as cited elsewhere (Cybulski 1990, 2006a; Cybulski et al. 2007) and in this compendium. As might be expected, there is sampling inconsistency in the types of data reported here (i.e., dental pathology versus isotopic readings versus postcranial metrics) as well as in the specifics of each type (e.g., estimates of stature versus limb proportions, or stable carbon versus nitrogen isotope readings). Subregional representations within each culture area, however, are not badly compromised.

Skeletal data from 245 designated archaeological sites or documented localities of museum collection entered into my analysis. Included were 141 coastal locations and 104 from the Plateau. The appendix reports the sites and references in 10 subregions for the Northwest Coast and 11 for the Plateau. The subregions were carefully constructed to consider potential regional variation within each culture area concomitant with ecological variation, as well as to demonstrate the broad representativeness of each culture area. Not all of the sites within each subregion could, however, be consistently included for all types of data.

As intimated in the introduction, there is temporal variation within the samples. Included are remains that I've described elsewhere as Ancient (ca. 10,000–5500 B.P.), Early (5500–3500 B.P.), Middle (3500 1800/1500 B.P.), or Late Prehistoric (1800/1500–175/150 B.P.), Prehistoric (where the more specific category is unknown), Historic, or unknown (Cybulski 2006a:532). The Northwest Coast sample has an overall better representation of Early and, especially, Middle period remains than that of the Plateau, while the Plateau sample is biased toward the Late Prehistoric and Historic periods. Most Ancient remains considered in this study, however, save one individual from Alaska, were discovered on the Plateau. Samples represented by each period are discussed in detail as appropriate.

Data Studied

Measurements were not specifically collected originally to address questions of potential variation in body size, limb proportions, and robusticity as considered in this paper. Rather, I used whatever raw measurements were available to assemble the indicators, given the long history of data collection, especially for the Northwest Coast. All measurements utilized in this study were of adults with no disparity evident between the two major cultural-environmental groupings in the distribution of young, middle, and old adults.

Estimates of stature from long-bone lengths were used to assess body size. Rather than combine estimates that had been reported by different authors with my own and thereby introduce uncertainties owing to the use of different formulas, I recalculated almost all statures investigated here from the original individual long bone lengths that had been reported using the formulas of Trotter and Gleser

(1958) for Mongoloid males and Trotter and Gleser (1952) for white females. While those regression formulas have been questioned for their application to ancestral North American Indian remains (e.g., Chatters 2000), they give, on the average, acceptable results for Northwest Coast skeletal remains of the Historic period. Skeletally derived means from that time level (ca. 175–75 B.P.) compare favorably with those calculated from the measured statures of living individuals that belonged to the same ethnic groups (Cybulski 1978). Table 4-1 summarizes the similarities. There is little reason to suspect that the formulas do not also apply to earlier time periods since there are no differences in body build and proportions as could be determined skeletally. For the most part, I avoided reliance on tibia length in the calculation of stature because of the different ways tibiae can be measured and reported by different authors (Jantz et al. 1994). The overall emphasis was placed on lower limb lengths utilizing the femur, or fibula or both.

Stature is not strictly an indicator of body size without body mass or weight being taken into account. As recommended by Auerbach and Ruff (2004), body mass was estimated from formulas involving measurements of the diameter of the femur head using the average of the three methods tested by those authors. Limb proportions were compared by means of the humero-femoral index (Olivier 1969:262). The sample sizes were small but larger than those that would be available if the more comprehensive intermembral index, which depends on the lengths of four limb bones (humerus, radius, femur, and tibia), had been used. Again, there is the problem of tibia length and how it might have been measured.

External measurements were available for the shafts of the femur and, to a lesser extent, the humerus (anteroposterior and mediolateral diameters in both instances). Those measurements were used to estimate long-bone strength or robusticity in lieu of cross section data, which are obtained radiographically but were generally unavailable. While cross section data might be considered more accurate for studies of robusticity, data based on external measurements are acceptable substitutes (e.g., Westcott 2006).

Beattie (1981) developed a "deltoid index" to quantify robusticity he recognized in the humeri of Northwest Coast skeletal remains. The index evaluates circumferential cross-sectional variation of the shaft without reference to the length of the bone ([circumference of shaft at deltoid / least shaft circumference] × 100). This index may be useful as an indicator of the qualitative difference sometimes observed between coastal and interior humeri in British Columbia (Cybulski 2005). A probably better quantifier of humeral robusticity that has been found useful in other geographic areas is "HSRI-2." This is a humeral shaft index that is calculated by the sum of the shaft's minimum and maximum diameters at midshaft as a proportion of maximum length (Heathcote et al. 2010). Both indices are reported here.

In the case of the femur, I investigated the index of platymeria, which measures the shape of the shaft at the subtrochanteric or upper level (Olivier 1969:263); the shape at midshaft (FMS = anteroposterior / mediolateral diameters, according to Westcott 2006); and femur midshaft robusticity (FMR = [(anteroposterior + mediolateral femur shaft diameters) × 100] / femur head diameter, according to Westcott 2006). In theory, higher values in all instances would indicate stronger femoral shafts which would sustain greater loading forces than those with

Table 4-1. *Mean Statures from Bones of Historic Period Northwest Coast Males Compared with Those of Males Living in the Late Nineteenth Century*

	Southern Kwakiutl		Coast Salish		Nootka	
	Living	Skeletal	Living	Skeletal	Living	Skeletal
Sample size	49	25	4	13	4	26
Mean	163.31	163.56	163.48	163.76	161.55	160.89
SD	5.20	4.18	2.67	4.87	6.69	3.86
Minimum	154.00	157.84	160.60	159.00	156.50	155.64
Maximum	174.90	176.01	166.30	174.43	171.10	170.89

Source: The "Living" data were calculated from individual measurements reported by Franz Boas (1892, 1895; Boas and Farrand 1899).

smaller values. In this connection, platycnemia of the tibia (Olivier 1969) was also investigated although smaller (platycnemic) values would be expected for stronger tibia shafts in the anteroposterior direction (Lovejoy et al. 1976).

All told, 684 adults' limb measurements and their derived characters or indices including stature were studied: 549 from the Northwest Coast and 135 were from the Plateau. Not every one of them, however, could contribute equally to the 9 characters presented here.

Stable Isotopes

Three hundred fifty-three samples were available for study of stable carbon isotope variation including 249 for the Northwest Coast and 104 for the Plateau. The vast majority of the samples came from adults with none used from any tested or reported individual below a physiological age of 12–16 years. Sixty percent of the delta values per mil were taken from the reported work of Chisholm in British Columbia, which involved direct readings (Chisholm 1986, 1996; Chisholm et al. 1983; Howe et al. 1994; Lovell et al. 1986). The rest were drawn from Ball (1981), Carlson and Hobler (1993), Chatters and Zweifel (1987), Condrashoff (1984), Curtin (1984:15), Cybulski (1993, 1994, 2008a), Cybulski and colleagues (1981), Dixon (1999), Green and colleagues (1986, 1998), Leonhardy and colleagues (1987), Oliver (1992a), Pennefather-O'Brien and Strezewski (2002), Pokotylo and colleagues (1987), Taylor (2001), Taylor and colleagues (1998), and Thom (1995:79). Almost all the latter values were by-products of radiocarbon dating.

Sixty-three bone samples were tested for both nitrogen and carbon isotopes. They were derived largely from isotopic values that accompanied radiocarbon dates from bone that I had tested for the British Columbia coast and Canadian Plateau. A few were reported by others also as radiocarbon dated derivatives (Brady 2005; Green et al. 1998).

Dental Pathology

The frequencies of dental caries and periapical abscess lesions were also studied as potential indicators of subsistence behavior. In all, the jaws of 889 individuals entered into the analyses, 791 for the Northwest Coast (79 sites or localities of collection) and 98 for the Plateau (52 sites or localities of collection). I personally studied 781 sets of teeth or jaws, the data for which are largely unpublished in terms of site reports but have been summarily included for pathology elsewhere (Cybulski 1990, 2006a; see also Cybulski 1973, 1980a, 1980b, 1982a, 1986, 1988a, 1989, 1991, 1992, 1993, 1994, 1997a, 2005, 2008b and Cybulski et al. 2007 for specific site treatments).

Data for the other teeth or jaws, not necessarily reported individually but often grouped, were assembled from: Breschini (1979); Chatters (1985, 2000); Curtin (1986, 1990); Curtin and Lawhead (1985); Gordon (1974); Hall and Haggerty (1981); Krantz in Boreson (1985); Krentz in Chatters and Zweifel (1987); Landis and colleagues (1984); Leonhardy and colleagues (1987); McLeod and Skinner (1987); Mulinski in Chance and colleagues (1977); Oliver (1992b, 1992c, 1995a, 1995b, 1997a, 1997b); Oliver and Crellin (1991); Pennefather-O'Brien and Strezewski (2002); Pokotylo and colleagues (1987); Richards (1982); Skinner (1987a); Sprague and Birkby (1970, 1973); Stewart (1950); Sumpter (1982).

Although the numbers of individuals studied have been reported here, the frequencies of dental pathology were, in fact, compiled per tooth in the case of carious lesions and per tooth socket in the case of abscess lesions following the methodology reported elsewhere (Cybulski 1992:129–134). Those sample sizes are shown in the Results section.

For all categories of data, mean differences between the two culture areas were tested for statistical significance using the two-sample T test in SYSTAT 6.1.2 for Windows.[1] Proportionate differences were tested using chi-square or chi-square with Yates's correction, or Fisher's exact test as appropriate.

Results

Subsistence Behavior

Both stable isotopes and dental pathology show significant differences between Northwest Coast and Plateau samples, complementing their respective consumers' different emphases on subsistence. At the same time, internal variation is apparent in the isotope data, most notably for the Plateau.

Stable Isotopes

Figure 4-1 demonstrates the variation in $\delta^{13}C$ values (for which most of the data are available in this genre) in seven sample groupings using box plots (Wilkinson et al. 1996:340–342). The boxes enclose the middle 50 percent range of the values for each group, and the vertical line within each group is the median.

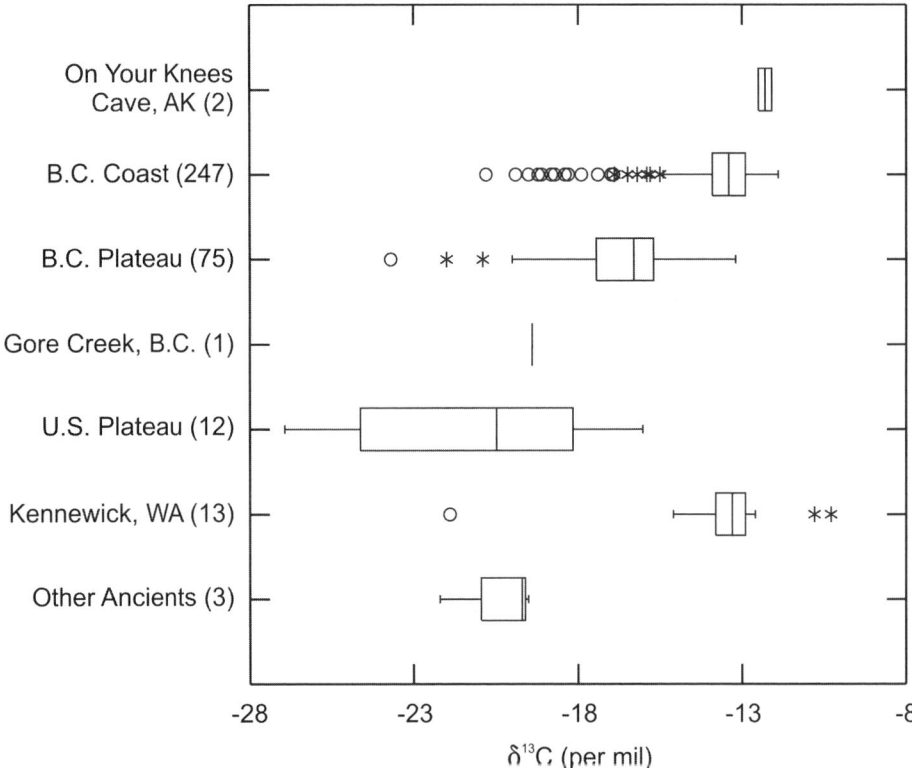

Figure 4-1. *Box plots of the distribution of stable carbon isotope ratios ($\delta^{13}C$ values per mil) in samples from northwestern North America including the Northwest Coast and Plateau (sources cited in the text). Numbers in parentheses represent the number of individuals tested except in the cases of On Your Knees Cave, Alaska, and Kennewick, Washington, where multiple tests of the same individual are indicated. Other Ancients include Clark's Fork, Buhl, and DeMoss in Idaho. (See text in Results section for explanation of the symbols.)*

The stems or whiskers, and the special symbols beyond, identify each group's range. The asterisks denote outliers and the small circles the extreme values in the ranges. The direction in which the data are skewed or nonsymmetrical for each group is also apparent.

In Figure 4-1, the carbon isotopic data are grouped for the British Columbia coast, the British Columbia Plateau, and the U.S. Plateau. They are reported separately for the Ancient human remains from On Your Knees Cave, Alaska (two readings from Dixon 1999); Gore Creek, British Columbia (one reading from Cybulski et al. 1981 and Chisholm and Nelson 1983); and Kennewick, Washington (13 readings from Taylor 2001 and Taylor et al. 1998). Three other Plateau Ancients are grouped, including Clark's Fork (Pennefather-O'Brien and Strezewski 2002), Buhl (Green et al. 1998), and DeMoss (Green et al. 1986), all sites in

Idaho. Buhl appears to be the oldest of this triad, radiocarbon dated to about 10,700 B.P., while Clark's Fork appears to be the youngest, dated at about 5900 B.P. Kennewick was originally dated at about 8400 B.P. (Chatters 2000) and is commensurate with the antiquity of Gore Creek at about 8300 B.P. (Cybulski et al. 1981). The DeMoss burial site, comparable in geochronology to Clark's Fork, comprised multiple individuals but only one sample was dated and tested for carbon isotopes.[2]

The coast samples show a tendency toward high or less negative $\delta^{13}C$ values whereas the Plateau values show a tendency toward low or more negative values. This appears to reflect differences in the relative consumption of marine and terrestrial protein, the latter traceable to C_3 grasses and plants. In a landmark study of stable carbon isotopes for the British Columbia coast, Chisholm and colleagues (1983; see also Chisholm et al. 1982) reported an average $\delta^{13}C$ value of -13.3 per mil for 37 samples they considered accurate and concluded from this figure that an average 90 percent of the dietary protein for coastal British Columbia consumers came from marine sources.

Later published research demonstrated more negative values for samples from the British Columbia interior (Lovell et al. 1986). In that instance, no overall average was discussed because meaningful environmental variation was found among the 21 localities studied. Those locations ranged from along the Fraser River, near the western boundary of the Canadian Plateau, to the Slocan Valley–Kootenay region near the eastern boundary. Selecting eight localities along the Thompson River in the western half of this area, covering a distance of about 200 km, the authors noted in the $\delta^{13}C$ values a "clear trend" for decreasing consumption of migratory anadromous salmon with increasing distance of the localities upstream. Another way of seeing this is that while an intake of marine protein via the consumption of anadromous salmon could be cited to account for less negative $\delta^{13}C$ values, more negative values are indicative of a greater Plateau culture dependence on terrestrial protein.

The data presented here further suggest a greater C_3-derived terrestrial protein dependence on the U.S. Plateau than on the Canadian Plateau, an exception being the data reported for Kennewick, which in the main simulate the $\delta^{13}C$ values of British Columbia coastal consumers. The authors of those figures interpreted them to indicate that the individual found at Kennewick, Washington, depended heavily on a marine diet presumed to have come from fishing anadromous salmon in the Columbia River (Taylor 2001; Taylor et al. 1998). Figure 4-1 clearly indicates that Kennewick would have been the only tested U.S. Plateau individual to have done so, either in the ancient or more recent prehistoric past. The other individuals have reported $\delta^{13}C$ values that hover around a median of -20.0 per mil, which according to Lovell and colleagues (1986) reflects a median 100 percent intake of C_3-derived terrestrial protein.

It has been noted that at around the time of Kennewick, stream flows were reduced in the Columbia Basin and fish runs, therefore, would have been severely limited (Chatters 1998:44). As discussed elsewhere, it is more likely that at 9000 to 8000 B.P., Kennewick Man hunted large herbivores, such as bison and elk, which, in the main, consumed C_4 grasses (Cybulski 2006a:543). Conditions in the Columbia Basin would have been right for such plant life at that time since the environment was warm and arid, in contrast to the northern Plateau which was relatively cooler and wetter (Chatters 1998) and, therefore, conducive to C_3-plant proliferation (hence, the more negative $\delta^{13}C$ value for Gore Creek). It is generally known that C_4-derived protein and marine protein give identical stable carbon isotope signatures in humans. Certainly, in body size, limb proportions, and long-bone strength, as revealed below, Kennewick Man resembled his later Plateau brethren rather than men of the Northwest Coast.

The extent of dietary variation on the Plateau is more succinctly shown in Figure 4-2, which illustrates the distribution of 63 bone samples (each a separate person) tested for both carbon and nitrogen isotope ratios. In this case, the addition of the N^{15}/N^{14} signatures gives more precise insight into the differences in protein consumption (Schoeninger et al. 1983) and time would seem to enter more critically into the dietary explanation equation. Unfortunately, no nitrogen information has yet been reported for Kennewick, DeMoss, or Clark's Fork, but it has been reported for Buhl.[3]

The coastal signatures are relatively homogeneous. In part, this may be the product of regional differences on the Northwest Coast. Most of the currently available data come from sites on the northern coast of British Columbia including Prince Rupert Harbour and the Nass River valley (Cybulski 1992, 2008a). Only two samples are from outside this region and they are from sites on the Strait of Georgia (Cybulski 2008a). It has been suggested that marine-protein consumption decreases among coastal consumers who have access to large lowland areas where, presumably, the hunting of C_3-plant-consuming deer is feasible (e.g., Chisholm 1986). Regional carbon isotope values, which entered into the production of Figure 4-1, identify this possibility for segments of the Strait of Georgia and for Haida Gwaii (the Queen Charlotte Islands). Variability in $\delta^{13}C$ values in those regions is greater than in others but still clearly directed toward the overall coastal norm.

It is nonetheless evident that the coastal individuals shown in Figure 4-2 depended heavily on marine-protein sources and that there was little variation. Five species of salmon were the mainstay of the Northwest Coast diet but other marine foods also played an important role depending on local circumstances (Ames and Maschner 1999:113–123). At least eight other kinds of food fish were taken from the sea, as were a variety of sea mammals, seaweeds and kelp, and a variety of waterfowl including shorebirds and migratory seabirds (Ames 2003; Suttles 1990). The maritime subsistence base was probably established, if not firmly entrenched, in some parts of the coast by about 6000 to 5500 B.P. and flexibly under way with land-game hunting during earlier Archaic period times (Ames and Maschner 1999:123–139; see also Hebda and Frederick 1990).

Plateau inhabitants were more dependent on large game animals and plant foods although the degree of that dependence varied according to habitat and proximity to major river drainages (Cybulski 2006a:542). People living along the Fraser and Thompson rivers in south-central British Columbia and along the middle Columbia River in Washington had ready access to anadromous salmon though not necessarily all species enjoyed by the people of the coast. Even in

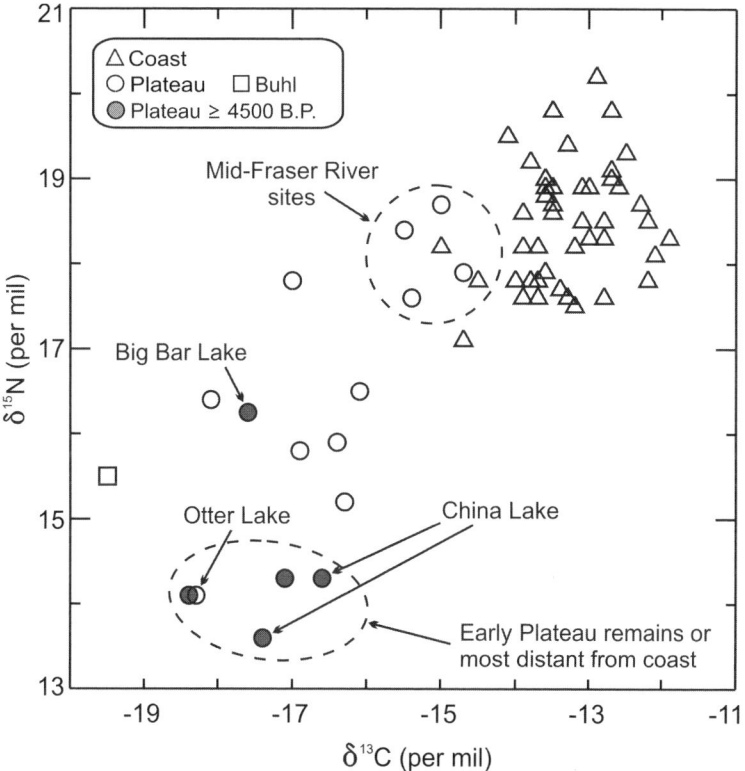

Figure 4-2. *Scatter plot of the distribution of stable nitrogen ($\delta^{15}N$ values per mil) versus stable carbon isotopes ($\delta^{13}C$ values per mil) for 63 Northwest Coast and Plateau individuals (descriptive details explained in the text).*

these locations, however, salmon need not have been the focus of subsistence that it was for the coast. The hunting of land mammals such as elk, deer, caribou, and even bison by people in the eastern part of the northern (Canadian) Plateau figured prominently, the process attaining ritual importance for some ethnographic groups such as the Shuswap (Ignace 1998). While fishing and the consumption of salmon has been emphasized in some contexts (e.g., Hewes 1998), beginning especially about 4000 B.P. (Kuijt 1989), and was certainly important to ethnographic groups such as the Lillooet and Lower Thompson, the much more flexible subsistence pattern involving land-derived resources, especially large ungulates, appears deeply rooted archaeologically (Ames et al. 1998; Chatters and Pokotylo 1998; Pokotylo and Mitchell 1998; Roll and Hackenberger 1998).

Of interest in the Figure 4-2 scatter diagram is the broadly spaced distribution of the Plateau samples particularly in contrast to the relatively tightly knit Northwest Coast samples. In part, the Plateau samples follow the "clear trend" by distance in $\delta^{13}C$ values associated with relative salmon accessibility as noted by Lovell and colleagues (1986). In this case, however, the data are more precise and time is an important player.

The four Plateau samples closest to those of the coast are from archaeological sites on the middle Fraser River near the towns of Lytton (EbRj-1) and Lillooet (EdRl-10) (Dawson 1892; Sanger 1963). They are late in time, dating from 1230 B.P. to 990 B.P. as shown by recently obtained radiocarbon dates (Cybulski 2008a). Their nearest Plateau neighbors in the scatter diagram are late as well (ca. 1000 B.P.) but are also farther removed geographically, located for the most part in the Kamloops Lake area or near the city of Kamloops, some 100 km to the east.

The Plateau samples most distinctive in comparison to the Northwest Coast samples tend to be the oldest. They are from China Lake (EiRm-7; $n = 2$), dating to around 5100 B.P. (McKendry 1983; Malhi et al. 2007), and Pritchard (EeQw-21; $n = 2$), dating to around 4500 B.P. (Brady 2005). While this group is joined by a much later skeleton from Otter Lake (EcQt-12), dating to 1280 B.P. (Cybulski 2005), that site—near Armstrong, British Columbia—is also the most geographically distant in the series from both the Thompson and Fraser rivers.

China Lake and Pritchard date from the Middle Prehistoric period of the Canadian Plateau, prior to the Late Prehistoric introduction of salmon abundance on the Fraser and Thompson rivers (Kuijt 1989). They are, however, also somewhat removed geographically from the Fraser and Thompson rivers and they are not riverine sites. One might presume that these people were not primarily fishers but that their subsistence practices revolved more around hunting land game and plant collecting.

One of their contemporaries, on the other hand, a woman from Big Bar Lake dated at around 5000 B.P. (Cybulski et al. 2007), seems to have depended more on salmon like her neighbors later in time in the Cariboo and Kamloops Lake regions. However, it is possible that this woman originated on and migrated from the coast. She was of mtDNA haplogroup A and her haplotype presented an identical match with a modern Nuu-chah-nulth (Nootka) individual from the west coast of Vancouver Island (Cybulski et al. 2007:70). A second Nuu-chah-nulth individual was found to be only one mutational step away from the Big Bar woman, a second possible indicator of genetic origin. In the report on her skeletal remains and burial recovery, the close genetic similarity was interpreted to indicate regional continuity across the Pacific Northwest over the last 5,000 years (Cybulski et al. 2007). Not entertained was an alternative explanation that she could have been a coastal migrant. As detailed below, her body build and limb proportions could be interpreted as being those of a coastal individual. Perhaps, the Big Bar woman migrated to the Plateau some ten years or more prior to her death, within a period sufficiently long enough to have witnessed partial turnover in her stable isotope composition (cf. Corr et al. 2009).

Dental Pathology

Skeletal remains from the Northwest Coast and Plateau commonly exhibit dental conditions associated with hunter-gatherers or fisher-gatherers (Cybulski 2006a:543–544). Both groups display high rates of occlusal attrition, few tooth caries (cavities), and frequent occurrences of periapical abscesses. Nonetheless, statistically significant differences in the frequencies of caries and abscesses can be seen between Northwest Coast and Plateau samples, likely reflecting differences in the relative consumption of traditional plant foods and their types.

As may be seen in Figure 4-3, clear differences occur in the frequencies of caries and abscesses between historic and prehistoric samples from the Northwest Coast. The historic sample shows more caries and fewer abscesses than the prehistoric, an inverse relationship that likely reflects a lower rate of occlusal attrition. The change following contact was likely the result of the introduction of soft European foods that were also high in carbohydrate content. The prevalence of caries was kept low, however, owing to a continued dependence on the traditional fish diet (Cybulski 1990:58).

The frequencies for the Plateau follow a similar visual pattern but in neither case is the difference statistically significant. Perhaps the samples are just too small to properly test the reality that may have existed, but the dental conditions could also be reflecting a more flexible, varied Plateau diet than that of the Northwest Coast.

What is clear is that there is a significantly higher frequency of caries in prehistoric and historic Plateau samples than in coeval samples from the Northwest Coast. For the state of Oregon, Hall and colleagues (1986) found the highest frequency of caries in samples of teeth studied from areas where dependence on plant resources, such as camas bulbs, appears to have been greatest. Indeed, ethnobotanical studies indicate that Plateau peoples depended much more heavily on plant foods in their diets than did people of the Northwest Coast. On the northern and southern Plateau, plant foods contributed more than 50 percent of the caloric intake (Hunn et al. 1998). Carbohydrates were plentiful in the guise of root plants such as balsamroot, bitterroot, yellow avalanche lily, and camas bulbs. Their consumption, often as a sweet dessert, likely accounted for the higher prevalence of caries in Plateau dentitions both in the prehistoric and historic periods.

Skeletal Metric Derivatives

Table 4-2 presents postcranial metrical derivatives. Statistically significant differences between the two culture areas are identified by greater than or less than symbols, which also indicate the direction of differences in the sample means and the level of probability. For nine comparisons, the differences were highly significant with a probability level of less than .01 utilizing a two-tailed t-test (less than .001 in seven instances). The measurements included stature, humerus robusticity (HSRI-2), the index of platymeria, and femur midshaft shape (FMS) for both sexes. Females were significantly different at the 5 percent level in the humero-femoral index, and males were significantly different at that level in the deltoid index and femur midshaft robusticity (FMR). There were no statistically significant differences between females in the means of the deltoid index and femur midshaft robusticity or in body mass and the index of platycnemia for both sexes (shown by equals signs in Table 4-2).

As shown by statistical significance, sex dimorphism was indicated for both culture areas in stature, body mass, and femur midshaft shape, and for the

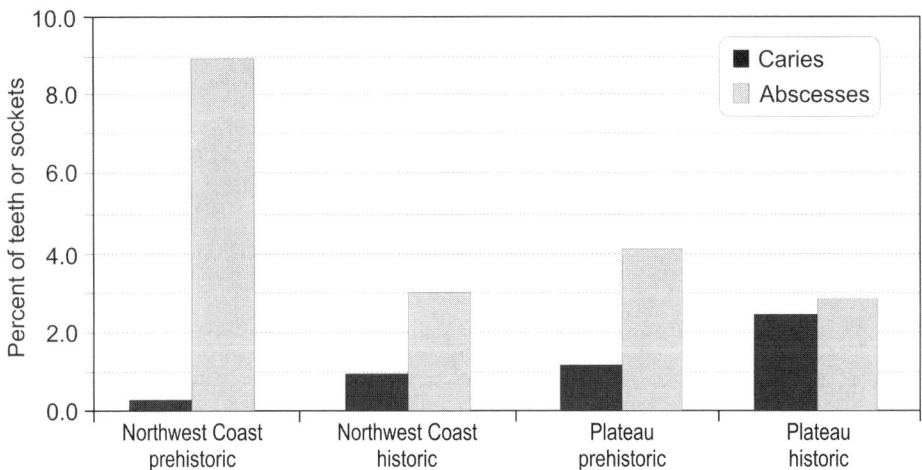

Figure 4-3. *Distribution of caries and abscesses in samples of Northwest Coast and Plateau teeth and alveolar sockets. Northwest Coast sample sizes are 7,435 teeth and 7,300 sockets for the prehistoric period; 6,925 teeth and 10,990 sockets for the historic period. Plateau sample sizes are 941 teeth and 1,149 sockets for the prehistoric period; 694 teeth and 839 sockets for the historic period.*

Northwest Coast alone in humero-femoral index, humerus robusticity (HSRI-2), femur midshaft robusticity (FMR), and the index of platycnemia.

Mean body mass differs between the sexes, as might be expected, but does not differ significantly between culture areas. Mean stature, however, does differ between the two areas as do upper and lower limb proportions as measured by the humero-femoral index. Plateau lower limbs are longer than those of coastal inhabitants relative to the upper limb insofar as the femur and humerus are concerned. This is not as clear-cut for the females, however, as it is for the males. When the underlying entry data are evaluated, the Plateau female sample shows significantly longer humeri as well as femora than the coastal female sample (294 mm vs. 286 mm for the humeri and 412 mm vs. 395 mm for the femora). In males, the humeri are identical in mean length at 312 mm, but Plateau males have significantly longer femora than coastal males (441 mm as opposed to 424 mm).

Coastal femur shafts are significantly flatter at the subtrochanteric level and rounder at midshaft (FMS) than Plateau femoral shafts in both sexes. Conversely, the emphasis is on increased anteroposterior depth of the Plateau shaft, an indication of stronger femora. The FMR data are statistically equivocal in that respect although the Plateau means are higher than those of the coast. Presumably, platymeria and the FMS data are more descriptive and reflective of femoral shaft strength.

With regard to humerus robusticity or strength, the sample sizes were notably smaller than those available for the femur shaft. What data there are indicate that humerus robusticity in the form of the HSRI-2 index is significantly higher for both sexes on the Northwest Coast than on the Plateau. As in the case of FMR, the deltoid index is a statistically equivocal indicator of stronger humeri on the coast although the means are higher than those for the Plateau.

Table 4-2. *Northwest Coast versus Plateau Metric Derivatives*

| Measurement | Sex | Northwest Coast | | | Difference[a] | Plateau | | |
		n	Mean	SD		n	Mean	SD
Stature (cm)	M	278	163.28	4.26	<<	68	167.78	4.04
	F	225	152.25	5.47	<<	49	156.21	4.55
Humero-femoral index	M	136	73.55	1.90	>>	25	70.80	1.83
	F	109	72.60	2.15	>	19	71.49	1.95
Humerus robusticity (HSRI-2)	M	72	13.68	1.11	>>	24	12.45	1.25
	F	55	12.50	1.04	>>	14	11.78	0.80
Deltoid index	M	97	118.18	5.94	>	12	113.18	5.91
	F	71	116.99	5.50	=	11	113.79	6.50
Index of platymeria	M	206	72.51	5.49	<<	59	80.71	8.18
	F	154	72.11	5.72	<<	38	80.99	6.06
Femur midshaft shape (FMS)	M	172	1.06	0.07	<<	56	1.13	.10
	F	126	1.01	0.08	<<	32	1.06	.07
Femur midshaft robusticity (FMR)	M	165	118.96	6.70	<	53	122.05	9.82
	F	120	120.78	7.32	=	26	123.95	7.74
Body mass (kg)	M	210	65.62	5.26	=	57	64.60	6.53
	F	150	56.16	4.65	=	31	54.74	3.22
Index of platycnemia	M	188	64.23	5.91	=	27	62.76	5.65
	F	140	66.81	5.87	=	16	66.13	5.24

[a]The greater than and less than symbols (> and <) indicate that the mean differences are statistically significant; a double symbol means at less than the 1% probability level and a single symbol at less than the 5% level. The equals symbol (=) indicates the difference is not statistically significant.

Ancient and Early Remains

It is of interest to look at the postcranial variability among the Ancient and Early humans included in the samples. In this case, the focus is on the Plateau representatives, especially Kennewick, China Lake, and Big Bar Lake, which were singled out in the section on isotopic variation. While Pritchard was also mentioned, there were no measurement data available that could be considered. Likewise, the partial skeleton from On Your Knees Cave in Alaska, a Northwest Coast location, lacks relevant measurements. Table 4-3 presents comparative data for all of the Plateau Ancient and Early remains for which measurement data were available.

Table 4-3. *Metric Derivatives of Ancient and Early Plateau Skeletons*

Skeleton	Estimated stature (cm)	Humero-femoral index	Humerus robusticity (HSRI-2)	Deltoid index	Index of platymeria	Femur midshaft shape (FMS)	Femur midshaft robusticity (FMR)
Males							
Clark's Fork	171.63 ± 4.25	—	12.00	—	77.67	—	—
Gore Creek	167.96 ± 3.18	—	—	122.58	65.71	1.15	126.09
Kennewick	174.79 ± 3.18	72.55	11.14	—	80.00	1.24	132.65
Braden 3	165.88 ± 3.80	—	—	—	80.00	1.14	126.32
Braden 5	173.24 ± 4.25	—	13.39	—	77.78	—	—
Braden 9	164.16 ± 3.80	70.66	12.04	—	77.42	1.16	122.73
China Lake 1	159.86 ± 3.80	67.81	11.59	106.12	80.00	1.26	110.47
China Lake 2	163.52 ± 3.80	69.98	10.47	114.58	80.00	1.16	108.14
Females							
Buhl	156.85 ± 3.72	—	—	—	92.59	.96	114.63
Big Bar Lake	154.26 ± 3.72	72.01	—	120.79	75.00	—	—
Braden 2	156.85 ± 3.72	71.22	12.12	—	86.67	1.04	142.50
Braden 8	161.57 ± 3.57	—	—	—	82.76	—	—

Kennewick. With the possible exception of the humero-femoral index, the metric data for Kennewick clearly fit the picture of a Plateau male: tall with strong lower limbs and low humeral robusticity. His estimated stature exceeded the coastal male average by over 10 cm and was the second highest estimate of all 68 Plateau males treated in this study.

China Lake. The two China Lake males fit the Plateau male pattern in almost all measurements. However, they appear unusually short, at or below the Northwest Coast male average. While at least three Plateau males were shorter than China Lake 1, the combination of short stature in the two men is a hint of a unique commonality in their body builds relative to other Plateau males. Figure 4-4 compares the lengths of their femora and humeri relative to midshaft perimeter and least circumference of the shaft respectively, traditional indicators of long-bone robusticity. In addition to short stature, the China Lake males have remarkably gracile long bones.[4] Both men, in their late teens or early 20s, were buried next to one another and also exhibited supratrochlear spur (also known as supracondylar process) in their left humeri, a rare morphological trait that has been figured at less than 1 percent in other archaeological skeletal samples (Saunders 1978). These common and unusual details suggest that the two young men could have been relatives, indeed brothers, the reality of which seems to be supported by mtDNA testing. Both tested as haplogroup M, unique among North American aboriginal populations (Malhi et al. 2007). The significance in the present context is that the two men alone would not typify Plateau adaptation in body build and stature. Yet, their isotopic signatures were as expected and the dental condition in China Lake 1 was characteristic of the Plateau pattern (Figure 4-5). The teeth and jaws of China Lake 2 were not available for study.

Big Bar Lake. Due to poor preservation, only three metric characters were available for the female skeleton from Big Bar Lake. Perhaps the most telling indicator of a possible coastal affiliation is its deltoid index. While this measurable character is not the best indicator of separation between the Northwest Coast and the Plateau, the value is notably high in the Big Bar Lake female at 120.79, well in excess of the Northwest Coast male average. It is, in fact, the second highest among Plateau females. The other two available indices, humero-femoral and platymeric, might also be interpreted as Northwest Coast characteristics. Overall, the isotope and mtDNA signatures (discussed previously) and the metric data are suggestive of a possible coastal origin.

Conclusions

Average body mass is the same for the Northwest Coast and Plateau at about 65 kg in males and 55 kg in females, perhaps a latitudinal optimum. People of the Northwest Coast have short bodies with strongly developed upper limbs and relatively short lower limbs, well suited for a marine-oriented subsistence base dependent on canoe travel. People of the Plateau, especially males,

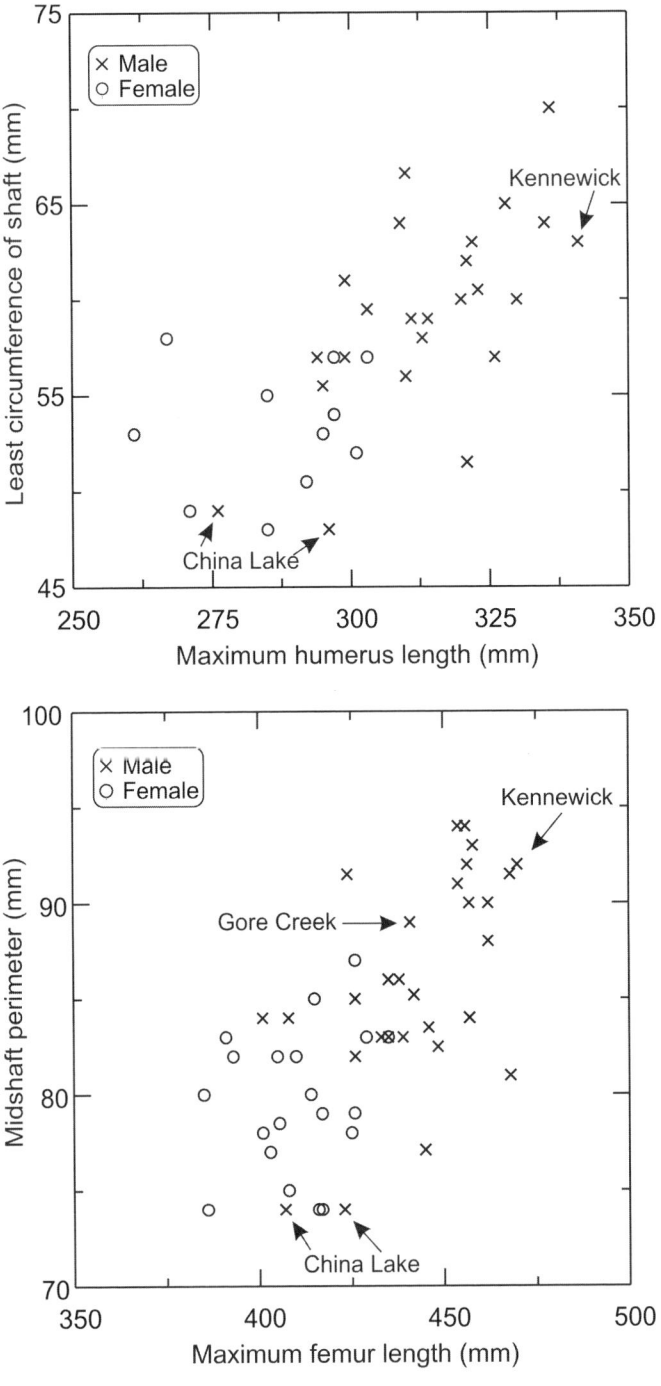

Figure 4-4. *Scatter plots of traditional indicators of humerus and femur robusticity in Plateau males and females. The measurements are in millimeters. Comparative details are explained in the text.*

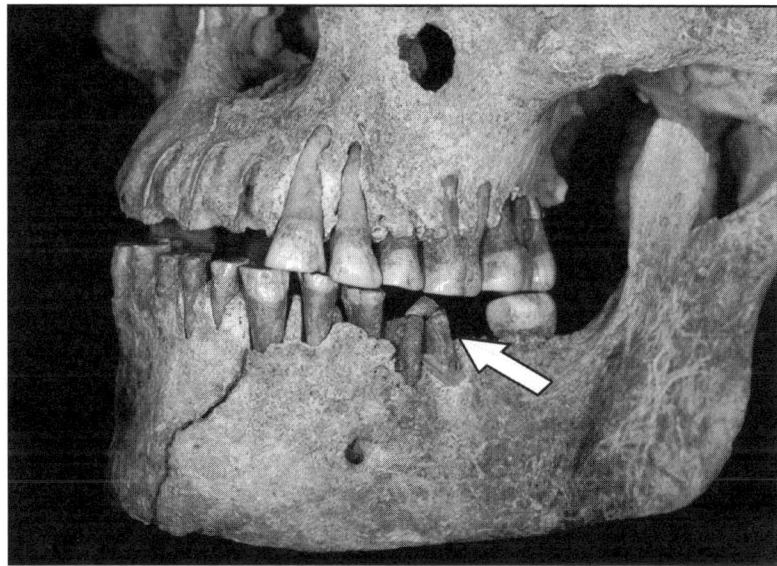

Figure 4-5. *Heavy occlusal wear and crown caries (arrow) in the teeth of China Lake 1 (EiRm-7:1), the skeleton of an 18-to-24-year-old male radiocarbon dated at 5100 B.P. (photo by the author).*

are taller and, therefore, leaner and have relatively longer and stronger lower limbs, which are more suitable for hunting and gathering. Hunting and gathering formed important elements of the economy, in addition to fishing later in prehistory, especially for those groups who lived away from rivers and streams conducive to anadromous salmon runs.

In terms of its physique, the ancient skeleton from Kennewick, Washington, was typically Plateau adapted. Its high carbon isotope signature was more likely from hunting and eating herbivores that consumed C_4 grasses more than 8,000 years ago in the Columbia Basin than from a dependence on a marine diet, via the Columbia River, apparently only meagerly available at that time.

Two males from the 5,000-year-old site of China Lake in the interior of British Columbia present isotopic data and dental conditions that are characteristic of Plateau adaptation but display unique elements of skeletal morphology that could not be considered typically Plateau. Their common genetic signatures, atypical in terms of North American Indians, suggest they were brothers. A contemporary female skeleton from the nearby Plateau site of Big Bar Lake displays physical features, isotopic signatures, and a genetic makeup that together could indicate emigration from a coastal origin.

Nutritional differences could very well have accounted for the differences in stature between coastal and interior individuals. The diet was much more flexible and varied for the Plateau than for the Northwest Coast peoples, and this is reflected in the distribution of stable isotopes and in differences in dental pathology. The caloric intake on the Plateau was significantly greater given Plateau dependence on plant foods. But would nutritional differences account for differences in limb proportions and long-bone strength? Clearly, stronger arms are advantageous for marine-oriented people, while longer and stronger legs are advantageous for those who would chase game. Selective pressures were likely at work over long periods of time to bring about some, if not most, of the human physical differences between the two natural and cultural areas.

Acknowledgments

I thank Arabella Bowman, Janet Young, Carrie Dan, and Andrew Hickok for sharing stable isotope data. Overall, this paper would not have been possible without the research efforts of those and other colleagues in archaeology and physical anthropology of the Pacific Northwest cited herein. I can only hope I've done their reported data, as well as my own, justice.

Funding for this and other of my projects was provided by the Canadian Museum of Civilization through its Physical Anthropology Program. Logistical assistance for various field projects during which some of my data were collected was provided by Indian Affairs Canada, the British Columbia Archaeology Branch, and the Hesquiat, Oowekeeno, Quatsino, Canoe Creek, High Bar, and Laxgalts'ap First Nations. I very much appreciate the interest of those First Nations as well as others over the years who have supported my research. My thanks must also be directed to the staff of the British Columbia Archaeology Branch and the Canadian Museum of Civilization who assisted with the identification of archaeological sites, literature research, and administrative chores. Finally, I thank Ben Auerbach for inviting my participation in the 25th Annual Visiting Scholar Conference at Southern Illinois University Carbondale, and Anne Katzenberg and Susan Pfeiffer for their very helpful comments on an earlier draft of this manuscript.

Appendix

Archaeological sites (including official Canadian or U.S. designations where available) and localities used in data collection by culture area and subregion of study; data references are also provided.

Northwest Coast

Coast of Alaska
> Old Tongass, Staney Creek (73Acn1341), On Your Knees Cave (49PET408)
> *References*: Cybulski 1967, 1974; Dixon 1999

Haida Gwaii (Queen Charlotte Islands)
> Blue Jackets Creek (FlUa-4), Chaatl, Cumshewa, Delkatla (GaUa-2), Gust Island (FhUb-1), Haans Island (FgTw-5), Kaidju, Kaisun, Ka-Yung, Kiusta, Kung (undesignated), Kung (GaUd-1), Naden Harbour, New Masset

(GaUa-3), Ninstints, North Island Cave, Old Masset, Skedans, Skidegate, Skincuttle, Tanu, Yaku (undesignated), Yaku (GbUg-2), Yan
References: Chisholm 1986; Cybulski 1967, 1973, 1975a, 1990; Cybulski and Pett 1981; Murray 1981; Skinner 1984

Nass River Valley
Greenville (GgTj-6, localities A and B), Ishkeenickh River Cave (GfTj-1)
References: Cybulski 1992, 1996, 2007, 2008a

Prince Rupert Harbour
Baldwin (GbTo-36), Boardwalk (GbTo-31), Dodge Island (GbTo-18), Garden Island (GbTo-23), Grassy Bay (GbTn-1), Lachane (GbTo-33), Lucy Island (GbTp-1), Parizeau Point (GbTo-30), Port Essington, Port Simpson, Ridley Island (GbTn-19)
References: Cybulski 1986, 1988b, 1990, 1992, 2001, 2008a

Central Mainland Coast of British Columbia
Anutlict (FaSu-6), Bella Bella, Bella Coola (FcSq-4?), Clatse Bay (FcSx-1), Elcho Harbour, Kimsquit (FeSr-10), Knight Inlet, K'nts Island (FaSu-9), Mc-Naughton Island (ElTb-10), Milbanke Sound, Namu (ElSx-1), Nugent Sound, Owikeno Lake (EkSp-13)
References: Chisholm 1986; Chisholm et al. 1983; Curtin 1984; Cybulski 1975b, 1995, 1997b, 1999, 2001

North and East Coasts of Vancouver Island
Bear Cove (EeSu-8), Bull Harbour, Cape Commerell, Cape Scott, Cayilth, Cluxewe River localities, Coal Harbour, Fort Rupert localities, Koskimo–Quatsino Sound localities, May's Place, Nimpkish River localities, O'Connor (EeSu-5), Peel Island rock-shelter, Quattishe East Gravehouse (EdSv-4), Salmon River, Shell Island gravehouse (EeSu-25), Wakashan localities
References: Carlson, C. 1979; Chapman 1972; Chisholm 1986; Cybulski 1967, 1997a, 1997b

West Coast of Vancouver Island
Clayoquot Sound, Hesquiat Harbour (DiSo-1, 5, 9, 11, 12, 13, 16, 18, 19, 20), Kyuquot Sound, Nuchatlitz Inlet, Whalers House (Nootka Sound), Yuquot
References: Chisholm 1986; Cybulski 1967, 1978, 1980a

Strait of Georgia mainland sites
Beach Grove (DgRs-1), Belcarra Park (DhRr-6), Boundary Bay, Crescent Beach (DgRr-1), Glenrose Cannery (DgRr-6), Locarno Beach (DhRt-6), Marpole (DhRs-1), St. Mungo Cannery (DgRr-2), Tsawwassen (DgRs-2), Whalen Farm (DfRs-3), White Rock (DgRq-18)
References: Ball 1981; Beattie 1981; Chisholm 1986; Chisholm et al. 1983; Conaty and Curtin 1984; Curtin 1980, 1991; Cybulski 1991, 1992, 2008a; Heglar 1958; Lazenby 1986; Seymour 1976; Styles 1976; Trace 1981

Strait of Georgia island sites

Bliss Landing (EaSe-2), Cape Mudge Village (EaSh-3), Clemclemalutz, Courtenay River (DkSf-1), Cowichan localities, Cox Island, Deep Bay (DiSe-7), Departure Bay (DhRx-16), Duke Point (DgRx-5), False Narrows (DgRw-4), Green Point (DeRv-151), Harbour House (DfRu-3), Helen Point (DfRu-8), Hill (DfRu-4), Kuiniekeu, Kulleet Bay (DgRw-17), Maple Bay (DeRv-12), North Saanich, Pender Canal (DeRt-1, DeRt-2), Qualicum Beach (DiSc-26), Quamitchan, Saanich, Somenos Creek (DeRw-18), South Saanich, Stevenson Cairn (DcRu-52), Victoria, Welbury Point (DfRu-42), Willows Beach (DcRt-10)
References: Beattie 1981; Brown 1996; Carlson 1986; Chisholm 1986; Condrashoff 1984; Curtin 1998; Curtin et al. 1991; Cybulski 1967, 1988a, 1991, 1992, 1997c, 2008a; Gordon 1974; Hall and Haggarty 1981; Howe et al. 1994; Monks 1977; Skinner 1986; Skinner and McKendry 1984; Thom 1995

Coast of Washington

Ozette Village (45CA024), Semiahmoo Spit, (45WH017), Walan Point (45JE016)
References: Aronstam 1975; Blukis Onat and Haversat 1977; Grabert et al. 1978

Plateau

Chilcotin/Carrier

Alexis Creek, Nechako-Nautley confluence (GaSd-3), Puntzi Lake, Swan Lake Park, Quesnel (FfRo-1)
References: Chisholm 1986; Cybulski 1982a; Lovell et al. 1986; Oliver 1997c, 1999; Skinner 1987a, 1987b

Cariboo (Northern Shuswap)

Big Bar Lake (EhRk-4), China Gulch (EiRn-1), Clinton (EgRk-2), China Lake (EiRm-7), Horsefly and Caribou Road Junction (FaRl-y)
References: Chisholm 1986; Cybulski 2006b, 2006c, 2008a; Cybulski et al. 2007; Malhi et al. 2007

Mid-Fraser River region

Fountain (EeRl-19), Jones Ranch (EdRl-10), Lillooet (EeRl-16, -80, -167, and -169), Lillooet Museum, Lillooet School Trail, Lytton (EbRj-1), Murray (EeRl-18), Seton Lake, Seton Substation (EeRl-192), West Fountain (EeRl-6)
References: O. Beattie in Stryd 1980; Chisholm 1986; Cybulski 1980b, 2006c, 2008a; Lazenby and McKendry 1984; Lovell et al. 1986; McLeod and Skinner 1987; Oliver 1995a, 1995c; Richards 1982; Smith 1899; Stryd and Baker 1968

Thompson River West

Basque (EdRh-26), Cache Creek (EeRh-1), Hat Creek (EdRk-3), Walhachin
References: Chisholm 1986; Curtin 1986; Cybulski 2006c; Lovell et al. 1986; Pokotylo et al. 1987

Kamloops Lake–Kamloops
> Brocklehurst (EeRc-8), Guerin Creek (EeRc-58), Kamloops (undesignated), Kamloops (EeRb-10, EeRc-42, and E*Rc-61), CNPR Stn 1046, CNPR Stn 1125, Savona, Savonas, Tranquille Centre (EeRd-24), Young Place (EeRc-57)
> *References*: Chisholm 1986; Cybulski 1982b, 2006c, 2008a, 2008b; Knowles 1916; Oliver 1992b, 1997a; Wilson 1972

North Thompson River
> Barriere (undesignated), Barriere (EhRa-39)
> *References*: Oliver 1998a, 1998b

South Thompson River–Shuswap Lakes
> Chase (EeQw-1), Chase T31 (EeQw-T31), Gore Creek (EeQw-48), Lee Creek (EfQv-1), Monte Creek (EdQx-20; EdQx-25), Neskonlith Reserve 1 (EeQw-75), O'Keefe Ranch (EcQt-13), Pinaus Lake (EcQv-2), Pritchard (EeQw-21), South Thompson, Squilax (EfQv-10; EgQw-66?), Wildlife Park (EdRa-y)
> *References*: Brady 2005; Carlson 2002, 2003; Chisholm 1986; Curtin and Lawhead 1985; Cybulski 2008a; Cybulski et al. 1981; Harrison 1963; Oliver 1989, 1992a, 1995b, 1997b

Slocan Valley–Kootenay
> Arrow Lakes (EaQl-10), Brilliant (DhQj-1), Canal Flats (EbPw-1), Creston (DgQd-y), Drimmie Creek, Grand Forks (DgQo-2), Koocanusa Lake, Vallican (DjQj-1)
> *References*: Chisholm 1986; Oliver 1992c, 1994; Sumpter 1982

Nicola
> Merritt/Spences Bridge (EdRh-y), Nicoamen River (EbRi-7), Nicola (EaRf-6), Quilchena Hotel (EaRd-14), Shackan IR 11 (EbRg-6), Spences Bridge (EdRh-53?)
> *References*: Chisholm 1986; Curtin 1990; Cybulski 1995; Lovell et al. 1986; Skinner and Copp 1986; Skinner and Thacker 1988

Okanagan
> Kelowna (EaQu-6), Kelowna (DlQu-y), Kopp (EbQt-19), Oliver (DhQv-86), Osoyoos (DgQu-13), Otter Lake (EcQt-12), Penticton (EjQv-65?)
> *References*: Chisholm 1986; Cybulski 1989, 2005; Kusmer and Lawhead 1987; Oliver and Crellin 1991; Skinner and Thacker 1989

Washington Plateau
> Banks Lake, Beebe Orchards (45CH218), Bonaparte Creek (45OK512), Chaudiere (45FE047), Chelan Station (45CH296), Congdon (45KL041), Deadhorse (45ST063), Ferguson (45WT055B), Freeland (45FE001), Kennewick (45BN052), Kettle Falls Railroad Bridge (45FE038), Marmes (45FR050), Mill Creek (45LI006), Nancy (45FE016), Narrows (45OK011), Nespelem (45OK197), Pakootas (45OK159), Palus Talus (45WT056), Sntl'exwenewixwtn (45OK355), Steamboat Rock Mass Grave (45GR098), Tonasket (45OK561)

References: Boreson 1985; Breschini 1979; Butler 1963; Chance et al. 1977; Chatters 1985, 2000; Chatters and Zweifel 1987; Krantz 1979; Landis et al. 1984; Sprague and Birkby 1973; Sprague and Mulinski 1980

Oregon Plateau
    Butte Creek Cave
    Reference: Stewart 1950

Idaho Plateau
    Braden (10WN117), Buhl (10TF1019), Clark's Fork, Cottonwood Creek (10NP182), DeMoss (10AM193)
    References: Green et al. 1986, 1998; Harten 1980; Leonhardy et al. 1987; Neves and Blum 2000; Pennefather-O'Brien and Strezewski 2002

Notes

1. © 1996, SPSS Inc., 444 North Michigan Avenue, Chicago, Illinois, U.S.A. 60611.

2. It is uncertain whether Buhl should be regarded as a *Plateau* Ancient. The site is at a gravel quarry just south of the Snake River in south-central Idaho, a region allocated to the Great Basin culture area (Murphy and Murphy 1986). However, Buhl is effectively on the culture area boundary and the nearby sites of DeMoss and Braden (Harten 1980), the long-bone measurements of which have been entered into the present study, are considered part of Eastern Plateau prehistory (Roll and Hackenberger 1998).

3. No nitrogen isotopic information has been reported for the human remains from On Your Knees Cave, Alaska, radiocarbon dated older than Kennewick at 9800 B.P. (Dixon 1999). However, there is little reason to suspect that this human did not depend heavily on a marine diet given the location of the find on Prince of Wales Island.

4. There is little doubt that the two skeletons were male. The pelvis was available in the case of China Lake 1 and partially available in the case of China Lake 2. Prior to my 2002 study of the remains, independent male sex assessments were made on recovery of the remains (McKendry 1983) and during a subsequent osteological investigation (Stijelja and Williams 1986).

References

Ames, Kenneth M.
    2003   The Northwest Coast. *Evolutionary Anthropology* 12:19–33.
Ames, Kenneth M., Don E. Dumond, Jerry R. Galm, and Rick Minor
    1998   Prehistory of the Southern Plateau. In *Plateau*, edited by Deward E. Walker Jr., pp. 103–119. Handbook of North American Indians, Vol. 12, William C. Sturtevant, general editor, Smithsonian Institution, Washington, D.C.

Ames, Kenneth M., and Herbert D. G. Maschner
   1999   *Peoples of the Northwest Coast: Their Archaeology and Prehistory.* Thames and Hudson, London.
Aronstam, Elliott Mark
   1975   Human Remains from the Ozette Village Archeological Site. Master's thesis, Department of Anthropology, Washington State University, Pullman.
Auerbach, Benjamin M., and Christopher B. Ruff
   2004   Human Body Mass Estimation: A Comparison of "Morphometric" and "Mechanical" Methods. *American Journal of Physical Anthropology* 125:331–342.
Ball, Bruce F.
   1981   Archaeological Evaluation of the Beach Grove Site, DgRs 1. *Midden* 13(1):3–7.
Beattie, Owen B.
   1981   An Analysis of Prehistoric Human Skeletal Material from the Gulf of Georgia Region of British Columbia. Ph.D. dissertation, Department of Archaeology, Simon Fraser University, Burnaby, British Columbia.
Blukis Onat, Astrida R., and Trudy Haversat
   1977   *Archaeological Excavations at Site 45JE16, Indian Island, Jefferson County, Washington, Burial Report.* Washington Archaeological Research Center, Project Report No. 61, Washington State University, Pullman.
Boas, Franz
   1892   Third Report on the Indians of British Columbia. *Report of the Meeting of the British Association for the Advancement of Science* 1891:408–449.
   1895 Fifth Report on the Indians of British Columbia. *Report of the Meeting of the British Association for the Advancement of Science* 1895:523–592.
Boas, Franz, and Livingston Farrand
   1899   Physical Characteristics of the Tribes of British Columbia. *Report of the Meeting of the British Association for the Advancement of Science* 1898:628–644.
Boreson, Keo
   1985   *The Burials at 45CH296, Chelan County, Washington.* Eastern Washington University Reports in Archaeology and History 100-40. Archaeological and Historical Services, Eastern Washington University, Cheney.
Boyd, Robert T.
   1990   Demographic History, 1774-1874. In *Northwest Coast*, edited by Wayne Suttles, pp. 135–148. Handbook of North American Indians, Vol. 7, William C. Sturtevant, general editor, Smithsonian Institution, Washington, D.C.
   1998   Demographic History until 1990. In *Plateau*, edited by Deward E. Walker Jr., pp. 467–83. Handbook of North American Indians, Vol. 12, William C. Sturtevant, general editor, Smithsonian Institution, Washington, D.C.
Brady, Karen
   2005   *Report on Archaeological Monitoring and Emergency Impact Management CP Rail Pritchard Track Expansion Project Mile 101.3 to 104.35 Shuswap Subdivision.* Canadian Pacific Railway Engineering Services, Calgary, Alberta. Also available from Golder Associates Limited, Kamloops, British Columbia.
Breschini, Gary S.
   1979   The Marmes Burial Casts. *Northwest Anthropological Research Notes* 13(2):111–158.
Brown, Douglas R.
   1996   Disposing of the Dead: A Shell Midden Cemetery in British Columbia's Gulf of Georgia Region. Master's thesis, Department of Anthropology and Sociology, University of British Columbia, Vancouver.

Butler, B. Robert
    1963   Further Notes on the Burials and the Physical Stratigraphy at the Congdon Site, a Multi-component Middle Period Site at The Dalles on the Lower Columbia River. *Tebiwa* 6(2):16–32.

Carlson, Catherine C.
    1979   *Preliminary Report on Excavations at Bear Cove (Site EeSu 8), Port Hardy, B.C.* Report on file (Permit No. 1978-3), British Columbia Archaeology Branch, Victoria.
    2002   *Report on Found Human Remains, Kamloops Wildlife Park, Campbell Creek, British Columbia.* Report on file, Permit 2002-118, British Columbia Archaeology Branch, Victoria.
    2003   *Report on Found Human Remains, Neskonlith Indian Reserve #1, British Columbia, Archaeological Site EeQw75.* Report on file, File 2002-18B, British Columbia Archaeology Branch, Victoria.

Carlson, Roy L.
    1986   *The 1985 Excavations at the Canal Site (DeRt 1 and DeRt 2).* Report on file, Permit No. 1985-10, British Columbia Archaeology Branch, Victoria.

Carlson, Roy L., and Philip M. Hobler
    1993   The Pender Canal Excavations and the Development of Coast Salish Culture. *BC Studies* 99:25–52.

Chance, David H., Jennifer V. Chance, and John L. Fagan
    1977   *Kettle Falls 1972; Salvage Excavations in Lake Roosevelt.* University of Idaho Anthropological Research Manuscript Series, Vol. 31, Laboratory of Anthropology, University of Idaho, Moscow.

Chapman, Margaret W.
    1972   Salvage Excavation at Two Coastal Middens. In *Salvage '71: Reports on Salvage Archaeology Undertaken in British Columbia in 1971,* edited by Roy L. Carlson, pp. 59–84. Department of Archaeology Publication 1, Simon Fraser University, Burnaby, British Columbia.

Chatters, James C.
    1985   *Forensic Analysis of a Prehistoric Interment, Bonaparte Creek, Okanogon County, Washington.* Central Washington Archaeological Survey, Archaeological Report No. 85-1, Central Washington University, Ellensburg. Available through U.S. Department of Commerce, National Technical Information Service, www.ntis.gov.
    1998   Environment. In *Plateau,* edited by Deward E. Walker Jr., pp. 29–48. Handbook of North American Indians, Vol. 12, William C. Sturtevant, general editor, Smithsonian Institution, Washington, D.C.
    2000   The Recovery and First Analysis of an Early Holocene Human Skeleton from Kennewick, Washington. *American Antiquity* 65:291–316.

Chatters, James C., and David L. Pokotylo
    1998   Prehistory: Introduction. In *Plateau,* edited by Deward E. Walker Jr., pp. 73–80. Handbook of North American Indians, Vol. 12, William C. Sturtevant, general editor, Smithsonian Institution, Washington, D.C.

Chatters, James C., and Matthew W. Zweifel
    1987   *The Cemetery at Sntl'exwenewixwtn, Okanagon County, Washington.* Central Washington Archaeological Survey, Archaeological Report No. 87-1, Central Washington University, Ellensburg.

Chisholm, Brian S.
    1986   Reconstruction of Prehistoric Diet in British Columbia Using Stable-Carbon Isotope Analysis. Ph.D. dissertation, Department of Archaeology, Simon Fraser University, Burnaby, British Columbia.

1996   Appendix B: The DeRw 18 (Somenos Creek) Burials Stable Isotope Analysis. In *Interim Report on Archaeological Investigations at the Somenos Creek Site (DeRw 18). B.C. Investigation Permit 1994-122*, edited by Douglas R. Brown, pp. 60–70. British Columbia Archaeology Branch, Victoria.

Chisholm, Brian S., and D. Erle Nelson
1983   An Early Human Skeleton from South Central British Columbia: Dietary Inference from Carbon Isotopic Evidence. *Canadian Journal of Archaeology* 7:85–86.

Chisholm, Brian S., D. Erle Nelson, and Henry P. Schwarcz
1982   Stable-Carbon Isotope Ratios as a Measure of Marine Versus Terrestrial Protein in Ancient Diets. *Science* 216:1131–1132.
1983   Marine and Terrestrial Protein in Prehistoric Diets on the British Columbia Coast. *Current Anthropology* 24:396–398.

Conaty, Gerald T., and A. Joanne Curtin
1984   *Crescent Beach Monitoring Programme: Final Report*. Report on file, Permit No. 1983-45, British Columbia Archaeology Branch, Victoria.

Condrashoff, Nancy
1984   *DcRu 52: A Salvaged Burial Cairn in Songhees Territory*. Report on file, DcRu-52, British Columbia Archaeology Branch, Victoria.

Corr, Lorna T., Michael P. Richards, Colin Grier, Alexander Mackie, Owen Beattie, Richard P. Evershed
2009   Probing Dietary Change of the Kwa˙da̱y Da̱ˑn Tsʼ̱ˑnchi̱ Individual, an Ancient Glacier Body from British Columbia: II. Deconvoluting Whole Skin and Bone Collagen $\delta^{13}C$ Values via Carbon Isotope Analysis of Individual Amino Acids. *Journal of Archaeological Science* 36:12–18.

Curtin, A. Joanne
1980   *Analysis of Burial 2 from the Beach Grove Site DgRs 1*. Report on file, Department of Archaeology, Simon Fraser University, Burnaby, British Columbia.
1984   Human Skeletal Remains from Namu (ElSx 1): A Descriptive Analysis. Master's thesis, Department of Archaeology, Simon Fraser University, Burnaby, British Columbia.
1986   *Osteological Analysis of Human Skeletal Material from the Clinton Museum, Clinton, B.C.* Report on file (Unnumbered), British Columbia Archaeology Branch, Victoria.
1990   *Analysis of Human Skeletal Remains from EbRg 6 near Merritt, B.C.* Report on file, Permit 1989-34, British Columbia Archaeology Branch, Victoria.
1991   *Archaeological Investigations at Tsawwassen, B.C., Volume III, Human Osteology*. Report on file, Permit 1989-41 and 1990-2, British Columbia Archaeology Branch, Victoria.
1998   Prehistoric Mortuary Variability on Gabriola Island, British Columbia. Ph.D. dissertation, Department of Anthropology, Ohio State University, Columbus.

Curtin, A. Joanne, Mark Finnis, and Morley Eldridge
1991   *A Human Burial from the Willows Beach Site, DcRt 10*. Report on file, Permit No. 1991-14, British Columbia Archaeology Branch, Victoria.

Curtin, A. Joanne, and Stephen Lawhead
1985   *Spallumcheen Heritage Inventory Project: Pinaus Lake Burial Salvage (1984)*. Report on file, Permit 1984-40, British Columbia Archaeology Branch, Victoria.

Cybulski, Jerome S.
1967   Osteological Data on Northwest Coast Skeletal Collections in the Field Museum of Natural History, Chicago, and the American Museum of Natural History, New York. Notes on file, Canadian Museum of Civilization, Gatineau.
1973   The Gust Island Burial Shelter: Physical Anthropology. In *Haida Burial Prac-*

tices: Three Archaeological Examples, edited by George F. MacDonald, pp. 60–113. Archaeological Survey of Canada, Mercury Series Paper No. 9, Canadian Museum of Civilization, Hull.

1974   *Identification and Analysis of a Human Skeleton from Prince of Wales Island, Alaska.* Library Archives, Archaeology Ms. 1032, Canadian Museum of Civilization, Gatineau.

1975a   *Skeletal Variability in British Columbia Coastal Populations: A Descriptive and Comparative Assessment of Cranial Morphology.* Archaeological Survey of Canada Mercury Series Paper No. 30, Canadian Museum of Civilization, Gatineau.

1975b   Physical Anthropology at Owikeno Lake, 1975. *Canadian Archaeological Association Bulletin* 7:201–210.

1978   *An Earlier Population of Hesquiat Harbour, British Columbia; A Contribution to Nootkan Osteology and Physical Anthropology.* Cultural Recovery Papers 1, Royal British Columbia Museum, Victoria.

1980a   Osteology of the Human Remains from Yuquot, British Columbia. *History and Archaeology* 43:175–194.

1980b   Skeletal Remains from Lillooet, British Columbia, with Observations for a Possible Diagnosis of Skull Trephination. *Syesis* 13:53–59.

1982a   *Human Skeletal Remains from Quesnel, British Columbia, Site FfRo 1.* Library Archives, Archaeology Ms. 2027, Canadian Museum of Civilization, Gatineau.

1982b   Osteological Data on Two Human Skeletons from Kamloops, EeRc-42. Notes on file, Canadian Museum of Civilization, Gatineau.

1986   *Human Remains from Lucy Island, British Columbia, Site GbTp 1, 1984/85.* Library Archives, Archaeology Ms. 2360, Canadian Museum of Civilization, Gatineau.

1988a   *Human Skeletal Remains from Site DeRu(v) 12, Maple Bay, British Columbia (April 1988).* Library Archives, Archaeology Ms. 3041, Canadian Museum of Civilization, Gatineau.

1988b   Osteological Data on Human Burials from Eight Shell Midden Sites in the Prince Rupert Harbour Region of British Columbia, 1968–1988. Notes on file, Canadian Museum of Civilization, Gatineau.

1989   *Human Skeletal Remains from Site DgQu 13, Osoyoos Indian Reserve No. 1, British Columbia.* Library Archives, Archaeology Ms. 3060, Canadian Museum of Civilization, Gatineau.

1990   Human biology. In *Northwest Coast,* edited by Wayne Suttles, pp. 52–59. Handbook of North American Indians, Vol. 7, William C. Sturtevant, general editor, Smithsonian Institution, Washington, D.C.

1991   *Human Remains from Duke Point, British Columbia, and Probable Evidence for Pre-Columbian Treponematosis.* Library Archives, Archaeology Ms. 3454, Canadian Museum of Civilization, Gatineau.

1992   *A Greenville Burial Ground: Human Remains and Mortuary Elements in British Columbia Coast Prehistory.* Archaeological Survey of Canada Mercury Series Paper No. 146. Canadian Museum of Civilization, Gatineau.

1993   *Human Remains from Somenos Creek (DeRw 18).* Library Archives, Archaeology Ms. 3679, Canadian Museum of Civilization, Gatineau.

1994   *Human Remains and Other Materials from Kulleet Bay (DgRw 17).* Library Archives, Archaeology Ms. 3720, Canadian Museum of Civilization, Gatineau.

1995   Osteological Data on British Columbia Collections in the Department of Archaeology, Simon Fraser University, and the Royal British Columbia Museum. Notes on file, Canadian Museum of Civilization, Gatineau.

1996   *Archaeological Human Remains and Associated Cultural Materials from Greenville, B.C., October 1995.* Library Archives, Archaeology Ms. 3947, Canadian Museum of Civilization, Gatineau.

1997a   *Fieldwork at Quattishe (EdSv 4), 1996.* Library Archives, Archaeology Ms. 3968, Canadian Museum of Civilization, Gatineau.

1997b   *Preliminary Report on Human Remains Collections from Quatsino Sound and Environs, Studied at the Royal British Columbia Museum, February 5-15, 1997.* Library Archives, Archaeology Ms. 3877, Canadian Museum of Civilization, Gatineau.

1997c   Osteological Data on Green Point Burials, DeRv-151. Notes on file, Canadian Museum of Civilization, Gatineau.

1999   *The Biological Affinities of Elcho Harbour: Preliminary Observations Based on Discrete Traits of the Skull.* Library Archives, Archaeology Ms. 4228, Canadian Museum of Civilization, Gatineau.

2001   Human Biological Relationships for the Northern Northwest Coast. In *Perspectives on Northern Northwest Coast Prehistory,* edited by Jerome S. Cybulski, pp. 107–144. Archaeological Survey of Canada Mercury Series Paper No. 160, Canadian Museum of Civilization, Gatineau.

2005   *A Human Skeleton (2004-7B) Found at Otter Lake Road, EcQt-12, near Armstrong, British Columbia.* Library Archives, Archaeology Ms. 4813, Canadian Museum of Civilization, Gatineau.

2006a   Skeletal Biology: Northwest Coast and Plateau. In *Environment, Origins, and Population,* edited by Douglas H. Ubelaker, pp. 532–547. Handbook of North American Indians, Vol. 3, William C. Sturtevant, general editor, Smithsonian Institution, Washington, D.C.

2006b   Osteological Data and Associated Notes on a Human Skeleton from the China Gulch Site, EiRn-1. Notes on file, Canadian Museum of Civilization, Gatineau.

2006c   Osteological Data on Interior British Columbia Collections in the Canadian Museum of Civilization. Notes on file, Canadian Museum of Civilization, Gatineau.

2007   Osteological Data from Locality B at Greenville Discovered in 2006. Notes on file, Canadian Museum of Civilization, Gatineau.

2008a   Radiocarbon Dates and Stable Isotope Ratios for Human Bone Samples from the Pacific Northwest, 1997–2008. Notes on file, Canadian Museum of Civilization, Gatineau.

2008b   *A Human Skeleton (2006-12B) from Lac du Bois Grasslands Provincial Park, near Tranquille, British Columbia (Site EeRd-24 "Mara Hill Burial").* Library Archives, Archaeology Ms. 4875, Canadian Museum of Civilization, Gatineau.

Cybulski, Jerome S., Donald E. Howes, James C. Haggarty, and Morley Eldridge
1981   An Early Human Skeleton from South-Central British Columbia: Dating and Bioarchaeological Inference. *Canadian Journal of Archaeology* 5:49–59.

Cybulski, Jerome S., Alan D. McMillan, Ripan Singh Malhi, Brian M. Kemp, Harold Harry, and Scott Cousins
2007   The Big Bar Lake Burial: Middle Period Human Remains from the Canadian Plateau. *Canadian Journal of Archaeology* 31:55–79.

Cybulski, Jerome S., and Bradley Pett
1981   Bone Changes Suggesting Multiple Myeloma and Metastatic Carcinoma in Two Early Historic Natives of the British Columbia Coast. In *Contributions to Physical Anthropology, 1978–1980,* edited by Jerome S. Cybulski, pp. 176–186. Archaeological Survey of Canada Mercury Series Paper No. 106, Canadian Museum of Civilization, Gatineau.

Dawson, George M.
  1892   Notes on the Shuswap People of British Columbia. *Transaction of the Royal Society of Canada for the Year 1891* 12(2):3–44.
Dixon, E. James
  1999   *Bones, Boats & Bison: Archeology and the First Colonization of Western North America*. University of New Mexico Press, Albuquerque.
Drucker, Philip
  1965   *Cultures of the North Pacific Coast*. Chandler Publishing Company, New York.
Gordon, Marjory E.
  1974   A Qualitative Analysis of Human Skeletal Remains from DgRw-4, Gabriola Island, British Columbia. Master's thesis, Department of Archaeology, University of Calgary, Alberta, Canada.
Grabert, Garland F., Jacki A. Cressman, and Anne L. Wolverton
  1978   *Prehistoric Archaeology at Semiahmoo Spit, Washington; A Report on Salvage Archaeology at 45WH17*. Reports in Archaeology No. 8, Department of Anthropology, Western Washington University, Bellingham.
Green, Thomas J., Bruce D. Cochran, Todd W. Fenton, James C. Woods, Gene L. Titmus, Larry Tieszen, Mary Anne Davis, and Susanne J. Miller
  1998   The Buhl Burial: A Paleoindian Woman from Southern Idaho. *American Antiquity* 63:437–456.
Green, Thomas J., Max G. Pavesic, James C. Woods, and Gene L. Titmus
  1986   The DeMoss Burial Locality: Preliminary Observations. *Idaho Archaeologist* 9(2):31–40.
Hall, Roberta L., and James C. Haggarty
  1981   Human Skeletal Remains and Associated Cultural Material from the Hill Site, DfRu 4, Saltspring Island, British Columbia. In *Contributions to Physical Anthropology, 1978–1980*, edited by Jerome S. Cybulski, pp. 64–105. Archaeological Survey of Canada Mercury Series Paper No. 106, Canadian Museum of Civilization, Gatineau.
Hall, Roberta L., Robert Morrow, and J. Henry Clarke
  1986   Dental Pathology of Prehistoric Residents of Oregon. *American Journal of Physical Anthropology* 69:325–334.
Harrison, Peter D.
  1963   *Report on a Study of the Osteological Collection of EeQw-1*. Library Archives, Archaeology Ms. 132, Canadian Museum of Civilization, Gatineau.
Harten, Lucille B.
  1980   The Osteology of the Human Skeletal Material from the Braden Site, 10-WN-117, in Western Idaho. In *Anthropological Papers in Memory of Earl H. Swanson, Jr.*, edited by Lucille B. Harten, Claude N. Warren, and Donald R. Tuohy, pp. 130–148. Idaho Museum of Natural History, Pocatello.
Heathcote, Gary M., Vincent P. Diego, Hajime Ishida, and Vincent J. Sava
  2010   Legendary Chamorro Strength: Skeletal Embodiment and the Boundaries of Interpretation. In *The Bioarchaeology of Individuals*, edited by Ann L.W. Stodder and Ann M. Palkovich. University Press of Florida, Gainesville, in press.
Hebda, Richard, and S. Gay Frederick
  1990   History of Marine Resources of the Northeast Pacific since the Last Glaciation. *Transactions of the Royal Society of Canada*, Sixth Series 1:319–342.
Heglar, Rodger
  1958   *A Report on Indian Skeletal Material from Locarno Beach Site (DhRt-6)*. Report on file, Laboratory of Archaeology, University of British Columbia, Vancouver.

Hewes, Gordon W.
  1998   Fishing. In *Plateau*, edited by Deward E. Walker Jr., pp. 620–640. Handbook of North American Indians, Vol. 12, William C. Sturtevant, general editor, Smithsonian Institution, Washington, D.C.

Howe, Geordie, Robert Muir, Randy Bouchard, Dana Lepofsky, Brian Chisholm, James Malcolm, and Rick Schulting
  1994   *Archaeological Investigations at the Departure Bay Site (DhRx 16), Nanaimo, B.C.* Report on file, Permit 1993-103, British Columbia Archaeology Branch, Victoria.

Hunn, Eugene S., Nancy J. Turner, and David M. French
  1998   Ethnobiology and Subsistence. In *Plateau*, edited by Deward E. Walker Jr., pp. 525–545. Handbook of North American Indians, Vol. 12, William C. Sturtevant, general editor, Smithsonian Institution, Washington, D.C.

Ignace, Marianne Boelscher
  1998   Shuswap. In *Plateau*, edited by Deward E. Walker Jr., pp. 203–219. Handbook of North American Indians, Vol. 12, William C. Sturtevant, general editor, Smithsonian Institution, Washington, D.C.

Jantz, Richard L., David R. Hunt, and Lee Meadows
  1994   Maximum Length of the Tibia: How Did Trotter Measure It? *American Journal of Physical Anthropology* 93:525–528.

Katzmarzyk, Peter T., and William R. Leonard
  1998   Climatic Influences on Human Body Size and Proportions: Ecological Adaptations and Secular Trends. *American Journal of Physical Anthropology* 106:483–503.

Knowles, Francis H. S.
  1916   Physical anthropology. In *Summary Report of the Geological Survey, Department of Mines, for the Calendar Year 1915*, Part III, pp. 278–283. Ottawa.

Krantz, Grover S.
  1979   Oldest Human Remains from the Marmes Site. *Northwest Anthropological Research Notes* 13(2):159–174.

Kroeber, Alfred L.
  1939   *Cultural and Natural Areas of Native North America*. University of California Press, Berkeley.

Kuijt, Ian
  1989   Subsistence Resource Variability and Culture Change During the Middle-Late Prehistoric Cultural Transition on the Canadian Plateau. *Canadian Journal of Archaeology* 13:97–118.

Kusmer, Karla, and Stephen Lawhead
  1987   *Recovery of Human Skeletal Remains from Site EcQt 12, Otter Lake, B.C.* Report on file (Unnumbered), British Columbia Archaeology Branch, Victoria.

Landis, Daniel G., Jerry R. Galm, and Harvey S. Rice
  1984   *Archaeological Monitoring of Highway Construction Through Beebe Orchards and Sites 45CH218, 45CH216, and 45CH289, Chelan County, Washington*. Archaeological and Historical Services, Eastern Washington University, Cheney.

Lazenby, Richard A.
  1986   *Case 85-25: An Aboriginal Skeleton from White Rock, B.C. (DgRq 18)*. Report on file, Permit No. 1985-4, British Columbia Archaeology Branch, Victoria.

Lazenby, Richard A., and Jean McKendry
  1984   *Results of Analysis of Human Skeletal Remains*. Report on file, Permit No. 1984-28, British Columbia Archaeology Branch, Victoria.

Leonhardy, Frank C., Bruce D. Cochran, and Raymond Carino
  1987   *An Analysis of Two Disturbed Ancestral Nez Perce Burials from the Cottonwood*

Creek Burial Site, 10-NP-182, Hells Canyon National Recreation Area, Idaho. Letter Report No. 87-2, Alfred W. Bowers Laboratory of Anthropology, University of Idaho, Moscow.

Lovejoy, C. Owen, Albert H. Burstein, and Kingsbury G. Heiple
1976   The Biomechanical Analysis of Bone Strength: A Method and Its Application to Platycnemia. *American Journal of Physical Anthropology* 44:489–506.

Lovell, Nancy C., Brian S. Chisholm, D. Erle Nelson, and Henry P. Schwarcz
1986   Prehistoric Salmon Consumption in Interior British Columbia. *Canadian Journal of Archaeology* 10:99–106.

McKendry, Jean
1983   *Summary Analysis of Archaeological Skeletal Remains No. 82-13.* Report on file, File 82-13, British Columbia Archaeology Branch, Victoria.

McLeod, Ann, and Mark Skinner
1987   *Analysis of Burial 86-6 from the Fountain Creek Site (EeRl 19), near Lillooet.* Report on file, Permit 1986-20, British Columbia Archaeology Branch, Victoria.

Malhi, Ripan S., Brian M. Kemp, Jason A. Eshleman, Jerome S. Cybulski,
David Glenn Smith, Scott Cousins, and Harold Harry
2007   Mitochondrial Haplogroup M Discovered in Prehistoric North Americans. *Journal of Archaeological Science* 34:642–648.

Monks, Gregory G.
1977   An Examination of Relationships between Artifact Classes and Food Resource Remains at Deep Bay, DiSe 7. Ph.D. dissertation, Department of Anthropology and Sociology, University of British Columbia, Vancouver.

Murphy, Robert F., and Yolanda Murphy
1986   Northern Shoshone and Bannock. In *Great Basin,* edited by Warren L. d'Azevedo, pp. 284–307. Handbook of North American Indians, Vol. 11, William C. Sturtevant, general editor, Smithsonian Institution, Washington, D.C.

Murray, Jeffrey S.
1981   Prehistoric Skeletons from Blue Jackets Creek (FlUa 4), Queen Charlotte Islands, British Columbia. In *Contributions to Physical Anthropology, 1978–1980,* edited by Jerome S. Cybulski, pp. 127–175. Archaeological Survey of Canada Mercury Series Paper No. 106, Canadian Museum of Civilization, Gatineau.

Neves, Walter, and Max Blum
2000   The Buhl Burial: A Comment on Green et al. *American Antiquity* 65:191–193

Oliver, Lindsay J.
1989   *Found Human Remains 89-9B Archaeological Site EeQw-T31 (1989) Chase, B.C.* Report on file, Permit 1989-34, British Columbia Archaeology Branch, Victoria.
1992a   *Found Human Remains Burial 92 16B, Diamond "L" Ranch, Highway 97, Vernon, B.C.* Report on file, File 92-16B, British Columbia Archaeology Branch, Victoria.
1992b   *Found Human Remains Burial 92-15B, 2574 Young Place, Kamloops, B.C.* Report on file, Permit No. 1992-1, File 92-15B, British Columbia Archaeology Branch, Victoria.
1992c   *Found Human Remains Burial 92-24B, Lake Koocanusa, nr. Jaffray, British Columbia.* Report on file, Permit No. 1992-1, File 92-24B, British Columbia Archaeology Branch, Victoria.
1994   *Found Human Remains 94-7B, Drimmie Creek, nr. Revelstoke, British Columbia.* Report on file, File 94-7B, British Columbia Archaeology Branch, Victoria.
1995a   *Found Human Remains Burial 95-3B, Lillooet Museum, Lillooet, British Columbia.* Report on file, File 95-3B, British Columbia Archaeology Branch, Victoria.

1995b   *Found Human Remains 94-29B, Lillooet, British Columbia.* Report on file, File 94-29B, British Columbia Archaeology Branch, Victoria.

1995c   *Found Human Remains Burial 95-4B, 2707 Anglemont Highway, Lee Creek [nr. Chase], British Columbia.* Report on file, File 95-4B, British Columbia Archaeology Branch, Victoria.

1997a   *Found Human Remains Burial 96-23B, Geurin Creek, Kamloops, B.C.* Report on file, File 96-23B, British Columbia Archaeology Branch, Victoria.

1997b   *Found Human Remains Burial 97-10B, Monte Creek, B.C.* Report on file, Ministerial Order 1997-003; File 97-10B, British Columbia Archaeology Branch, Victoria.

1997c   *Found Human Remains Burial 96-17B Puntzi Lake, nr. Alexis Creek, B.C.* Report on file, File 96-17B, British Columbia Archaeology Branch, Victoria.

1998a   *Found Human Remains Burial 98-4B, Barriere, B.C.* Report on file, Ministerial Order 1998-002; File 98-4B, British Columbia Archaeology Branch, Victoria.

1998b   *Found Human Remains Burial 98-8B, Barriere, B.C.* Report on file, Ministerial Order 1998-002; File 98-8B, British Columbia Archaeology Branch, Victoria.

1999   *Re: Branch Burial No. 99-7B, Swan Lake Provincial Park, nr. Dawson Creek, B.C.* Report on file, File 99-7B, British Columbia Archaeology Branch, Victoria.

Oliver, Lindsay J., and Dave Crellin

1991   *Found Human Remains Burial 90-17B, Oliver, B.C., Archaeological Site DhQv-T26 (-86).* Report on file, Permit No. 1990-1, British Columbia Archaeology Branch, Victoria.

Olivier, Georges

1969   *Practical Anthropology.* Charles C. Thomas, Springfield, Illinois.

Pennefather-O'Brien, Elizabeth E., and Michael Strezewski

2002   An Initial Description of the Archaeology and Morphology of the Clark's Fork Skeletal Material, Bonner County, Idaho. *North American Archaeologist* 23:101–115.

Pokotylo, David L., Marian E. Binkley, and A. Joanne Curtin

1987   *The Cache Creek Burial Site (EeRh 1), British Columbia.* Contributions to Human History No. 1, pp. 1–14. Royal British Columbia Museum, Victoria.

Pokotylo, David L., and Donald Mitchell

1998   Prehistory of the Northern Plateau. In *Plateau,* edited by Deward E. Walker Jr., pp. 81–102. Handbook of North American Indians, Vol. 12, William C. Sturtevant, general editor, Smithsonian Institution, Washington, D.C.

Richards, Thomas H.

1982   *Salvage Excavation of a Historic Cairn Burial, Site EeRl 169, Near Lillooet, British Columbia.* Report on file, Permit Report 1976-9, British Columbia Archaeology Branch, Victoria.

Richards, Thomas H., and Michael K. Rousseau

1987   *Late Prehistoric Cultural Horizons on the Canadian Plateau.* Publication No. 16, Department of Archaeology, Simon Fraser University, Burnaby, British Columbia.

Roll, Tom E., and Steven Hackenberger

1998   Prehistory of the Eastern Plateau. In *Plateau,* edited by Deward E. Walker Jr., pp. 120–137. Handbook of North American Indians, Vol. 12, William C. Sturtevant, general editor, Smithsonian Institution, Washington, D.C.

Ruff, Christopher B.

1994   Morphological Adaptation to Climate in Modern and Fossil Hominids. *Yearbook of Physical Anthropology* 37:65–107.

2002   Variation in Human Body Size and Shape. *Annual Review of Anthropology* 31:211–232.

Sanger, David
1963   *EdRl-10, an Unusual Burial Site near Lillooet, British Columbia*. Library Archives, Archaeology Ms. 756, Canadian Museum of Civilization, Gatineau.
Saunders, Shelley R.
1978   *The Development and Distribution of Discontinuous Morphological Variation of the Human Infracranial Skeleton*. Archaeological Survey of Canada Mercury Series Paper No. 81. Canadian Museum of Civilization, Gatineau.
Schoeninger, Margaret J., Michael J. DeNiro, and Henrik Tauber
1983   Stable Nitrogen Isotope Ratios of Bone Collagen Reflect Marine and Terrestrial Components of Prehistoric Human Diet. *Science* 220:1381–1383.
Seymour, Brian D.
1976   1972 Salvage Excavations at DfRs3, the Whalen Farm Site. In *Current Research Reports*, edited by Roy L. Carlson, pp. 83–98. Department of Archaeology Publication 3, Simon Fraser University, Burnaby, British Columbia.
Skinner, Mark F.
1984   *Ancient Native Human Remains Discovered at Masset Trailer Court (Dec. 8, 1983) Masset, Queen Charlotte Islands, B.C.* Report on file, Permit No. 1983-46, British Columbia Archaeology Branch, Victoria.
1986   *Analysis of Human Skeletal Remains, Cape Mudge Village, Quadra Island, B.C., EaSh 3*. Report on file, Permit No. 1985-4, British Columbia Archaeology Branch, Victoria.
1987a   *Analysis of Human Skeletal Remains (86-16) from the Nechako/Nautley River Confluence (GaSd3)*. Report on file, Permit No. 1987-2, British Columbia Archaeology Branch, Victoria.
1987b   *Analysis of Found Human Remains 514 Front Street (Behind Masonic Hall) Quesnel, B.C. 86 09 06*. Report on file, Permit No. 1986-20, British Columbia Archaeology Branch, Victoria.
Skinner, Mark F., and Stanley Copp
1986   *The Nicoamen River Burial Site (EbRi 7), near Lytton, British Columbia*. Report on file, Permit 1985-4, British Columbia Archaeology Branch, Victoria.
Skinner, Mark F., and Jean McKendry
1984   *Human Remains Salvage Report, DfRu-42-Burial 1 (Saltspring Island, B.C.)*. Report on file, Permit No. 1983-46, British Columbia Archaeology Branch, Victoria.
Skinner, Mark F., and Ron Thacker
1988   *The Quilchena Hotel Burial Site (EaRd 14), Nicola Lake, B.C.* Report on file, Permit 1987-2, British Columbia Archaeology Branch, Victoria.
1989   *Analysis of Human Skeletal Remains: The Kopp Burial Site (EbQt 19), Vernon, British Columbia*. Report on file, Permit No. 1988-1, British Columbia Archaeology Branch, Victoria.
Smith, Harlan I.
1899   Archaeology of Lytton, British Columbia. In *Memoirs of the American Museum of Natural History,* Vol. 2, Pt. 3, pp. 129–161. New York.
Sprague, Roderick, and Walter H. Birkby
1970   Miscellaneous Columbia Plateau Burials. *Tebiwa* 13(1):1–32.
1973   *Burials Recovered from the Narrows Site (45OK11), Columbia River, Washington*. University of Idaho Anthropological Research Manuscript Series No. 11, Moscow.
Sprague, Roderick, and Thomas M. J. Mulinski
1980   *Ancestral Burial Relocations, Chief Joseph Dam, 1979*. University of Idaho Anthropological Research Manuscript Series No. 63, Moscow.

Stewart, T. Dale
    1950   Report on Skeleton from the Butte Creek Cave, John Day River Region, Oregon. *Proceedings of the American Philosophical Society* 94:385–387.
Stijelja, Maryanne, and Todd Williams
    1986   An Analysis of the Skeletal Remains of EiRm-7. Report on file, Department of Archaeology, Simon Fraser University, Burnaby, British Columbia.
Stock, Jay T.
    2006   Hunter-Gatherer Postcranial Robusticity Relative to Patterns of Mobility, Climatic Adaptation, and Selection for Tissue Economy. *American Journal of Physical Anthropology* 131:194–204.
Stryd, Arnoud H.
    1980   A Review of the Recent Activities Undertaken by the Lillooet Archaeological Project. *Midden* 12 (2):5–20.
Stryd, Arnoud H., and James Baker
    1968   Salvage Excavations at Lillooet, British Columbia. *Syesis* 1:47–56.
Styles, Norla
    1976   Preliminary Report on the Burials at Glenrose. In *The Glenrose Cannery Site*, edited by R. G. Matson, pp. 203–213. Archaeological Survey of Canada Mercury Series Paper No. 52, Canadian Museum of Civilization, Gatineau.
Sumpter, Ian D.
    1982   *Analysis of Human Skeletal Remains and Associated Skeletal Material from Site DjQj 1, Vallican, British Columbia*. Report on file (Unnumbered), British Columbia Archaeology Branch, Victoria.
Suttles, Wayne
    1990   Environment. In *Northwest Coast*, edited by Wayne Suttles, pp. 16–29. Handbook of North American Indians, Vol. 7, William C. Sturtevant, general editor, Smithsonian Institution, Washington, D.C.
Suttles, Wayne (editor)
    1990   *Northwest Coast*. In Handbook of North American Indians, Vol. 7, William C. Sturtevant, general editor, Smithsonian Institution, Washington, D.C.
Taylor, Royal E.
    2001   Amino Acid Composition and Stable Carbon Isotope Values on Kennewick Skeleton Bone. Archeology and Ethnography Program, U.S. National Parks Service. Electronic document, http://www.cr.nps.gov/aad/kennewick/Taylor2, accessed April 16, 2002.
Taylor, Royal E., Donna L. Kirner, John R. Southon, and James C. Chatters
    1998   Radiocarbon Dates of Kennewick Man. *Science* 280:1171–1172.
Thom, Brian D.
    1995   The Dead and the Living: Burial Mounds and Cairns and the Development of Social Classes in the Gulf of Georgia Region. Master's thesis, Department of Anthropology and Sociology, University of British Columbia, Vancouver.
Trace, Andrew A.
    1981   An Examination of the Locarno Beach Phase as Represented at the Crescent Beach Site, DgRr 1, British Columbia. Master's thesis, Department of Archaeology, Simon Fraser University, Burnaby, British Columbia.
Trotter, Mildred, and Goldine C. Gleser
    1952   Estimation of Stature from Long Bones of American Whites and Negroes. *American Journal of Physical Anthropology* 10:463–514.
    1958   A Re-evaluation of Estimation of Stature Based on Measurements of Stature

Taken During Life and of Long Bones after Death. *American Journal of Physical Anthropology* 16:79–123.

Turner, Nancy J.
1997   *Food Plants of Interior First Peoples.* Royal British Columbia Museum Handbook Series, University of British Columbia Press, Vancouver.

Walker, Deward E. (editor)
1998   *Plateau.* Handbook of North American Indians, Vol. 12, William C. Sturtevant, general editor, Smithsonian Institution, Washington, D.C.

Westcott, Daniel J.
2006   Effect of Mobility on Femur Midshaft External Shape and Robusticity. *American Journal of Physical Anthropology* 130:201–213.

Wilkinson, Leland, Grant Blank, and Christian Gruber
1996   *Desktop Data Analysis with SYSTAT.* Prentice Hall, Upper Saddle River, New Jersey.

Wilson, Robert
1972   *Report on Brocklehurst Burial Site EeRc 8 Kamloops, British Columbia.* Report on file, Permit No. 1971-47, British Columbia Archaeology Branch, Victoria.

5. | Discrete Dental Trait Evidence of Migration Patterns in the Northern Southwest

Kathy R. Durand, Meradeth Snow,
David Glenn Smith, and Stephen R. Durand

Summary Statement: This chapter uses discrete dental trait evidence to explore intraregional migration in the U.S. Southwest. Migration patterns are the focus of much current research in the Southwest but are usually identified using indirect methods, such as documenting changes in ceramic styles or architecture. This study more directly tests two migration models in the prehistoric northern Southwest by using discrete dental traits. These traits are under genetic control and provide a nondestructive means of documenting genetic relatedness among regional populations. We have documented patterns in the dentition of samples from the Middle San Juan region, allowing us to estimate biodistances among Great House and small-site inhabitants in this region. The models we test through this research include the timing of the southward migration from the Mesa Verde region and the possible movement of groups from Chaco Canyon into the Middle San Juan region in the early twelfth century. Despite the usefulness of discrete dental traits for addressing the issue of migration, they have been underutilized. This study helps bring these data to the forefront to document migrations in the prehistoric Southwest.

Introduction

Archaeologists have long grappled with the issue of regional migrations, searching for evidence of migration processes and using them to explain unusual patterns in the archaeological record (e.g., Haury 1958; Morris 1938). The

Human Variation in the Americas: The Integration of Archaeology and Biological Anthropology, edited by Benjamin M. Auerbach. Center for Archaeological Investigations, Occasional Paper No. 38. © 2010 by the Board of Trustees, Southern Illinois University. All rights reserved. ISBN 978-0-88104-095-1.

northern Southwest, or Ancestral Puebloan region, is no exception to this trend, with migrations the focus of many recent studies (e.g., Ahlstrom et al. 1995; Cordell et al. 2007; Duff and Wilshusen 2000; Lekson and Cameron 1995). Typically, episodes of migration are documented through the material culture that people left behind. However, this is an indirect means that can be inaccurate when migrants quickly adopt local styles or local styles change due to culture contact with no migration (Cordell 1995; Willey et al. 1956). A more direct method of tracing migration patterns is available through DNA analysis of skeletal remains (Carlyle et al. 2000; Jones 2003; Kaestle and Smith 2001; Malhi et al. 2003; Mulligan 2006; Schurr 2004), but it is rarely possible to use this destructive technique on Native American remains. The use of discrete dental trait data provides a completely nondestructive alternative to DNA analysis that can provide estimates of biological affinity among populations (e.g., Durand and Wheelbarger 2007; Haydenblit 1996; Scott and Turner 1997; Sofaer et al. 1986).

We use discrete dental traits to test two models of migration in the Middle San Juan region of the northern Southwest during the late prehistoric period. Although this region may have been an important center in prehistory (McKenna and Toll 1992), it is often considered to have been peripheral to the larger centers of Mesa Verde to the north or Chaco Canyon to the south. Two hypotheses have been proposed regarding the movement of people into the Middle San Juan region that can be tested with discrete dental trait data. First, we test Lekson's (1999) proposal that the elites from the Great House of Pueblo Bonito, in Chaco Canyon, moved north to Aztec Ruin after the Great House communities in Chaco Canyon were abandoned in the early twelfth century. Second, we look for evidence regarding Duff and Wilshusen's (2000) hypothesis that the abandonment of the Mesa Verde region actually began between 800 and 725 B.P. (A.D. 1150 and 1225), earlier than approximately 650 B.P. (A.D. 1300) as previously believed (Lekson and Cameron 1995; Rohn 1983, 1989; Varien et al. 1996). Our results for the former hypothesis are stronger, as we have samples from both the source and destination populations for this hypothesized migration from Chaco Canyon. For the latter hypothesis, we have samples only from the destination communities. Nevertheless, our results suggest that during the twelfth century, the Middle San Juan region received migrants from both Chaco Canyon in the south and from another source, possibly the Mesa Verde region to the north.

Discrete Dental Traits

Discrete dental traits are nonmetric aspects of a tooth's morphology, such as shoveling of the incisors and lower molar cusp number. These traits are drawn from teeth in all parts of the dental arcade, including anterior and posterior teeth in both the maxilla and mandible. Discrete dental traits are under genetic control (Nichol 1990; Scott 1973; Scott and Turner 1997) and can be used to estimate genetic relationships among populations (Coppa et al. 2007; Haydenblit 1996; Howell and Kintigh 1996; Irish 2005, 2006; Scott and Turner 1988, 2006; Sofaer et al. 1986). Discrete dental trait data can be collected from living populations using dental casts, as well as from skeletal samples, thus facilitating comparisons through time.

While previous studies of discrete dental traits have been conducted in the U.S. Southwest (Scott and Dahlberg 1982; Scott and Turner 2006; Turner 1993), these studies either included only modern samples or did not examine variation within cultural regions. For example, Scott and Dahlberg (1982) found that the Zuni were the most distinctive population of the modern Native American groups they sampled, while the Navajo (Diné) exhibited the most similarity to the other groups in their study. This pattern for Navajo dentition is consistent with the results of mtDNA studies showing that the Navajo are highly admixed with other Southwest populations (e.g., Malhi et al. 2003), providing another example of the strong genetic basis of discrete dental traits. Because all the samples in Scott and Dahlberg's (1982) study were modern, they could not trace these regional patterns back in time. Turner's (1993) study included both modern and prehistoric samples, but because he was interested in biodistances between populations on a continent-wide basis, his data were pooled such that it is not possible to compare particular sites within a given region.

Haydenblit (1996) examined regional variation among skeletons from sites in southern Mexico and compared these results with prehistoric and modern samples across the New World and eastern Asia. She found that the sample from Tlatilco, the oldest site in her study, was distinctively different from the other Mesoamerican samples. Interestingly, Haydenblit (1996:237) also discovered that all the Mesoamerican samples displayed the southern Asian dental pattern, which exhibits a simplified expression of eight key traits[1] versus the more complex dental pattern found among northern Asian populations (Hanihara 1991, 1992; Hillson 1996; Turner 1983, 1990). The typical dental pattern found in studies of U.S. prehistoric samples, including those in the U.S. Southwest, is the northern Asian pattern (Turner 1979, 1985, 1987, 1990), rather than the southern Asian pattern that Haydenblit (1996) found for her Mesoamerican samples. This finding suggests that the Mesoamerican samples may represent a different ancestral population than the more Northeast Asian ancestral group suggested by other New World dentitions.

Biodistance estimates based on discrete dental traits also have been made for Old World regions. Coppa and colleagues (2007) found that their pre-Neolithic sample in prehistoric Italy was different from Neolithic and later samples. Their results support the model of immigrants entering Italy during the Neolithic period, but they were able to show that there was some admixture with the indigenous population, suggesting that there was not complete replacement of the indigenous population. Studies by Irish (2006) in northern Africa have shown continuity through time among 15 samples from Egypt, but a case of population discontinuity farther south in Nubia (Irish 2005). Irish's data allowed him to test several different models of regional population movement in northern Africa.

Roler (1992) found that discrete dental traits from population samples in the Near East were more similar to one another during the Bronze Age, when regional trade networks and, apparently, gene flow flourished in the region. Modern Near Eastern population samples exhibited greater heterogeneity between samples, reflecting the immigration of diverse groups to the region and the lack of mate exchange between many of these groups.

Thus, a number of studies from around the world have shown that discrete dental traits are useful for studying biological relationships among regional populations. These studies have documented isolated, stable populations in some regions and extensive admixture in others, helping give definitive answers to questions about prehistoric migrations. Thus, a regional study of these traits should prove useful to document the migration of populations in the later prehistoric periods of the northern Southwest.

While material culture remains (Clark 2004; Webster and Loma'omvaya 2004) and craniometric studies (Akins 1986; Buikstra et al. 1990; Schillaci 2003; Schillaci et al. 2001; Schillaci and Stojanowski 2002) have been used to document prehistoric Southwest migrations, we know from other regions that the results of the latter do not always match those obtained from mtDNA (e.g., Bolnick and Smith 2007). Further, craniometric studies of Chaco Canyon samples have produced diametrically opposed conclusions. Akins (1986, 2003) has found that "the Pueblo Bonito burials represent closely related groups rather than accumulations from a wider region brought together on ceremonial occasions" (Akins 2003:101). Schillaci (2003:239), on the other hand, argues that a model in which Chaco Canyon populations develop in situ from an original Basketmaker population "is inconsistent with [the] high level of genetic heterogeneity and differentiation observed for Chaco Canyon populations."

Regional Context

The Chaco region of the U.S. Southwest was centered in Chaco Canyon in the San Juan basin, approximately 150 kilometers northwest of Albuquerque, New Mexico. Chaco Canyon is home to 12 large Great House structures within a 15 km stretch, as well as numerous other smaller structures. Moving beyond Chaco Canyon, a large number of outlying Great Houses, often with smaller community structures surrounding them, are found across and beyond the San Juan basin (Kantner and Mahoney 2000; Lekson 1991; Marshall et al. 1979; Powers et al. 1983). For decades, this area has been recognized as an interconnected culture region that may have had some kind of administrative and economic ties across its span (e.g., Judge 1989; Lekson et al. 1988; Vivian 1990). The Chaco Phenomenon (Irwin-Williams 1972)—one of the earlier terms applied to the region—still captures the connected yet enigmatic nature of its prehistoric populations.

While over a decade ago the main focus of research in the Chaco region was on defining the nature of the Chaco Phenomenon (Was it a centrally organized form of complex society, or did the similarities across the region arise from some form of emulation?), today that focus is shifting in part to tracing patterns of migration within the region (Clark 2004; Duff 1998; Hegmon 2000) with a whole session devoted to this topic at the 20th Anniversary Southwest Symposium. Though the establishment of Great House communities around the San Juan basin and beyond was undoubtedly associated with the movement of people, it has been somewhat difficult to identify migrations during the Chaco fluorescence (during the Pueblo II period from 1050–800 B.P. [A.D. 900–1150]). Archaeologists have evidence to indicate that, on a temporary basis, people came into the center at Chaco Canyon periodically for religious or other activities (e.g., Judge 1989; Lekson et al. 1988; Toll 2001). On a more permanent basis, people moved into the central San Juan basin in the late Pueblo I period (1250–1050 B.P. [A.D. 700–900]) prior to the fluorescence of Chaco Canyon (Wilshusen and Van Dyke 2006; Windes 2007), and people may have moved out of Chaco Canyon to the peripheral areas during Chaco's fluorescence to establish new outlying Great House communities during the Pueblo II period. Elsewhere in the Southwest, Duff and Lekson (2006:328) suggest that there were "significant migrations or expansions from the Zuni-Acoma areas into the Mogollon Highlands . . . during the Chaco era."

Regional migrations continued in the subsequent Pueblo III period (800–650 B.P. [A.D. 1150–1300]), during which many people left Chaco Canyon permanently. Archaeologists have long speculated about where these emigrants went (e.g., Cordell 1984; Lekson 1999). Finally, sometime during the late Pueblo III and Pueblo IV periods (650–450 B.P. [A.D. 1300–1500]), a large influx of people moved to the Rio Grande region (Ahlstrom et al. 1995; Cordell 1995; Lekson and Cameron 1995). There may well have been more regional migration events than cited above, but those listed are agreed upon by most scholars working in the northern Southwest. In part to document the origins and destinations of the migrants, a variety of studies are under way to source artifacts (Cameron 1984, 1997, 2001; Cordell et al. 2008; Hull et al. 2008; Mathien 1986, 2001; Toll 2006). Of course, one of the difficulties with this approach to migration is that while the goods may be found to have been constructed in or obtained from another region, it is unclear whether they were brought into the new region by migrants or through trade. If migrations at this small scale leave sufficient genetic traces, studying the biological evidence from the human remains themselves could clarify whether changes in the material culture record were related to migration or were caused by one of various cultural practices, such as trade, emulation, or simply changes in styles over time.

There are two proposed intraregional migrations that can readily be tested with samples of Ancestral Puebloan discrete dental traits. The first concerns the timing of migrants coming from the Mesa Verde region in the north. Did people gradually begin to leave the Mesa Verde region between 800 and 725 B.P. (A.D. 1150 and 1225), as Duff and Wilshusen (2000) propose, rather than migrating around 650 B.P. (A.D. 1300) as previously supposed (Lekson and Cameron 1995; Rohn 1983, 1989; Varien et al. 1996)? If these migrants passed through the Middle San Juan region, a logical first step in their southeastward migration, there should be evidence of their presence in the form of a change in discrete dental trait frequencies between Pueblo II and early Pueblo III skeletal samples in the region. If the migrations occurred at a later date, there should be more continuity through time in the dental traits from the Pueblo II to the Pueblo III period. We have regional small-site samples from both periods, which allowed us to test this model.

A second prehistoric migration model that can be tested with discrete dental traits involves the movement of populations into the Middle San Juan region from the south. Lekson (1999) proposes that the elites from the Chaco Canyon Great Houses moved from Chaco Canyon up to Aztec Ruin after the abandonment of the Chaco Canyon Great Houses in the early twelfth century. This proposal has gained support from recent studies of material culture from Aztec Ruin and Pueblo Bonito (Brown et al. 2008; Webster 2008). We have tested this model with dental data from Aztec Ruin and Pueblo Bonito, the largest Great House in Chaco Canyon. Our expectation was that if there was a migration of elites from Chaco Canyon to Aztec Ruin, the discrete dental trait frequencies from Aztec Ruin should be very similar to those from Pueblo Bonito, while the Aztec trait frequencies should be more distinctive in comparison to the local small-site population.

Study Samples

The samples collected for this study come from seven archaeological sites located in the San Juan basin in the U.S. Southwest (Figure 5-1). All the data from these assemblages was collected by K. Durand. In addition, we compare our samples with seven modern population samples collected by Scott (Scott 1973; Scott and Dahlberg 1982) from plaster casts of modern Native Americans (Figure 5-2). These modern population samples were drawn from the San Carlos Apache, Hopi, Mohave, Navajo, Tohono O'odham (formerly the Papago), Yuma, and Zuni groups.

With the exception of the Pueblo Bonito sample, from Chaco Canyon, the prehistoric Southwest samples (Table 5-1) come from a particular part of the Chaco region, known as the Middle San Juan, or Totah, region (McKenna and Toll 1992). This region, centered at Farmington, New Mexico, was named for the confluence of the San Juan, Animas, and La Plata rivers. It is a relatively lush region for the desert Southwest that would have provided ample farmland and wild game, as well as a continuous source of freshwater to its inhabitants. Two of the population samples from this region come from the Chacoan Great Houses of Aztec Ruins (along the Animas River) and Salmon Ruins (along the San Juan River). Both of these samples date to the Pueblo III or post–Chacoan period.

The other four Southwest samples come from Chacoan small-house or residential structures. These include several sites along the La Plata River (Table 5-1), excavated by the Office of Archaeological Studies (Martin et al. 2001) from 1988 to 1991, pooled here by occupation period (individuals from seven sites were pooled for the Pueblo II sample and individuals from three sites were pooled for the Pueblo III sample). The other two prehistoric samples come from two sites, the Tommy site and the Mine Canyon site (Figure 5-1), located near the confluence of the three rivers. Both of these sites are on the B-Square Ranch, a working ranch of over 12,000 acres that runs along the San Juan River and makes up the southern edge of Farmington. We also conducted a study of the mtDNA from these two sites (Snow et al. 2010). According to Wheelbarger (2008:221–225) the ceramics from the Tommy site predominantly date to the Pueblo II period (1050–800 B.P. [A.D. 900–1150]), while the ceramics from the Mine Canyon site suggest it was occupied in the Pueblo III period (800–650 B.P. [A.D. 1150–1300]).

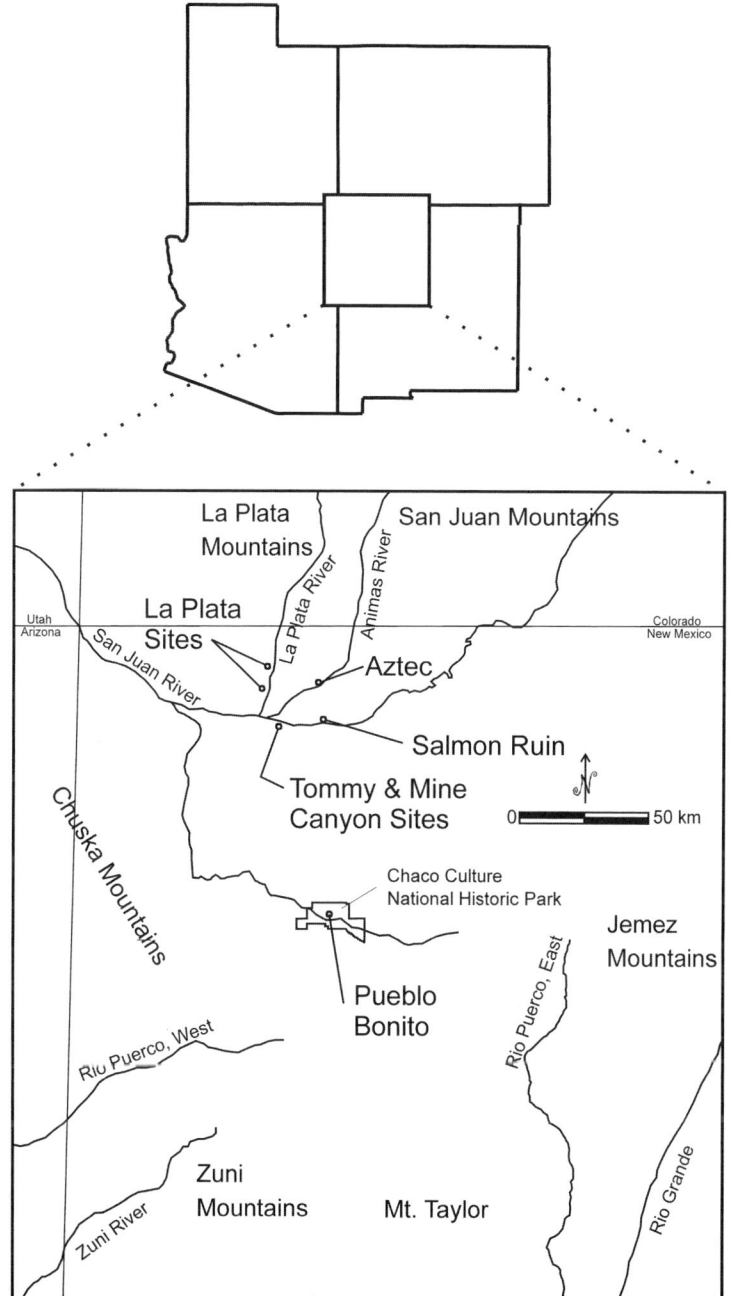

Figure 5-1. *Map of northwestern New Mexico and the Four Corners region showing the locations of the prehistoric sites sampled in this study. The La Plata samples were derived from seven sites that are located in the two site clusters depicted on the map.*

Figure 5-2. *Map of the American Southwest and Mexico showing the general locations of the modern groups sampled by Scott and Dahlberg (1982) and the prehistoric sites in Mexico sampled by Haydenblit (1996).*

Table 5-1. *Samples Collected for This Study*

Sample	N	Current Location
Aztec Ruin	22	American Museum of Natural History
Pueblo II La Plata[a]	15	Office of Archaeological Studies, Santa Fe, New Mexico
Pueblo III La Plata[b]	12	Office of Archaeological Studies, Santa Fe, New Mexico
Mine Canyon	10	B-Square Ranch, Farmington, New Mexico
Pueblo Bonito	42	National Museum of Natural History
Salmon Ruin	18	Salmon Ruins Museum, Bloomfield, New Mexico
Tommy Site	26	B-Square Ranch, Farmington, New Mexico

[a]This sample includes skeletons from LA 37592, LA 37593, LA 37595, LA 37599, LA 37600, LA 37601, and LA 65030.
[b]This sample includes skeletons from LA 37592, LA 37601, LA 65030.

Results of Discrete Dental Traits Analysis

The traits utilized for this study are given in Table 5-2. They were collected using the descriptions and plaques from the Arizona State University Dental Anthropology System (ASUDAS, Turner et al. 1991). All discrete dental trait data for the prehistoric U.S. Southwest samples was collected by K. Durand, thus eliminating any interobserver error. In addition to the discrete dental traits, data concerning the presence or absence of and wear for each tooth, as well as overall dental health (including caries, enamel hypoplasias, and abscesses) of all individuals, were collected, although these data were not used in the following analysis.

The discrete dental trait data were collected on an interval scale and then converted to presence/absence data using the cutoff points provided in Table 5-2. The mean measure of divergence (MMD, Berry and Berry 1967), as modified by Harris and Sjøvold (2004), was used to calculate the distances between sample populations.[2] Average linkage clustering was then used to create the dendrograms shown here, as it seemed to be the most discriminating solution, avoiding chaining problems.

Ancestral Puebloan Samples

The dendrogram in Figure 5-3 is based on 21 traits (Table 5-2) and the samples fall into two main clusters. The top cluster consists of three of the small sites in the Middle San Juan region: the Pueblo II Tommy site sample, the Pueblo III Mine Canyon site sample, and the Pueblo III La Plata Valley sample. Although this is a reasonable relationship to expect based on the spatial proximity of these samples, it is inconsistent with our analysis of the mtDNA, in which the Tommy site and Mine Canyon site samples have very different haplogroup frequencies (Table 5-3). The Pueblo III La Plata sample, while grouping with these two sites in the dendrogram, is the most distinctive of the population samples. The Pueblo III La Plata sample is the last to join or form a cluster; all the other samples join with another sample earlier (at a lower MMD value) than this La Plata sample does.

The second cluster contains the three Great House samples and one of the regional small-site samples. The samples from the Great Houses of Aztec Ruin and Pueblo Bonito form a subgroup within this cluster, indicating that they are more similar to one another than either is to the Great House of Salmon Ruin. The Salmon Ruin sample, on the other hand, groups first with the local small-house Pueblo II sample from the La Plata Valley before joining with the other Great Houses.

The clustering of the Aztec Ruin and Pueblo Bonito samples is intriguing, particularly because these samples are not close together in space or time. As discussed below, this pattern provides support for Lekson's (1999) model of a migration event between Pueblo Bonito and Aztec Ruin at the Pueblo II/III boundary.

Table 5-2. *Discrete Dental Traits for Prehistoric Southwest Samples*

Traits (Breakpoints)		Tooth	AR	LP PII	LP PIII	MC	PB	SR	TS
Winging	n	UI1	16	4	7	6	25	2	4
(+ = ASU 1)	%		18.8	0	42.9	83.3	32.0	0	75.0
Palatine torus	n		22	3	8	10	40	5	10
(+ = ASU 2–3)	%		13.6	0	0	10.0	0	20.0	40.0
Shovel-shape	n	UI1	11	9	6	10	27	14	21
(+ = ASU 3–7)	%		27.3	33.3	83.3	40.0	40.7	50.0	61.9
Double shovel	n	UI1	18	8	7	9	27	15	17
(+ = ASU 2–6)	%		0	87.5	71.4	55.6	40.7	33.3	70.6
Interruption groove	n	UI2	13	7	6	7	31	9	17
(+ = ASU +)	%		92.3	100.0	66.7	71.4	64.5	77.8	52.9
Tuberculum dentale	n	UI2	12	7	7	9	32	9	18
(+ = ASU 2–6)	%		25.0	14.3	0	22.2	9.4	66.7	22.2
Peg-reduced	n	UI2	18	9	8	3	34	14	11
(+ = ASU P or R)	%		0	0	12.5	33.3	2.9	7.1	18.2
Distal accessory ridge	n	UC	3	7	2	6	11	7	13
(+ = ASU 2–5)	%		66.7	71.4	100.0	66.7	45.5	42.9	53.9
Cusp 5	n	UM1	8	7	7	5	30	15	23
(+ = ASU 2–5)	%		12.5	57.1	14.3	60.0	20.0	33.3	13.0
Carabelli's cusp	n	UM1	14	7	7	9	42	18	26
(+ = ASU 2–7)	%		21.4	42.9	85.7	22.2	35.7	44.4	46.2
Parastyle	n	UM1	14	8	4	4	23	6	9
(+ = ASU 1–5)	%		7.1	12.5	25.0	25.0	8.7	33.3	0
Enamel extension	n	UM1	17	3	6	4	36	13	14
(+ = ASU 1–3)	%		11.8	0	66.7	50.0	50.0	46.2	50.0
Hypocone	n	UM2	13	7	6	0	33	11	2
(+ = ASU 3–5)	%		76.9	85.7	83.3	—	63.6	100.0	50.0
Absence	n	UM3	20	8	6	1	32	10	5
(+ = ASU 1)	%		5.0	0	16.7	0	0	0	40.0
Anterior fovea	n	LM1	4	3	4	5	17	11	19
(+ = ASU 2–4)	%		25.0	66.7	75.0	60.0	58.8	63.6	68.4
Cusp number	n	LM1	8	5	6	6	21	12	21
(+ = ASU 6+)	%		12.5	60.0	33.3	50.0	23.8	58.3	52.4
Deflecting wrinkle	n	LM1	4	3	5	2	13	10	18
(+ = ASU 2–3)	%		75.0	100.0	60.0	100.0	61.5	70.0	66.7
Protostylid	n	LM1	10	6	8	6	32	17	15
(+ = ASU 1–6)	%		60.0	33.3	37.5	33.3	78.1	52.9	60.0
Cusp 7	n	LM1	13	4	6	6	27	17	22
(+ = ASU 2–4)	%		0	25.0	33.3	0	7.4	5.9	4.6
Groove Pattern	n	LM2	12	7	5	5	25	11	13
(+ = ASU Y)	%		8.3	42.9	20.0	60.0	20.0	27.3	38.5
Rocker Jaw	n		19	2	3	6	30	8	7
(+ = ASU 1–2)	%		26.3	0	0	16.7	26.7	12.5	28.6

Sources: Breakpoints for traits are listed according to their value in the Arizona State University (ASU) Dental Anthropology System and are drawn from Irish (2005) and Scott and Turner (1997).
Note: AR – Aztec Ruin; LP PII – La Plata sites, Pueblo II Assemblages; LP PIII – La Plata sites, Pueblo III Assemblages; MC – Mine Canyon; PB – Pueblo Bonito; SR – Salmon Ruin; TS – Tommy site.

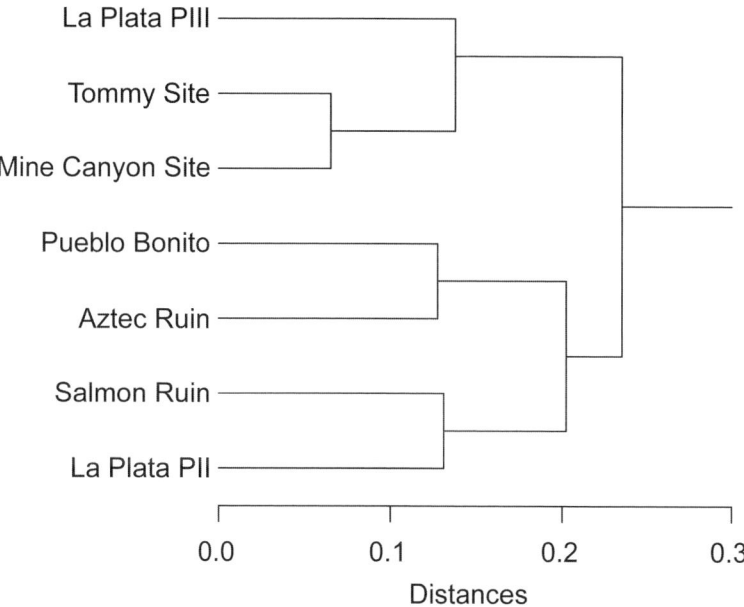

Figure 5-3. *Cluster dendrogram showing the distance between prehistoric southwestern samples based on 21 discrete dental traits. The input to the average linkage clustering analysis was the mean measure of divergence (MMD).*

Table 5-3. *mtDNA Haplogroup Frequencies for Tommy and Mine Canyon Sites*

Haplogroups	A	B	C	D	Total
Tommy site	1 (.03)	25 (.69)	5 (.14)	5 (.14)	36
Mine Canyon site	7 (.58)	4 (.33)	1 (.08)	0	12

Source: Data taken from Snow and colleagues (2010:1639).
Note: Proportions shown in parenthesis.

Modern and Prehistoric Southwest Samples

In Figure 5-4, we compare our prehistoric Southwest samples with those from modern Southwest populations. This analysis is based on nine traits (UI1 shoveling, UI2 tuberculum dentale, UC distal accessory ridge, UM1 cusp 5, UM1 Carabelli's cusp, LM1 protostylid, LM1 cusp 6, LM1 cusp 7, and LM2 groove pattern) taken from Scott (1973) and Scott and Dahlberg (1982). Here two main cluster groups are present, one consisting largely of modern samples and the other made up of Ancestral Puebloan samples. In this comparison, most of the Ancestral Puebloan samples cluster in the same groupings as they did in

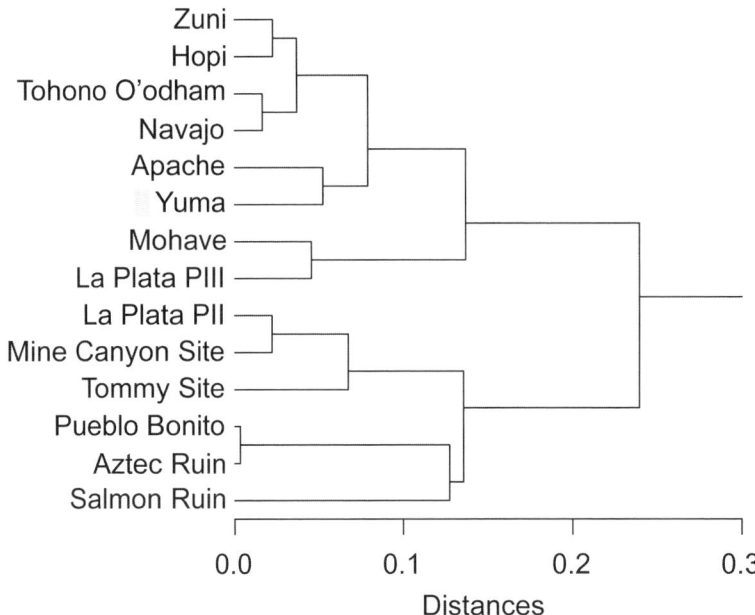

Figure 5-4. *Cluster dendrogram showing the distance between prehistoric and modern southwestern samples based on 9 discrete dental traits. The input to the average linkage clustering analysis was the mean measure of divergence (MMD).*

the 21 trait analysis. The primary differences are in the cluster groups for the La Plata Valley samples. For the 21 trait solution, the Pueblo II La Plata sample clustered with the Great Houses, while the Pueblo III sample grouped with the Tommy and Mine Canyon samples. For Figure 5-4, with a reduction in the number of traits and the addition of more comparative samples, the Pueblo II La Plata sample now groups with the local small-site samples, while the Pueblo III La Plata sample groups with the cluster of modern samples. Obviously, the elimination of 12 traits affected the MMD values for these samples. Just as sequencing of mtDNA provides a clearer picture of a population than simply categorizing according to blood types does, using a larger number of dental traits provides a more representative sample of the underlying DNA.

Although they do not provide a dendrogram to illustrate their results, it should be noted that the modern samples in Figure 5-4 cluster differently than they would have in Scott and Dahlberg's (1982) original analysis. While we are using some of Scott and Dahlberg's (1982) data, our MMD values are different from theirs for several reasons. First, we used breakpoints for the presence/absence dichotomy commonly used today (e.g., Coppa et al. 2007; Haydenblit 1996; Irish 2005, 2006; Scott and Turner 1997) that are different from those used by Scott and Dahlberg (1982), resulting in higher frequencies for some traits (shoveling and protostylid) but lower frequencies for another (Carabelli's cusp).

This was done to ensure that all the data included in our study was comparable to that in other recent studies. Second, we utilized some traits that were not included in Scott and Dahlberg's (1982) published analysis but that are available in Scott's (1973) dissertation (which also was our source to calculate new breakpoints). These traits were included to provide a larger number of traits for our site samples.

Thus, it is not surprising that our results are somewhat different from theirs. Further, this reanalysis of their data is more consistent with the results of recent DNA studies of Southwest populations (Malhi et al. 2003). For example, the Apache and Navajo samples do not fall into the same cluster using the new traits and breakpoints. Although both populations are Athapaskan in origin, recent DNA research has shown that while modern Apache and Navajo populations have had extensive admixture it has been with different groups, and consequently they exhibit "significantly different haplogroup frequencies" (Malhi et al. 2003:120). The Apache have admixed with the Yuma and the Pima, while the Navajo have admixed with Pueblo groups (Malhi et al. 2003:120). Hence, in Figure 5-4 the Navajo cluster with Puebloan samples (and the Tohono O'odham, who share particular genetic markers with the Zuni), while the Apache cluster with the Yuma (with whom they share a high frequency of the gene Albumin Mexico [AL*Mexico], which is found in only low frequency among the Navajo [Smith et al. 2000]).

Among the Ancestral Puebloan samples, the Great Houses continue to cluster together, although Aztec Ruin and Pueblo Bonito are still much more similar to each other than either is to Salmon Ruin. Indeed, for this smaller set of traits, the Aztec Ruin and Pueblo Bonito samples are nearly indistinguishable from one another. The sample from the Pueblo III La Plata Valley sites was the only prehistoric sample to cluster with the modern samples, forming an isolated cluster with the Mohave. Surprisingly, although we have found specific mtDNA mutations common to individuals from both the Tommy and Mine Canyon sites and to modern Zuni (Snow et al. 2010), we do not see a connection between these samples among the dental data. Why did none of the other Ancestral Puebloan samples cluster with any of the modern Southwest dental samples? The answer may lie in the amount of admixture that has occurred between Southwest populations in the intervening centuries. Most of the mtDNA studies are based on modern populations, as is Scott and Dahlberg's (1982) dental analysis. Dental and mtDNA data from samples dating to the period between the modern and Ancestral Puebloan samples might help to clarify the biodistances among these groups. It also is possible that the differences between the modern and prehistoric dental trait samples are largely due to interobserver error.

Furthermore, it must be recalled that MMD values are distance measures. Smaller MMD values mean that samples are more similar, while larger values indicate that two samples are more dissimilar or divergent from one another. Overall, the MMD values for all Southwest samples in Figure 5-4 are small. This indicates that, while there are some differences among the Southwest samples, there is a great deal of similarity among them.

Conclusions

Several conclusions can be drawn from these results that inform us about prehistoric Southwest migrations. First, we have found support for a close biological affinity between Aztec Ruin and Chaco Canyon. Both phases of our analysis revealed a close connection between the Pueblo Bonito and Aztec Ruin Great Houses, supporting Lekson's (1999) model of migration from Pueblo Bonito to Aztec Ruin at the end of the Pueblo II period. The dentition of the inhabitants of Aztec Ruin clearly did not have the same trait frequencies as the local Middle San Juan region population. On the other hand, the dentition of the inhabitants of the Salmon Ruin Great House looked more like a blend of the local population and the other Great Houses. This is not unexpected as the Salmon sample dates to the Pueblo III period during a time in which Salmon loses its distinctive Chacoan character and looks more like local sites in terms of, for example, the use of space (subdividing large Chacoan rooms), the proliferation of intramural kivas, and the shift to local and northern ceramic traditions. Irwin-Williams (1980) referred to this period as the Mesa Verdean/Secondary occupation (versus the Chacoan/Primary occupation) and Reed (2006:296) called it the San Juan period.

Our results suggest that it would be possible to test a second component of Lekson's (1999) model if the proper samples were obtained. Lekson (1999) proposes that the inhabitants of Aztec Ruin later migrated southward to Paquimé (Casas Grandes) in what is now northern Mexico. Certainly, there is abundant evidence of Mesoamerican cultural traits in the northern Southwest (e.g,. Lekson 1999; Lekson et al. 2007; McGuire 1980; McGuire et al. 1994; Mathien 1986; Nelson 2006). Future analyses of discrete dental traits from the southern Southwest (northern Mexico) could provide evidence of the movement (or lack thereof) of people across this larger region. If the Paquimé burial population had the southern Asian dental pattern Haydenblit (1996) found for Mesoamerican dentitions to the south of Paquimé, immigrants with the Ancestral Puebloan dental pattern would be dramatically different, making immigrants from Aztec Ruin easy to detect. Turner's (1993) study of Southwest dentition suggests that the Paquimé dentitions are different from those in the Puebloan Southwest, but a more specific comparison of Pueblo Bonito, Aztec Ruin, and post-650 B.P. (A.D. 1300) Paquimé dental frequencies will be required to test Lekson's model. If Paquimé residents had dentition that was more of a blend of the two types (due to their intermediate location near the border of the two regions), it should still be possible to detect the presence of migrants from the Chaco region through their dental trait frequencies.

The evidence for immigrants coming into the Middle San Juan region from the Mesa Verde region is not as clear in our analysis. The most distinctive of the dental samples is the Pueblo III La Plata Valley sample, which was the last to group with other prehistoric Southwest samples and was the only prehistoric sample to group with the modern samples. This is intriguing, as this is the exact place and time in which migrants are predicted to have entered the region from the north (Duff and Wilshusen 2000). Both the patterns in the dental data and our preliminary mtDNA data indicate the presence of migrants in the region during the Pueblo III period. While this supports the argument that immigrants came to the region at this time, without dental and mtDNA data from potential source populations to the north we are at present unable to identify the origin of the immigrants.

Based on our analyses, discrete dental trait data do support the two models we have tested. We find support for Lekson's (1999) model regarding the migration of some elites from Chaco Canyon to Aztec Ruin after the abandonment of the Great Houses in Chaco Canyon. Further, we have identified another migration to the Middle San Juan region during the Pueblo III period that may represent a population from Mesa Verde moving south. With further discrete dental trait data from the Mesa Verde region, this may support Duff and Wilshusen's (2000) model of an earlier migration out of the Mesa Verde region. We believe these results demonstrate the efficacy of using discrete dental trait data to document the movement of populations in prehistory.

Acknowledgments

We are very grateful to Tommy Bolack for providing funding and permission to study the Tommy site and Mine Canyon site assemblages. David Hunt was invaluable at the Smithsonian's National Museum of Natural History in facilitating the study of the Chacoan skeletons, as were David Hurst Thomas, Giselle Garcia, and Gary Sawyer at the American Museum of Natural History. A big thank you also goes to Larry Baker for allowing us to study Salmon Ruin's skeletal remains. We are indebted to Nancy Akins for much assistance in the La Plata data collection and for allowing K. Durand to tag along on the AMNH trip. Finally, we would like to thank our editor, Benjamin Auerbach, as well as Stephen Lekson and an anonymous reviewer for helpful suggestions on an earlier version of this paper.

Notes

1. The northern Asian dental pattern (often summarized by the term *Sinodont dentition*) is defined as having "higher frequencies of incisor shoveling and double shoveling, 1-rooted upper first premolars, upper first molar enamel extensions, missing-pegged-reduced upper third molars, lower first molar deflecting wrinkles, and 3-rooted lower first molars, as well as a lower frequency of 4-cusped lower second molars" (Scott and Turner 1997:270). Southern Asians have a lower frequency of these same traits (Hanihara 1991), and this pattern has been categorized as *Sundadont dentition*.

2. Smith's mean measure of divergence (MMD) is a dissimilarity or distance measure and is based on the proportion of discrete traits between pairs of samples (Harris and Sjøvold 2004). There is no standard computational algorithm or commercially available program to compute MMD and following Harris and Sjøvold (2004), we feel it necessary to be explicit regarding the exact method of computa-

tion. We used a spreadsheet program to accomplish the calculations. MMD requires converting the trait frequencies into angular measures and we used Equation 2 as described in Harris and Sjøvold (2004:85), called the Freeman-Tukey transformation. We also used the correction formula attributed to Freeman and Tukey (Equation 7, Harris and Sjøvold 2004:87) to correct for sample size in comparing the transformed trait frequencies. When sample sizes are small, the Freeman-Tukey correction can result in a negative value for the divergence between samples for a given trait. That is, the distance between two samples for a given trait can be smaller than the correction factor. In these cases we set the calculated value to zero (Harris and Sjøvold 2004:88–89). This solution is intuitively pleasing in that a small divergence between samples could easily be the result of sampling problems and should not be considered a true difference. It also eliminates the confusion that would ensue if one used negative divergences in the computation of the overall mean measure of divergence. We urge readers who are interested in applying this technique to consult Harris and Sjøvold (2004), the most comprehensive discussion of the application and computation of MMD that we have found.

References

Ahlstrom, Richard V. N., Carla R. Van West, and Jeffrey S. Dean
  1995   Environmental and Chronological Factors in the Mesa Verde–Northern Rio Grande Migration. *Journal of Anthropological Archaeology* 14:125–142.
Akins, Nancy J.
  1986   *A Biocultural Approach to Human Burials from Chaco Canyon, New Mexico*. Reports of the Chaco Center No. 9. Branch of Cultural Research, National Park Service, Santa Fe.
  2003   The Burials of Pueblo Bonito. In *Pueblo Bonito: Center of the Chacoan World*, edited by Jill E. Neitzel, pp. 94–106. Smithsonian Books, Washington, D.C.
Berry, A. Caroline, and R. J. Berry
  1967   Epigenetic Variation in the Human Cranium. *Journal of Anatomy* 101:361–379.
Bolnick, Deborah A., and David Glenn Smith
  2007   Migration and Social Structure among the Hopewell: Evidence from Ancient DNA. *American Antiquity* 72:627–644.
Brown, Gary M., Thomas C. Windes, and Peter J. McKenna
  2008   Animas Anamnesis: Aztec Ruins or Anasazi Capital? In *Chaco's Northern Prodigies: Salmon, Aztec, and the Ascendancy of the Middle San Juan Region after AD 1100*, edited by Paul F. Reed, pp. 231–250. University of Utah Press, Salt Lake City.
Buikstra, Jane E., Susan R. Frankenberg, and Lyle W. Konigsberg
  1990   Skeletal Biological Distance Studies in American Physical Anthropology: Recent Trends. *American Journal of Physical Anthropology* 82:1–7.
Cameron, Catherine M.
  1984   A Regional View of Chipped Stone Raw Material Use in Chaco Canyon. In *Recent Research on Chaco Prehistory*, edited by W. James Judge and John D. Schelberg, pp. 137–152. Reports of the Chaco Center, No. 8. Division of Cultural Research, National Park Service, Albuquerque, New Mexico.
  1997   The Chipped Stone of Chaco Canyon, New Mexico. In *Ceramics, Lithics, and Ornaments of Chaco Canyon*, Vol. 2, edited by Frances J. Mathien, pp. 531–658. Pub-

lications in Archeology 18G, Chaco Canyon Studies. National Park Service, U.S. Department of the Interior, Santa Fe.

2001   Pink Chert, Projectile Points, and the Chacoan Regional System. *American Antiquity* 66:79–102.

Carlyle, Shawn W., Ryan L. Parr, M. Geoffrey Hayes, and Dennis H. O'Rourke

2000   Context of Maternal Lineages in the Greater Southwest. *American Journal of Physical Anthropology* 113:85–101.

Clark, Jeffery J.

2004   Tracking Cultural Affiliation: Enculturation and Ethnicity. In *Identity, Feasting, and the Archaeology of the Greater Southwest*, edited by Barbara J. Mills, pp. 42–73. University Press of Colorado, Boulder.

Coppa, Alfredo, Andrea Cucina, Michaela Lucci, Domenico Mancinelli, and Rita Vargiu

2007   Origins and Spread of Agriculture in Italy: A Nonmetric Dental Analysis. *American Journal of Physical Anthropology* 133:918–930.

Cordell, Linda S.

1984   *Prehistory of the Southwest*. Academic Press, San Diego, California.

1995 Tracing Migration: Pathways from the Receiving End. *Journal of Anthropological Archaeology* 14:203–211.

Cordell, Linda S., H. Wolcott Toll, Mollie S. Toll, and Thomas C. Windes

2008   Archaeological Corn from Pueblo Bonito, Chaco Canyon, New Mexico: Dates, Contexts, Sources. *American Antiquity* 73:491–511.

Cordell, Linda S., Carla R. Van West, Jeffrey S. Dean, and Deborah A. Muenchrath

2007   Mesa Verde Settlement History and Relocation: Climate Change, Social Networks, and Ancestral Pueblo Migration. *Kiva* 72:379–405.

Duff, Andrew I.

1998   The Process of Migration in the Late Prehistoric Southwest. In *Migration and Reorganization: The Pueblo IV Period in the American Southwest*, edited by Katherine A. Spielmann, pp. 31–52. Anthropological Research Papers No. 51. Arizona State University, Tempe.

Duff, Andrew I., and Stephen H. Lekson

2006   Notes from the South. In *The Archaeology of Chaco Canyon, an Eleventh-Century Pueblo Regional Center*, edited by Stephen H. Lekson, pp. 315–337. School of American Research Press, Santa Fe.

Duff, Andrew I., and Richard H. Wilshusen

2000   Prehistoric Population Dynamics in the Northern San Juan Region, A.D. 950–1300. *Kiva* 66:167–190.

Durand, Kathy Roler, and Linda Wheelbarger

2007   The Point Community: Life in a Chacoan Small House Community. Paper presented at the 72nd Annual Meeting of the Society for American Archaeology, Austin, Texas.

Hanihara, Kazuro

1991   Dual Structure Model for the Formation of the Japanese Population. *Japan Review* 2:1–33.

1992   Dual Structure Model for the Formation of the Japanese Population. In *International Symposium on Japanese as a Member of the Asian and Pacific Populations*, edited by Kazuro Hanihara, pp. 244–251. International Research Center for Japanese Studies, Kyoto.

Harris, Edward F., and Torstein Sjøvold

2004   Calculation of Smith's Mean Measure of Divergence for Intergroup Comparisons Using Nonmetric Data. *Dental Anthropology* 17:83–93.

Haury, Emil W.

  1958   Evidence at Point of Pines for a Prehistoric Migration from Northern Arizona.
          In *Migrations in New World Culture History*, edited by Raymond H. Thompson, pp.
          1–6. University of Arizona Social Science Bulletin 27, Tucson.

Haydenblit, Rebeca

  1996   Dental Variation among Four Prehispanic Mexican Populations. *American Journal of Physical Anthropology* 100:225–246.

Hegmon, Michelle (editor)

  2000   *The Archaeology of Regional Interaction: Religion, Warfare, and Exchange Across the American Southwest and Beyond*. University Press of Colorado, Boulder.

Hillson, Simon

  1996   *Dental Anthropology*. Cambridge University Press, New York.

Howell, Todd L., and Keith W. Kintigh

  1996   Archaeological Identification of Kin Groups Using Mortuary and Biological Data: An Example from the American Southwest. *American Antiquity* 61:537–554.

Hull, Sharon, Mostafa Fayek, Frances J. Mathien, Phillip Shelley, and Kathy R. Durand

  2008   A New Approach to Determining the Geological Provenance of Turquoise Artifacts Using Hydrogen and Copper Stable Isotopes. *Journal of Archaeological Science* 35:1355–1369.

Irish, Joel D.

  2005   Population Continuity vs. Discontinuity Revisited: Dental Affinities among Late Paleolithic Through Christian-Era Nubians. *American Journal of Physical Anthropology* 128:520–535.

  2006   Who Were the Ancient Egyptians? Dental Affinities among Neolithic Through Postdynastic Peoples. *American Journal of Physical Anthropology* 129:529–543.

Irwin-Williams, Cynthia

  1972   *The Structure of Chacoan Society in the Northern Southwest. Investigations at the Salmon Site, 1972*. Eastern New Mexico University Contributions in Anthropology, Vol. 4, No. 3, Portales.

  1980   The San Juan Valley Archaeological Program Investigations at Salmon Ruin. *In Investigations at the Salmon Site: The Structure of Chacoan Society in the Northern Southwest*, edited by Cynthia Irwin-Williams and Phillip H. Shelley, Pt. 1, Vol. 1, pp. 3–104. Eastern New Mexico University Printing Services, Portales.

Jones, Martin

  2003   Ancient DNA in Pre-Columbian Archaeology: A Review. *Journal of Archaeological Science* 30:629–635.

Judge, W. James

  1989   Chaco Canyon–San Juan Basin. In *Dynamics of Southwest Prehistory*, edited by Linda S. Cordell and George J. Gumerman, pp. 209–262. Smithsonian Institution Press, Washington, D.C.

Kaestle, Frederika A., and David Glenn Smith

  2001   Ancient Mitochondrial DNA Evidence for Prehistoric Population Movement: The Numic Expansion. *American Journal of Physical Anthropology* 115:1–12.

Kantner, John, and Nancy M. Mahoney (editors)

  2000   *Great House Communities Across the Chacoan Landscape*. University of Arizona Press, Tucson.

Lekson, Stephen H.

  1991   Settlement Patterns and the Chaco Region. In *Chaco and Hohokam: Prehistoric Regional Systems in the American Southwest*, edited by Patricia L. Crown and W. James Judge, pp. 31–55. School of American Research Press, Santa Fe.

1999   *The Chaco Meridian: Centers of Political Power in the Ancient Southwest.* AltaMira Press, Walnut Creek, California.

Lekson, Stephen H., and Catherine M. Cameron
1995   The Abandonment of Chaco Canyon, the Mesa Verde Migrations, and the Reorganization of the Pueblo World. *Journal of Anthropological Archaeology* 14:184–202.

Lekson, Stephen H., Thomas C. Windes, and Patricia Fournier
2007   The Changing Faces of Chetro Ketl. In *The Architecture of Chaco Canyon, New Mexico*, edited by Stephen H. Lekson, pp. 155–178. University of Utah Press, Salt Lake City.

Lekson, Stephen H., Thomas C. Windes, John R. Stein, and W. James Judge
1988   The Chaco Community. *Scientific American* 259(1):100–109.

McGuire, Randall H.
1980   The Mesoamerican Connection in the Southwest. *Kiva* 46:3–38.

McGuire, Randall, E. Charles Adams, Ben A. Nelson, and Katherine A. Spielmann
1994   Drawing the Southwest to Scale: Perspectives on Macroregional Relations. In *Themes in Southwest Prehistory*, edited by George J. Gumerman, pp. 239–265. School of American Research Press, Santa Fe.

McKenna, Peter J., and H. Wolcott Toll
1992   Regional Patterns of Great House Development among the Totah Anasazi, New Mexico. In *Anasazi Regional Organization and the Chaco System*, edited by David E. Doyel, pp. 133–143. Maxwell Museum of Anthropology, Anthropological Papers No. 5, Albuquerque.

Malhi, Ripan S., Holly M. Mortensen, Jason A. Eshleman, Brian M. Kemp,
Joseph G. Lorenz, Frederika A. Kaestle, John R. Johnson, Clara Gorodezky, and
David Glenn Smith
2003   Native American mtDNA Prehistory in the American Southwest. *American Journal of Physical Anthropology* 120:108–124.

Marshall, Michael P., John R. Stein, Richard W. Loose, and Judith E. Novotny
1979   *Anasazi Communities of the San Juan Basin.* Public Service Company of New Mexico, Albuquerque.

Martin, Debra L., Nancy J. Akins, Alan H. Goodman, H. Wolcott Toll, and
Alan C. Swedlund
2001   Totah Time and the Rivers Flowing: Excavations in the La Plata Valley, Vol. 5. In *Harmony and Discord: Bioarchaeology of the La Plata Valley*. Archaeology Notes 242. Office of Archaeological Studies. Museum of New Mexico Press, Santa Fe.

Mathien, Frances J.
1986   External Contacts and the Chaco Anasazi. In *Ripples in the Chichimec Sea: New Considerations of Southwestern-Mesoamerican Interactions*, edited by Frances J. Mathien and Randall H. McGuire, pp. 220–242. Southern Illinois University Press, Carbondale.
2001   The Organization of Turquoise Production and Consumption by the Prehistoric Chacoans. *American Antiquity* 66:103–118.

Morris, Earl
1938   Mummy Cave. *Natural History* 42:127–138.

Mulligan, Connie J.
2006   Anthropological Applications of Ancient DNA: Problems and Prospects. *American Antiquity* 71:365–380.

Nelson, Ben A.
2006   Mesoamerican Objects and Symbols in Chaco Canyon Contexts. In *The Archaeology of Chaco Canyon, an Eleventh-Century Pueblo Regional Center*, edited by Stephen H. Lekson, pp. 339–371. School of American Research Press, Santa Fe.

Nichol, Christopher R.

1990 Dental Genetics and Biological Relationships of the Pima Indians of Arizona. Unpublished Ph.D. dissertation, Department of Anthropology, Arizona State University, Tempe.

Powers, Robert P., William B. Gillespie, and Stephen H. Lekson

1983 *The Outlier Survey: A Regional View of Settlement in the San Juan Basin*. Reports of the Chaco Center, No. 3. Division of Cultural Research, National Park Service, Albuquerque.

Reed, Paul F. (Contributions by Rex K. Adams)

2006 Chronology of Salmon Pueblo. In *Introduction, Architecture, Chronology, and Conclusions*, edited by Paul F. Reed, pp. 287–296. Thirty-Five Years of Archaeological Research at Salmon Ruins, New Mexico, Vol. 1, Center for Desert Archaeology, Tucson, Arizona, and Salmon Ruins Museum, Bloomfield, New Mexico.

Rohn, Arthur H.

1983 Budding Urban Settlements in the Northern San Juan. In *Proceedings of the Anasazi Symposium, 1981*, edited by Jack E. Smith, pp. 75–80. Mesa Verde Museum Association, Mesa Verde National Park, Colorado.

1989 Northern San Juan Prehistory. In *Dynamics of Southwest Prehistory*, edited by Linda S. Cordell and George J. Gumerman, pp. 149–177. Smithsonian Institution Press, Washington, D.C.

Roler, Kathy Lynne

1992 Near Eastern Dental Variation Past and Present. Unpublished Master's thesis, Department of Anthropology, Arizona State University, Tempe.

Schillaci, Michael A.

2003 The Development of Population Diversity at Chaco Canyon. *Kiva* 68:221–245.

Schillaci, Michael A., Erik G. Ozolins, and Thomas C. Windes

2001 Multivariate Assessment of Biological Relationships among Prehistoric Southwest Amerindian Populations. In *Following Through: Papers in Honor of Phyllis S. Davis*, Vol. 27, edited by Regge N. Wiseman, Thomas C. O'Laughlin, and Cordelia T. Snow, pp. 133–149. The Archaeological Society of New Mexico, Albuquerque.

Schillaci, Michael A., and Christopher M. Stojanowski

2002 A Reassessment of Matrilocality in Chacoan Culture. *American Antiquity* 67:343–356.

Schurr, Theodore G.

2004 The Peopling of the New World: Perspectives from Molecular Anthropology. *Annual Review of Anthropology* 33:551–583.

Scott, G. Richard

1973 Dental Morphology: A Genetic Study of American White Families and Variation in Living Southwest Indians. Unpublished Ph.D. dissertation, Department of Anthropology, Arizona State University, Tempe.

Scott, G. Richard, and Albert A. Dahlberg

1982 Microdifferentiation in Tooth Crown Morphology among Indians of the American Southwest. In *Teeth: Form, Function, and Evolution*, edited by Björn Kurtén, pp. 259–291. Columbia University Press, New York.

Scott, G. Richard, and Christy G. Turner II

1988 Dental Anthropology. *Annual Review of Anthropology* 17:99–126.

1997 *The Anthropology of Modern Human Teeth: Dental Morphology and Its Variation in Recent Human Populations*. Cambridge University Press, Cambridge.

2006 Dentition. In *Environment, Origins, and Population*, edited by Douglas H. Ubelaker, pp. 645–660. Handbook of North American Indians, Vol. 3, William C. Sturtevant, general editor, Smithsonian Institution, Washington, D.C.

Smith, David G., Joseph Lorenz, Becky K. Rolfs, Robert L. Bettinger, Brian Green,
Jason Eshleman, Beth Schultz, and Ripan Malhi

2000   Implications of the Distribution of Albumin Naskapi and Albumin Mexico for
New World Prehistory. *American Journal of Physical Anthropology* 111:557–572.

Snow, Meradeth, David G. Smith, and Kathy R. Durand

2010   Ancestral Puebloan mtDNA in the Context of the Greater Southwest. *Journal of
Archaeological Science* 37:1635–1645.

Sofaer, Jeffrey A., Patricia Smith, and Edith Kaye

1986   Affinities Between Contemporary and Skeletal Jewish and Non-Jewish Groups
Based on Tooth Morphology. *American Journal of Physical Anthropology* 70:265–275.

Toll, H. Wolcott

2001   Making and Breaking Pots in the Chaco World. *American Antiquity* 66:56–78.

2006   Organization of Production. In *The Archaeology of Chaco Canyon, an Eleventh-
Century Pueblo Regional Center*, edited by Stephen H. Lekson, pp. 117–151. School of
American Research Press, Santa Fe.

Turner, Christy G., II

1979   Dental Anthropological Indications of Agriculture among the Jomon People of
Central Japan. *American Journal of Physical Anthropology* 51:619–636.

1983   Sinodonty and Sundadonty: A Dental Anthropological View of Mongoloid Microevolution, Origin, and Dispersal into the Pacific Basin, Siberia, and the Americas. In *Late Pleistocene and Early Holocene Cultural Connections of Asia and America*,
edited by R. S. Vasilievsky, pp. 72–76. USSR Academy of Sciences, Siberian Branch,
Novosibirsk.

1985   Dental Evidence for the Peopling of the Americas. *National Geographic Society
Research Reports* 19:573–596.

1987   Late Pleistocene and Holocene Population History of East Asia Based on Dental Variation. *American Journal of Physical Anthropology* 73:305–321.

1990   Major Features of Sundadonty and Sinodonty, Including Suggestions about
East Asian Microevolution, Population History, and Late Pleistocene Relationships
with Australian Aboriginals. *American Journal of Physical Anthropology* 82:295–317.

1993   Southwest Indians: Prehistory Through Dentition. *National Geographic Research
and Exploration* 9:32–53.

Turner, Christy G., II, Christopher R. Nichol, and G. Richard Scott

1991   Scoring Procedures for Key Morphological Traits of the Permanent Dentition:
The Arizona State University Dental Anthropology System. In *Advances in Dental
Anthropology*, edited by Marc A. Kelley and Clark Spencer Larson, pp. 13–31. Wiley,
New York.

Varien, Mark D., William D. Lipe, Michael A. Adler, Ian M. Thompson, and
Bruce A. Bradley

1996   Southwestern Colorado and Southeastern Utah Settlement Patterns: A.D. 1100
to 1300. In *The Prehistoric Pueblo World A.D. 1150–1350*, edited by Michael A. Adler,
pp. 86–113. University of Arizona Press, Tucson.

Vivian, R. Gwinn

1990   *The Chacoan Prehistory of the San Juan Basin*. Academic Press, San Diego.

Webster, Laurie D.

2008   An Initial Assessment of Perishable Relationships among Salmon, Aztec, and
Chaco Canyon. In *Chaco's Northern Prodigies: Salmon, Aztec, and the Ascendancy of the*

Middle San Juan Region after A.D. *1100*, edited by Paul F. Reed, pp. 167–189. University of Utah Press, Salt Lake City.

Webster, Laurie D., and Micah Loma'omvaya
  2004   Textiles, Baskets, and Hopi Cultural Identity. In *Identity, Feasting, and the Archaeology of the Greater Southwest*, edited by Barbara J. Mills, pp. 74–92. University Press of Colorado, Boulder.

Wheelbarger, Linda
  2008   Puebloan Communities on the South Side of the Middle San Juan River. In *Chaco's Northern Prodigies: Salmon, Aztec, and the Ascendancy of the Middle San Juan Region after* A.D. *1100*, edited by Paul F. Reed, pp. 209–230. University of Utah Press, Salt Lake City.

Willey, Gordon R., Charles C. Dipeso, William A. Ritchie, Irving Rouse, John H. Rowe, and Donald W. Lathrap
  1956   Archaeological Classification of Culture Contact Situations. In *Seminars in Archaeology: 1955. Memoirs of the Society for American Archaeology*, No. 11, pp. 1–30, edited by Robert Wauchope. Society for American Archaeology, Salt Lake City, Utah.

Wilshusen, Richard H., and Ruth M. Van Dyke
  2006   Chaco's Beginnings. In *The Archaeology of Chaco Canyon, an Eleventh-Century Pueblo Regional Center*, edited by Stephen H. Lekson, pp. 211–259. School of American Research Press, Santa Fe.

Windes, Thomas C.
  2007   Gearing Up and Piling On: Early Great Houses in the Interior San Juan Basin. In *The Architecture of Chaco Canyon, New Mexico*, edited by Stephen H. Lekson, pp. 45–92. University of Utah Press, Salt Lake City.

6. The Introduction of Agriculture and the Foundation of Biological Variation in the Southern Southwest

James T. Watson

Summary Statement: The transition to agriculture has long been the subject of debate among archaeologists in the U.S. Southwest–Northwest Mexico. Arguments suggest that cultigens and agricultural technology were either brought in by migrating agriculturalists from Mesoamerica or diffused through an interconnected network of local foraging groups. The results of either of these disparate processes would have had a significant effect on the foundation of human biological variation in the region as groups began to farm and populations expanded. Here I examine several lines of anthropological evidence to test the hypothesis that agriculture was introduced into the Sonoran Desert by migrating farmers from Mesoamerica. Linguistic models tend to support migrations of maize-bearing proto–Uto-Aztecan peoples from central Mexico. Molecular analyses (modern mtDNA and aDNA) support diffusion of language, plants, and technology and perhaps limited migration. Studies in skeletal biology of early farmers in the area support a gradual integration of agriculture into local foraging groups. Although evidence for migration appears to be limited, we still lack crucial data that would provide conclusive evidence to the contrary. It is important to consider the introduction of agriculture as a process that contributed significantly to the foundation of biological variation among later complex Formative period cultures of the region.

Introduction

Does variation in the archaeological record coincide with biological variation in the Americas? As the central theme of this volume, many of the authors have focused on the nature of human variability and how the movement of peoples

Human Variation in the Americas: The Integration of Archaeology and Biological Anthropology, edited by Benjamin M. Auerbach. Center for Archaeological Investigations, Occasional Paper No. 38. © 2010 by the Board of Trustees, Southern Illinois University. All rights reserved. ISBN 978-0-88104-095-1.

across varied landscapes has shaped the archaeological record. This is a particularly relevant issue in the U.S. Southwest–Northwest Mexico (Figure 6-1) where preservation is generally good, the region has been intensively studied for over a century, and there are a wealth of modern Native American cultures to help contextualize the past and its progression to the present. The region boasts one of the oldest continuously occupied settlements in North America and one of the last native groups to arrive in the Americas. However, deciphering the relationship between variation in the archaeological record and the biology of the diverse peoples of the region requires a solid understanding of the foundation of that biological variation.

Although the ultimate foundation of biological variation in the U.S. Southwest–Northwest Mexico extends as far back as the earliest Paleoindian groups, I postulate that much of the biological variation distributed across modern native groups of the region is first defined and canalized (genetically fixed) with the introduction of agriculture. The adoption of cultigens and agricultural technology led to permanent settlements, greater investment in and defense of set resources, larger populations, greater social complexity, and the formation of Formative societies that have long been the focus of archaeological studies in the region—the Hohokam, Mogollon, and Ancestral Puebloans. Archaeological studies that focus on deciphering population relationships of late Formative groups and/or of historic native ethnogenesis hold their comparative prehistoric "biological baseline" at the beginning of the Formative cultural sequence (Clark 2004; Gregory and Wilcox 2007; Haury 1958; Hill et al. 2004; Lindsay 1983; Longacre 1973; Lyons 2003). The primary purpose of this paper therefore is to examine the evidence for the foundation of biological variation in the Southwest associated with the introduction of agriculture, specifically focusing on recent research from the Sonoran Desert of southern Arizona and northern Sonora.

The identification of biological variation associated with the introduction of agriculture has been of intense interest to anthropologists since the beginning of the discipline (Smith 1998). A classic example is the Neolithic transition in Western Europe and whether it was facilitated by migrations of Levantine farmers or by cultural diffusion (e.g., Childe 1965; Braidwood 1967; Ammerman and Cavalli-Sforza 1984; Renfrew 1987). Childe (1925) originally proposed a model of demic diffusion whereby unchecked population growth triggered economic and social problems among Near Eastern groups and forced farmers to search for new lands. Alternately, a model of cultural diffusion suggests that Europeans adopted agriculture by imitating Neolithic practitioners they encountered through trade or other interactions (Smith 1998). A recent explosion in molecular techniques has added new vigor to these arguments (Diamond and Bellwood 2003; Fix 2005).

Similarly, archaeologists have long debated the nature of the introduction and adoption of agriculture, and in particular maize, across the U.S. Southwest–Northwest Mexico, and several theories have been postulated to explain how it entered the area from its origins in Mesoamerica. Models are grouped within two general categories: (1) those that argue that domesticated crops were brought in by migrating groups of cultivators; and (2) those based on the idea that the diffusion of plants and technology from central Mexico led to an indigenous adoption of cultivars. Haury (1962) originally proposed that the Mesoamerican crop

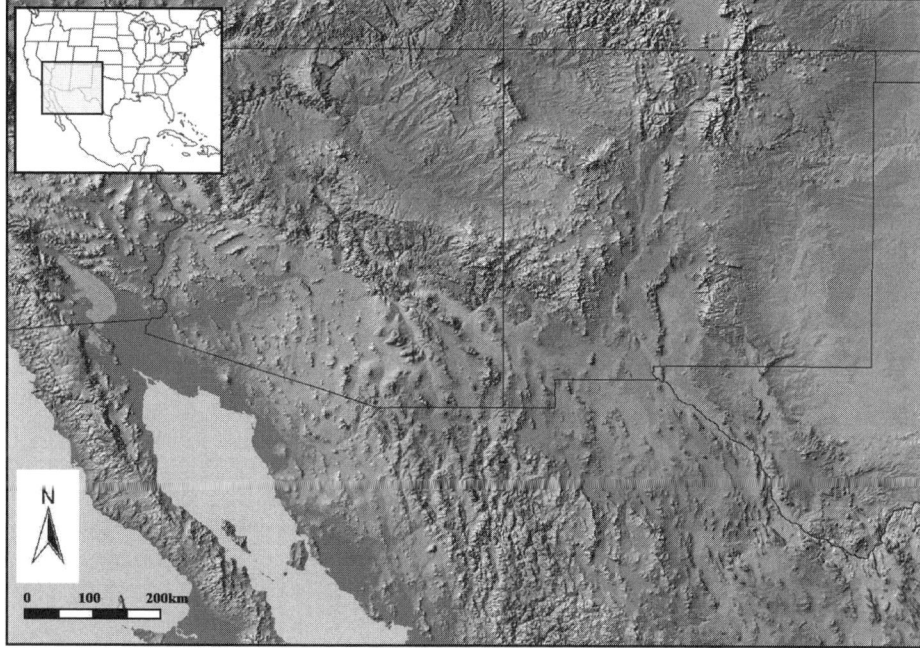

Figure 6-1. *Map of U.S. Southwest–Northwest Mexico culture area.*

complex (corn, beans, squash) likely entered the Southwest through a "highland corridor" in the Sierra Madre Occidental by 4500 B.P. but was confined to the Mogollon Highlands until approximately 2500 B.P. Berry and Berry (1986) envisioned agriculture coming into the Southwest by colonization, with farming groups moving into the area through the process of "punctuated equilibrium" with new technology and domesticated cultigens. Huckell (1990, 1995) pushes the date back earlier and associates the beginnings of the San Pedro phase (ca. 3500 B.P.) with the arrival of maize-bearing peoples from Mexico. Matson (1991) postulates analogous processes for the introduction of maize onto the Colorado Plateau, where he perceives similar assemblages between the San Pedro phase and Western Basketmaker II, suggesting that the latter descended from San Pedro or Cienega phase peoples who migrated onto the plateau following the development of cold-tolerant races of maize.

In contrast, models of cultural diffusion postulate that maize was adopted by indigenous foraging groups in the region for a number of reasons. Hard (1986) and Wills (1988, 1990, 1992, 1995) believe that local groups adopted the new cultigens to offset increasing environmental uncertainty, possibly the result of increasing population size and resource imbalance. Fish and colleagues (1986) and Roth (1996) are of the opinion that cultivation of domesticates was incorporated into seasonal cycles by semisedentary foragers occupying diverse ecotones where maize could easily be exploited. Matson (1991) provides a mechanism for diffusion by identifying a broad cultural continuum throughout the U.S. Southwest–Northwest Mexico and highland Mesoamerica at the beginning of the Late Archaic (related to the distribution of Gypsum Cave points and similar forms) that would have facilitated the rapid spread of cultigens northward. He also postulates that maize horticulture fits easily into the biannual cycle of the lowland Sonoran inhabitants.

These arguments are especially relevant to interpreting biological variation of the region and whether the introduction of agriculture was accompanied by an influx of new peoples (genotypes) or simply new plants (domesticates) and technology (farming). These competing theories can be similarly viewed as presenting opposing views for the foundation of biological variation in the region as populations of two distinct genetic origins: a native foraging population and a migrant agriculturalist population. In reality, these processes would exist along a continuum, in some combination of these polar definites, but it illustrates the importance of considering the relationship between variation in archaeology and human biology. Alan Fix (2005) sums it up best when addressing the interpretation of microevolution and biological variation in past human populations:

> The fundamental problem for all reconstructions that assume migration as a cause . . . is how to distinguish past population movements from other processes that lead to the same distributional outcomes. More particularly, how do present distributions of artifacts, languages, or gene "markers" signify movement of people as opposed to other processes such as cultural diffusion or transformation (in the case of genes or natural selection)? For this reason, the definitive demonstration of any prehistoric movement must depend on multiple converging lines of evidence [Fix 2005:xii].

Given competing arguments regarding the introduction of agriculture into the U.S. Southwest–Northwest Mexico and the idea that it defined social and biological structures within the region for the following several millennia, here I examine several lines of anthropological evidence to test the hypothesis that agriculture was introduced into the Sonoran Desert by migrating farmers from Mesoamerica. This hypothesis carries the assumption that the biological foundation resulting from the agricultural transition identifies some form of influx of biological diversity that could reflect gene flow and/or genetic drift in these early agricultural populations.

Information is presented in the subsequent four sections to describe and explore our current understanding of archaeology, language, genes, and skeletal biology as it pertains to the primary goal and central hypothesis of this paper. Each section describes information from studies that contribute to a better understanding of the introduction of agriculture and the foundation of biological variation in the Sonoran Desert. This material is subsequently synthesized and its significance considered in light of the theme of this volume.

Archaeology

The archaeological record provides the original impetus for the debate about the introduction of agriculture into the U.S. Southwest–Northwest Mexico. Early on, researchers recognized the importance of maize agriculture to the foundation of the later Formative cultures in the region (Haury 1962). Research over the past several decades has considerably improved our understanding of the cultural sequence after the arrival of maize in the region, but we are still lacking a great deal of information about the circumstances of its arrival and perhaps the time period just prior to its arrival. The Archaic and Early Agricultural archaeological periods are of greatest interest to the discussion of the introduction and adoption of agriculture in the Sonoran Desert. Figure 6-2 illustrates the cultural chronology for the Sonoran Desert from the late Archaic to the Historic period. The Archaic period extends from circa 10,500 to 3500 B.P. and is divided into Early, Middle, and Late subdivisions based on climatic differences first proposed by Antevs (1955). Archaeological remains from the Middle Archaic (7500–4500 B.P.) and Late Archaic (4500–3500 B.P.), most relevant to the introduction of agriculture, are found in a wide variety of contexts in the Sonoran Desert. The majority of sites assigned to these time periods are done so based on the presence of projectile point styles (Chiricahua, Cortaro, Gypsum Cave, Pinto) that have broad, poorly defined chronological associations (Mabry 1998a). Huckell and Roth (1992) identify that Cortaro points differ from other point types of similar temporality in that they are geographically restricted to the Sonoran Desert. They offer several hypotheses to account for its appearance: the Cortaro point served a different function than other points; its introduction is a sign of rapid cultural change; it shows multiple sociocultural entities perhaps occupied overlapping geographic space; or some combination of these. The timing and distribution of Cortaro points could have important implications for the argument of migration versus diffusion of agriculture into the Sonoran Desert.

Archaic sites are also recorded in a variety of settings (Huckell 2006; Roth 1996) and site size and type can vary according to location. Specialized sites for collecting acorns and piñon nuts are found in the mountains of southern Arizona (Huckell 1990). Sites from the *bajada* (piedmont) vary from hunting and gathering camps on the upper portions (Roth 1996) to plant collection and processing camps closer to the floodplain (Roth 1989). Excavations at the site of Los Pozos in southern Arizona provide evidence of short-term, episodic use of the Santa Cruz River floodplain during the Middle/Late Archaic (Gregory 1999). Artifacts and features include storage pits, a shallow hearth, and a variety of artifact assemblages that dated between 4700 and 3900 B.P. (Gregory 1999). In addition, a single direct date on a maize cupule from these deposits circa 4000 B.P. clearly indicates that maize was present during the late Middle and Late Archaic (Freeman 1998; Gregory 1999). The Los Pozos date adds to numerous other early dates on maize from floodplain sites along the Santa Cruz River that range from 4000 to 3500 B.P. (Kohler et al. 2008; Mabry 2005). Gregory (1999) identifies that the presence of these sites on the floodplain indicate a pattern of repeated site use and "tethering" to resources that provided conditions to support lower mobility during the Middle Archaic that were amplified during the Late Archaic and finally the Early Agricultural period. Maize recovered from McEuen Cave, dated to 4000 B.P., also demonstrates that it was present at locations other than on the floodplain (Huckell 2006; Huckell and Huckell 1999).

Although somewhat limited, it is Middle and Late Archaic occupations on the floodplains that provide some of the most intriguing evidence for the introduction of maize into the area. This ecological zone contains fertile soil, a shallow water

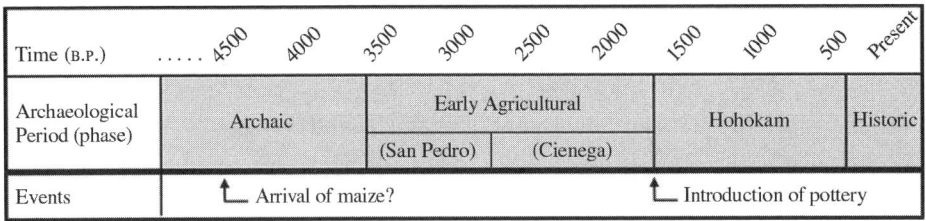

Figure 6-2. *Cultural sequence for the Sonoran Desert.*

table, abundant plant and animal resources, and is adjacent to active and perennial water flows. The sparse recovery of Archaic remains from these contexts has been variably attributed to their excessive depth in the thick alluvium that characterizes the floodplains of the Sonoran Desert (Huckell 1995) and to periods of floodplain incision that are ascribed to climatic degradation (Waters and Haynes 2001).

The Early Agricultural Period

The timing of the "Neolithic Revolution" (Childe 1925) varies by location around the globe, but changes in subsistence patterns associated with the adoption of agriculture are not identified archaeologically in the Sonoran Desert until approximately 3500 B.P., marking the beginning of the Early Agricultural period (3500–1800 B.P.). This period is divided into two archaeological phases based on changes in the material culture that reflect trends in site composition, artifact assemblages, and increasing population sizes: the San Pedro phase (ca. 3500–2800 B.P.) and the Cienega phase (2800–1800 B.P.).

San Pedro phase sites are found in river floodplains, along benches, on piedmonts, and in the uplands (Diehl 2005; Huckell et al. 1995; Roth 1996; Wellman 2000). They are generally characterized by the presence of small shallow domestic depressions ("pit" structures), large extramural storage pits, ubiquitous groundstone, an expedient lithic technology, San Pedro dart points, and the presence of maize (Huckell 1995; Huckell et al. 1995; Roth and Wellman 2001; Wellman 2000). In addition, some clay figurines and early ceramic sherds have been recovered from San Pedro contexts (Carpenter et al. 1997, 1999; Gregory 2001; Huckell 1995; Mabry 1999; Roth and Wellman 2001). Evidence for the earliest irrigation canals in the North American Desert West has also been documented at several sites in the area by about 3200 B.P. (Ezzo and Deaver 1998; Doolittle and Mabry 2006; Mabry 1999, 2005). Although the true extent of utilization of irrigation canals during this time period is not well understood, it minimally demonstrates that San Pedro groups were using technologies designed to intensify plant productivity.

The Cienega phase represents a greater variety and increased complexity of cultural characteristics from those defined for the San Pedro phase. This phase is characterized by larger villages, increased technological complexity, the establishment of both local and long distance commerce networks, deeper and larger subterranean houses, and the Cienega point type (Mabry 1997). The Cienega point likely represents experimentation with projectile-point technology and the introduction of the bow and arrow (Ochoa 2004; Sliva 2000). During the Cienega phase there is an elaboration of groundstone manufacture, the development of a shell ornament production industry, and the first formal ceramic tradition termed *Incipient Plainware* (Heidke 1999).

Huckell (1995) asserts that sometime between 3500 and 3000 B.P. a mixed subsistence system based on both agriculture and wild plant collection appears in southern Arizona, in which maize becomes an equally critical part of the whole system and that, by the Cienega phase, subsistence was primarily based in maize agriculture with surplus stored during different times of the year, as evidenced by the vast quantities of storage pits. Other researchers (Doelle and Fish 1988; Mabry 1999, 2005) cite the characteristics related to sedentism— including the ubiquity of maize and the presence of large middens, cemeteries, and irrigation canals—to suggest that this phase represents groups fully reliant on agriculture. Archaeobotanical remains additionally document multiseasonal residence as early as the San Pedro phase (Huckell et al. 1995). Diehl (2005) suggest that subsistence strategies remained a stable, mixed economy throughout the Early Agricultural period because pit storage technology was inadequate and made maize cultivation extremely risky. However, from storage features including maize remains, seeds, grasses, cacti, and mesquite at the Los Pozos site, Schurr and Gregory (2002) were able to identify a dramatic rise in all plant resources during a 200 year period of the Cienega phase.

Data from the archaeological record have been used by various researchers to argue in support of both migration and diffusion models. There is a clear continuity in site locations and areas of resource exploitation from at least the Middle Archaic through the Early Agricultural periods. In addition, researchers are continually pushing back the arrival of cultigens with earlier direct accelerator mass spectrometry (AMS) radiocarbon dates on maize from sites in various locations. Supporters of the migration model propose that the archaeological record demonstrates a rapid change from previous foraging practices to one largely reliant on cultivated crops (Carpenter et al. 2002, 2005; Huckell 1990, 1995; Mabry 2005). The apparent "sudden" appearance of crude ceramics and irrigation could be argued to reflect this transition. Minimally, the archaeological record demonstrates that sometime after approximately 4000 B.P. people settled along the rivers and began exploiting cultigens on the floodplains of the Sonoran Desert. These remains represent groups that form the foundation of biological variation for subsequent Formative cultures in the area. Ultimately, evidence from the archaeological record functions to both support and refute the migration hypothesis.

Language

Language models have long been used to address the question of migration and diffusion associated with the spread of agriculture (e.g., Bellwood 1997; Renfrew 1987). Research into the distribution of the Uto-Aztecan language family (Table 6-1) provides alternatives to explain the spread of maize agriculture into the U.S. Southwest–Northwest Mexico (Hill 1999, 2000, 2001). Western North America, from Mesoamerica to the Great Basin, is largely composed of languages that belong to the Uto-Aztecan language family (Figure 6-3). Based on previous

Table 6-1. *Uto-Aztecan Language Family*

Northern Uto-Aztecan	Southern Uto-Aztecan
Hopi	Tepiman
	Pima-Tohono O'odham (Papago)
Numic	*Pima*: Lower Pima, Mountain Pima
Western: Mono, Northern Paiute	*Tepehuan*: Northern, Southern
Central: Comanche, Gosiute, Shoshone, Tiimpisha Shoshone	Tepecano
Southern: Chemehuevi, Kawaiisu, Southern Paiute, Ute	Taracahitan
	Opata
Tübatulabal	Eudeve
	Tarahumara
Takic	Guarijio
Cupan: Cahuilla, Cupeno, Luiseno Serrano, Gabrielino-Fernandeno	Yaqui-Mayo (Cahitan)
	Tubar
	Corachol-Aztecan
	Corachol: Cora, Huichol
	Aztecan: Nahua, Pochutec, Pipil

Source: Adapted from Hill 2001

work by Bellwood (1997) and compared lexical terms for maize cultivation, Hill (2001) postulates that Proto–Uto-Aztecans were among the first maize cultivators in Mesoamerica and that they spread northward into the present range due to demographic pressures associated with cultivation. She attributes this expansion and the subsequent construction of the language family to demic diffusion and "leapfrogging" between approximately 4500 and 3500 B.P. Hill's hypothesis is very similar to that championed by Childe (1965) for Neolithic Europe.

A few archaeologists have embraced these linguistic arguments and further tie the movement of Uto-Aztecan speakers and maize cultivation to Middle Holocene climatic events in the Desert West (Carpenter et al. 1997, 2002; Mabry 2005). These models offer a more complex theory for migration of maize-bearing peoples into the region and the introduction of agriculture. They still rely on the acceptance of evidence for population movement into the area; evidence that is contested by the diffusionists. In addition, several aspects of the linguistic side of the argument

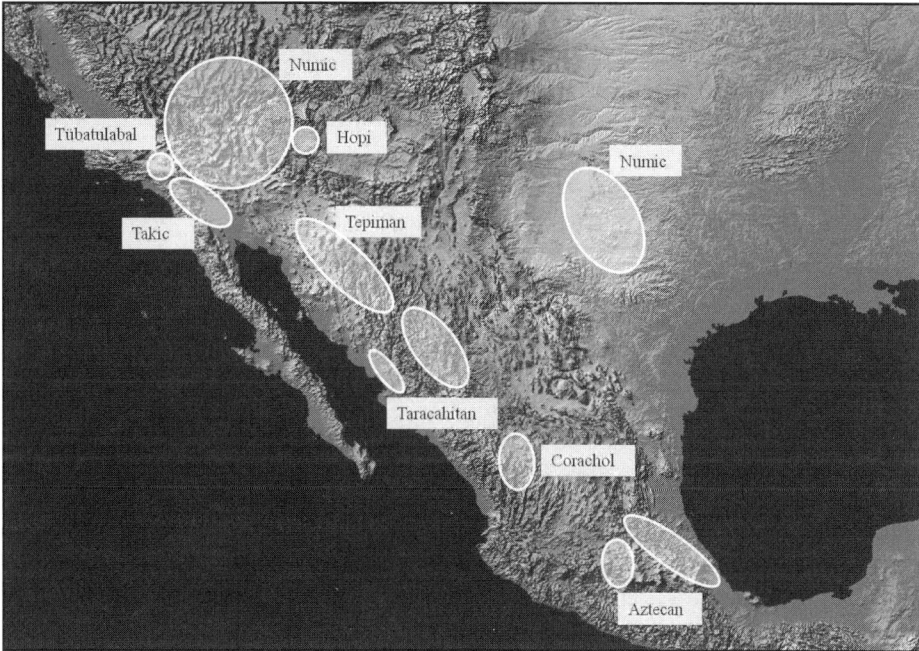

Figure 6-3. *Distribution of the Uto-Aztecan language family and the modern locations of its major branches.*

are equally contentious. The origin of the Uto-Aztecan language family, or the Proto–Uto-Aztecan (PUA) homeland, is the subject of some debate among scholars (Bellwood 1997; Fowler 1983, 1994; Lamb 1958; Miller 1984; Nichols 1981). Fowler (1983, 1994) identifies the PUA homeland in the mixed woodland/grassland zone of the northern foothills of the Sierra Madre based on the comparison of the modern distributions of plants and animals to 27 cognates presumed to be part of the PUA folk biological knowledge. Bellwood (1997) and Hill (1999, 2001) are convinced that the historical-linguistic evidence indicates that Uto-Aztecan peoples migrated out of Mesoamerica and into their historic distributions.

Linguistic Divergence

The timing of the diversification/divergence of the Proto–Uto-Aztecan linguistic group is equally important to understand cultural-linguistic and population relationships in the North American Desert West. Glottochronological analyses have estimated anywhere between 8000 and 3000 B.P. to account for the degree of linguistic divergence observed among Uto-Aztecan languages (Carpenter et al. 1997, 2002). Romney (1957) proposed the youngest dates at 3000 B.P., Hale (1958, 1959) and Hill (2000) agree on 4000 B.P., Fowler (1983) and Miller (1984) proposed 5000 B.P., and Miller (1986) proposed the latest date of about 6000 B.P. to explain the historic degree of language diversity. During this time it is believed that the Uto-Aztecan language family split into "Northern" and

"Southern" subgroups (Hale 1958) composed of Shoshonean, Sonoran, and Na-huatlan language branches. The Shoshonean represents the northern subgroup and the Nahuatlan represents the southern subgroup. The placement of the So-noran language branch has been somewhat disputed; however, many place it in the Southern subgroup (Fowler 1983; Hill 2001).

Situating the divergence of the PUA group into a chronological framework is important to Southwest archaeologists for several reasons. In fact, several argue that it offers the best explanation for population movement and the adoption of domesticated cultigens. The timing and distribution of the PUA divergence could place people migrating into the region with agricultural technology at or just prior to the period considered here. Hill (2000, 2001) suggests that the Proto-Southern Uto-Aztecan (PSUA) began to break up by about 4000 B.P. and that maize cultivation had spread throughout the PSUA region before diversification but that language divergence occurred prior to the introduction of pottery. This would place the breakup after 3000 B.P. and before 1900 B.P. (Carpenter et al. 1997, 2002), which constitutes the duration of the Cienega phase in the archaeo-logical record. Hill (2000) further postulates that the language group had differ-entiated into the historically observed languages by 1200 B.P. This information indicates that sometime after 3000 B.P., language and domesticated agriculture spread throughout the U.S. Southwest–Northwest Mexico and suggests that by 1200 B.P. much of the region had been settled into social constructions constitut-ing the origins of Formative archaeological cultures, which subsequently devel-oped into the historically observed native groups of the Southwest.

Migrating Farmers

The linguistic evidence provides a potential explanation for the mode and tempo of the introduction of maize agriculture into the Desert West. The lin-guistically approximated time period, from 3000 to 1200 B.P., represents a large segment of the cultural development of the Early Agricultural period. Carpenter and colleagues (1997, 2002) believe the correlation between the breakup of the PUA during the Early Agricultural period and the existence of a "Uto-Aztecan cultural-linguistic continuum," in combination with climatic influences, provides the simplest explanation for a "rapid" spread of agriculture across the Southwest. Mabry (2005) offers an alternative hypothesis that identifies Uto-Aztecan speak-ers as repopulating the desert lowlands at the beginning of the Late Holocene from highland Altithermal refuges; then later, agriculture and pottery technology from Mesoamerica diffused rapidly across a linguistic and cultural continuum. Based on excavations at the site of Las Capas (Mabry 1999, 2007) in Tucson and the discovery of early irrigation canals (ca. 3200–3100 B.P.), Hill (2001) proposes a model in which maize cultivators moved up the west Mexican coast "leapfrog-ging" between river valleys looking for favorable environments to cultivate.

These models of human and/or cultural-linguistic expansion make appro-priate use of the linguistic evidence but fall short of considering the internal dy-namics and complexity inherent in the diffusion or movement of any part of culture. Hill (2000) notes that the movement or stability of linguistic innovation varies across human groups. Working with the spread of Indo-European languages in Europe, Johanna Nichols (1997) distinguishes "residual zones" and "spread zones" as two opposing ends of a continuum in which humans utilize language. Residual zones represent those areas where groups exhibit greater linguistic diversity, greater antiquity, and locational stability. Spread zones represent those areas where groups exhibit low linguistic diversity, relatively recent origin, and geographic mobility. The historically observed languages of the Southwest and their relatedness as PUA would indicate a spread zone (Hill 2000).

Hill (2000) points out that spread zones are often associated with or inferred as human migrations but that this may not always be an accurate interpretation. She presents a similar model in which human groups use language to interact and control their culturally bounded environment through either a "localist strategy" or a "distributed strategy." These concepts are similar in appearance to Nichols's zones but differ in that they reflect the movement and stability of language over human groups without migration. Hill's (2000) research suggests that localist versus distributed sociolinguistic stances would be associated with a group's concepts about rights and access to resources and that these differences can develop without group migration. Groups that employ a localist strategy maintain socially closed systems/networks with strong internal ties and develop distinct linguistic differences to identify and control local resources. Groups that employ a distributed strategy maintain socially open systems/networks with "weak" internal ties and develop linguistic similarities to provide social and geographical mobility. All of these groups can use identical technologies.

Hill's (2000) model can be applied to long- or short-term linguistic development. When applied to the movement of Uto-Aztecan languages and agricultural technologies, it appears to support population stability on the landscape and the transmission of technology through diffusion and language differentiation through localist strategies. However, it is likely more appropriate to view the divergence of the Northern and Southern subgroups as distinct distributed strategies associated with population movements at the beginning of the Late Holocene. Subsequent internal language divergence—within these respective subgroups—then reflects the employment of localist strategies once agricultural technologies spread throughout the region. Based on this reinterpretation, language-based arguments function to both support and refute the migration hypothesis.

Genes

Recent molecular studies from modern native groups of the U.S. Southwest–Northwest Mexico and some ancient materials have added a new dimension to help address the potential for a farmer migration into the region (Kemp 2006; LeBlanc et al. 2007; Malhi et al. 2003). In a large study of structure of diversity within a sample of modern native mtDNA haplogroups, Malhi and colleagues (2002) postulate that they were able to identify several major prehistoric population events in North America. Their results identified the Southwest, with high frequencies of haplogroup B, as one area that experienced prehistoric population expansions. Examination of the haplotype network further supported their findings with the identification of high levels of reticulations and high frequencies of nodal haplotypes, in this case suggesting multiple founding lineages of haplogroup B. However, population expansions occurred at disparate times in prehistory and can be tied to particular events, such as the expansion of Clovis and the spread of agriculture throughout the region (Malhi et al. 2002). This is not indicative of a farmer migration from Mesoamerica but rather of regional homogeneity after the spread of farming within the region. The authors point out that individuals with a founding haplotype B exhibiting a particular mutation (np 16261C) are exclusively found in the Southwest. This may lend some credence to the argument that biological diversity in that region has been maintained in foraging groups far back into antiquity.

Several studies of mtDNA diversity in native North Americans have demonstrated that common ancestry among modern tribal groups may extend well into prehistory. Indeed several researchers hypothesize that "tribalization" of Native Americans occurred early in prehistory (Lorenz and Smith 1996; Torroni et al. 1993). These authors use the term *tribalization* to refer to the foundation of genetic variation inherent to modern Native American tribal groups. Torroni and colleagues (1993) recorded high incidence of private mtDNA polymorphisms and a limited distribution of shared mtDNA mutations to support the idea of early tribalization (Malhi et al. 2002). Lorenz and Smith (1996) reached similar conclusions on the antiquity of tribalization based on the observation of a greater level of intratribal homogeneity in mtDNA haplogroup frequencies than intraregion homogeneity in North America. Their study also focused on trying to decipher the roles geography and language groups or families played in the distribution of haplogroup frequencies. Most native groups in the Southwest display very high frequencies of haplogroup B, with the exception of Navajo and Apache, the Na-Dene groups in the region. Lorenz and Smith (1996) were able to reconstruct the nature of historic patterns of gene flow between these groups and at the same time demonstrate that the underlying genetic composition of the original inhabitants of the region is largely based in geographic association. Although approached from a much wider scope and scale, the results of these studies suggest that geographic proximity, extending well back into prehistory, contributed greatly to the modern distribution of native groups in the Southwest. In other words, geography, not language, was the greater determining factor in the modern distribution and frequencies of haplogroups in the Southwest. It is possible to extend this postulation back to the beginning of the Early Agricultural period when the earliest farmers were first settling into permanent places on the landscape.

A few studies have specifically tested the farmer migration hypothesis using mtDNA in the southern Southwest (Kemp 2006; Kemp et al. 2010; Malhi et al. 2003). These studies examined mtDNA haplogroup frequencies of a diverse group of modern Native Americans from the Southwest to compare within the region—by language group and geography—and with other groups in the Americas (Figure 6-4). They identify a high frequency of haplogroup B within Southwest groups, but there are differences in diversity between northern and southern groups. Puebloans generally exhibit low diversity within the haplogroup, whereas Yuma and Pimas exhibit a large amount of diversity within the

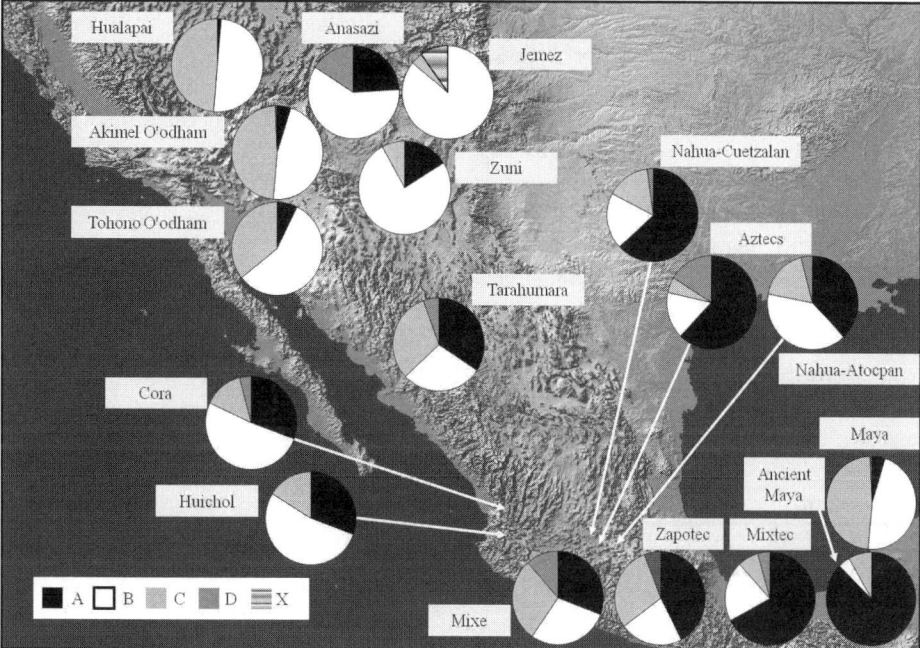

Figure 6-4. *Study populations and haplogroup frequencies (modified from Kemp 2006).*

haplogroup and in haplogroup frequencies. They postulate this is a reflection of dissimilar population histories guided by differences in archaeologically defined pre-Columbian cultural traditions (i.e., Ancestral Puebloan versus Hohokam). Their results further indicate that language was not a likely causal factor in structuring gene flow among populations within the Southwest. They contend that the distribution of mtDNA haplogroups and haplotypes among Uto-Aztecan–speaking groups in the Southwest and in central Mexico does not follow a pattern indicative of a population expansion northward whereby Uto-Aztecan spread in association with the development of maize cultivation (Malhi et al. 2003). The one caveat to these conclusions is the most recent discovery of a discord in genetic associations between the Southwest and Mesoamerica between mtDNA and Y-chromosome DNA variation (Kemp et al. 2010). Although the authors reject the idea of an agricultural migration based on the molecular evidence, they identify that if such an event did occur, it would have been largely biased to males.

Three factors would also have had profound consequences for the genetic composition and biological variation inherent in groups after farming and permanent settlements took hold in the Southwest, regardless of arguments for migration or diffusion of agriculture into the region. First, archaeologists have documented significant population movements throughout much of the Southwest from the thirteenth through fifteenth centuries (Adler et al. 1996; Clark 2004; Hill et al. 2004; Wilcox et al. 2008) where genetic mixing among previous separate populations was highly probable. Several hundred years of regional contraction and population consolidation likely resulted in a great deal of admixture of the founding populations of the Southwest. Second, there is little archaeological evidence of native groups inhabiting much of the Sonoran Desert for the period from the fifteenth to seventeenth century (Hill et al. 2004). There is evidence that suggests instead that the post–Classic period resulted in movement to and population consolidations at locations of modern Puebloan groups such as at Hopi, Zuni, and along the Rio Grande (Clark 2004; Hill et al. 2004). The resulting demographic shift greatly reduced the archaeological visibility of groups who may have remained in the area into the post-contact period. This would have minimally resulted in alterations to the distributions of numerous genetic markers such as mtDNA haplogroups. Last, many molecular studies using modern groups often do not consider the tremendous demographic upheaval caused by European contact, beginning with waves of epidemic disease and massive die-offs and continuing with colonization and subjugation of the native populace through to the present (Dobyns 1983, 1989, 1993; Ramenofsky 1987; Thomas 1989). These processes of historical population decline and reorganization likely caused episodes of founder effect, genetic drift, or in extreme cases natural selection. Stannard's (1992:57–95) estimate of the native death toll that resulted from Spanish colonialism is 60–80 million. This type of tremendous demographic upheaval could have so thoroughly altered genetic variability in the region that haplogroup frequencies among modern tribal groups cannot be assumed to approximate the pre-Columbian situation.

Examining haplogroup frequencies among modern populations of Southwest Amerindians could be of limited utility for testing the migration scenario. A recent study, designed to specifically test the assumptions associated with inferring into the genetic past from modern distributions of native haplogroups identified that, at least in a small sample on the eastern seacoast, both modern and ancient DNA haplogroup frequencies were indistinguishable (Halverson and Bolnick 2008). Other recent research similarly suggests that genetic drift has not had a large effect on mitochondrial haplogroup frequencies in recent North American prehistory (O'Rourke et al. 2000). Regardless of specific applicability, the important question to address is how haplogroup frequencies of the earliest farmers in the Southwest compare with those of the foragers who occupied the region for the millennia prior to the agricultural transition. By comparing the ancient mtDNA of the first farmers to that of earlier foragers, as well as to later groups across the region, researchers can avoid the problems inherent in modern samples and thereby provide a more realistic test of the farmer-migration hypothesis.

Ancient DNA

Very few studies have considered ancient DNA to address the farmer-migration hypothesis. LeBlanc and colleagues (2007) were able to extract ancient DNA from quids and aprons containing menstrual blood from Basketmaker caves in the northern Southwest. They compared the distribution of haplotypes across geographic regions—by language family and by time—and concluded that geography was the greatest contributing factor to the distribution of haplotype frequencies (Table 6-2). Although this study does not support a farmer-

Table 6-2. *Comparison of mtDNA Haplogroup Frequencies by Time, Geographic Region, and Language Family*

Population	N	Haplogroups (%)				
		A	B	C	D	X
Prehistoric Southwest:						
BM / Anasazi	58	12.7	73.7	12.2	4.3	0
Modern Southwest						
Uto-Aztecan	188	5.6	51.9	41.9	0	0
Not Uto-Aztecan	197	8.7	70.6	20.3	0	9.9
Modern Northwest Mexico						
Uto-Aztecan	207	31.8	44.5	20.5	4.8	0
Prehistoric Central Mexico						
Uto-Aztecan	23	65.2	13.0	4.4	17.4	0
Modern Central Mexico						
Uto-Aztecan	96	50.5	29.8	16.6	3.1	0
Not Uto-Aztecan	204	46.8	24.0	21.9	7.3	0

Source: Modified from LeBlanc and colleagues 2007

migration model associated with the spread of the Uto-Aztecan language family as they proposed, it does function to highlight a significant geographical division when considering this model in the greater Southwest region. The difference between cultural events and biological relatedness in the northern Southwest (on the Colorado Plateau) versus the southern Southwest (in the lowland deserts) can be considerable. This study functions to additionally underscore the need for early skeletal samples from the southern Southwest to effectively address the farmer-migration hypothesis.

A couple of related studies were able to identify some degree of similarity of ancient mtDNA haplogroup frequencies recovered from samples of ancient Anasazi from the "Four-Corners" area and ancient Fremont from the Salt Lake area (O'Rourke et al. 1996; Parr et al. 1996). Carlyle and associates (2000) expanded on these earlier studies to include more Anasazi groups and compare ancient mtDNA haplogroup frequencies to modern Southwest Amerindians, all of which were shown to share similar frequencies across the Southwest. In an unrelated study, using ancient mtDNA from western Nevada to address the question of the late migration of Numic peoples throughout the Great Basin, Kaestle and Smith (2001) present data that identify a close genetic relationship between mtDNA haplogroup frequencies from ancient Fremont samples and modern Southwest and Baja groups. Short Cavalli-Sforza and Edwards (1967) chord distances and clustering on phylogenetic dendrograms demonstrate that haplogroup frequencies from the samples grouped closest by language family and geographic proximity. Yet, they could not rule out the possibility that observed differences in haplogroup frequencies are instead the result of long-term microevolutionary change within a single population. As a result, the authors revisit the issue in a recent publication (Cabana et al. 2008) using computer simulations designed to test this specific problem. They determined that relatively low levels of gene flow and random genetic drift can produce sufficient degrees of genetic differences between population samples (Cabana et al. 2008). This is significant for arguments of migration using molecular evidence when different demographic circumstances have the potential to produce similar patterns.

The data indicate that much of the prehistoric Southwest had high frequencies of mtDNA haplogroup B, a pattern observed throughout many of the modern groups in the region, especially those belonging to the Uto-Aztecan language family. This supports the supposition that after the introduction of agriculture and the settlement of the region there was a great deal of genetic homogeneity across distinctive Formative cultures. Subsequent episodes of prehistoric migration, epidemics, colonialism, and social reorganization functioned to largely strengthen or maintain the biological foundation of the region and perhaps reduce overall genetic variability. However, the question again comes back to whether or not the foundation of biological variation in the region was brought in by migrant farmers or it constitutes an even more ancient heritage passed down from a large network of desert foragers. We still have yet to recover ancient DNA from preagricultural foragers in the area. Given the specific considerations that Malhi, Kemp, and their colleagues have given this particular question, most of the genetic evidence fails to support a migration hypothesis.

Skeletal Biology

Patterns in the skeletal biology of past human groups are commonly used to address questions about the transition from foraging to farming and as evidence of the consequences of migration across the globe (e.g., Cohen and Armelagos 1984; Larsen 1997; Schillaci and Stojanowski 2005; Turner 1986). Over 360 human inhumations have been recovered from Early Agricultural period (3600–1800 B.P.) sites in the Sonoran Desert to date and these skeletons represent the largest sample of the earliest pre-Columbian inhabitants of the region. There are very few burials that predate the Early Agricultural sample, not just within the Sonoran Desert, but throughout the North American Desert West (Mabry 1998b). Three inhumations have been identified that are possibly associated with the Middle Archaic period, estimated to date between circa 5800 and 3200 B.P., but the associations are dubious. During the terminal part of the Early Agricultural period a major shift in southern Southwest burial ritual—whereby cremation became more common during the subsequent ceramic period (Mabry 1998b; Watson and Cerezo-Román 2010)—resulted in fewer inhumations until the Hohokam Classic period (Reid and Whittlesey 1997), when this burial practice again regained some popularity.

Early Agricultural period skeletal samples are, therefore, extremely important for a number of reasons. First and foremost, as the largest sample of early inhabitants of the Sonoran Desert, they are an important resource for reconstructing population composition, movement, and relationship to subsequent groups documented to have made a living in the area. Second, as they represent the earliest farmers and the first people to settle into permanent villages in the region, these samples are important for addressing major questions about human populations—with regard to their adaptations, diet, and health—during the transition to agriculture. Lastly, given the dearth of inhumations from the Sonoran Desert, the samples are an important source of information about the people who called this desert their home and they constitute the only samples available to address the central goal of this paper and to test the central hypothesis.

To date, a few more than 12 Early Agricultural period sites from southern Arizona and northern Sonora have produced approximately 368 human inhumations (Table 6-3). Radiocarbon dates on human bone (Villalpando and Carpenter 2004) or associated with burials (Mabry 1997) range between 5100 and 1800 B.P.; these span the entire duration of both the San Pedro and the Cienega phases. Several other sites that date to, or contain components of, the Early Agricultural period have produced some fragmented remains, but these are very limited and not useful for reconstructing a larger picture of the demography, health, and diet of these early farmers. In addition, many of these sites have similarly produced several cremation burials that can be attributed to the terminal end of the Early Agricultural period, but again, the fragmented nature of these burials precludes their use in comparisons with the more numerous and temporally expansive inhumation samples. Despite the dramatic differences in numbers and the clustered, but widely geographically distributed, nature of these burial samples (Figure 6-5), many similarities have been identified between each of these sites as well as between the skeletal material recovered from them. Different from subsequent occupations and the archaeologically defined cultures in the Sonoran Desert, the Early Agricultural inhabitants appear in most ways to share both cultural and biological traits across this expanse. The burial samples from all of the Early Agricultural period sites are considered here as one skeletal assemblage representing a biologically and culturally related population encompassing a vast majority of its inherent variability.

Biological Distance Analyses

Since the beginning of bioarchaeology in the Southwest and Hooton's work at Pecos Pueblo (1930), understanding biological variation has been an integral part of research in prehistoric skeletal biology in the region. An increase in the use of multivariate statistics, beginning in the 1970s, fostered a cottage industry of testing biological continuity in skeletal samples from the U.S. Southwest–Northwest Mexico. The vast majority of these studies largely focused on either craniometrics or discrete dental traits (i.e., Bennett 1973; Corruccini 1972; El-Najjar 1978; Turner 1986), and for the most part, this continues to be the case

Table 6-3. *Inhumations Recovered from Early Agricultural Period Sites*

Site Name	ASM Site No.	N = 368	Phase	Citation
Clearwater	AZ BB:13:6	2	Cienega	Diehl 1997
Coffee Camp	AZ AA:6:19	3	Cienega	Dongoske 1993
Donaldson	AZ EE:2:30	5	Cienega	Minturn & Lincoln-Babb 1995
La Playa	SON F:10:3	267	SP (79+)/C (82+)	Watson 2005
Las Capas	AZ AA:12:111	15	San Pedro	McClelland 2005
Los Morteros	AZ AA:12:57	3	Cienega	Wallace 1995
Los Ojitos	AZ EE:2:37	10	Cienega	Minturn & Lincoln-Babb 1995
Los Pozos	AZ AA:12:91	23	Cienega	Minturn & Lincoln-Babb 2001
Pantano	AZ EE:2:50	3	Cienega	McClelland 2005
Rillito Fan	AZ AA:12:788	2	Cienega	Wöcherl 2003
Santa Cruz Bend	AZ AA:12:746	8	Cienega	Minturn et al. 1998
Stone Pipe	AZ AA:12:745	1	Cienega	Minturn et al. 1998
Valley Farms	AZ AA:12:736	2	San Pedro	Wellman 2000
Wetlands	AZ AA:12:90	24	Cienega	Guthrie & Lincoln-Babb 1998

today (LeBlanc et al. 2008; Schillaci and Stojanowski 2005). However, few of these studies have specifically tried to test the farmer-migration hypothesis, although the lack of skeletal remains predating the Early Agricultural period is a major limiting factor.

LeBlanc and colleagues (2008) is the one notable recent exception: a study designed specifically to test the farmer migration hypothesis. LeBlanc and colleagues compared discrete dental traits in samples from throughout the North American Desert West to eastern and western Basketmaker dentitions in the northern Southwest. They identify that the eastern Basketmakers were very divergent from the rest of the samples and hypothesize that these eastern groups display biological inheritance from indigenous foragers, whereas western groups display biological inheritance from migrant Uto-Aztecan farmers from Mesoamerica. The Basketmaker sample size is admittedly small and the comparisons are made with noncontemporaneous Formative groups making their interpretations somewhat unconvincing. A major limiting factor to appropriately addressing the migrant-farmer hypothesis in the region has been, until recently, a lack of early skeletal material from the Sonoran Desert.

Although not specifically designed to address the migration hypothesis, in a recent study I examined crown dimensions of a small set of molars ($n = 23$) in the

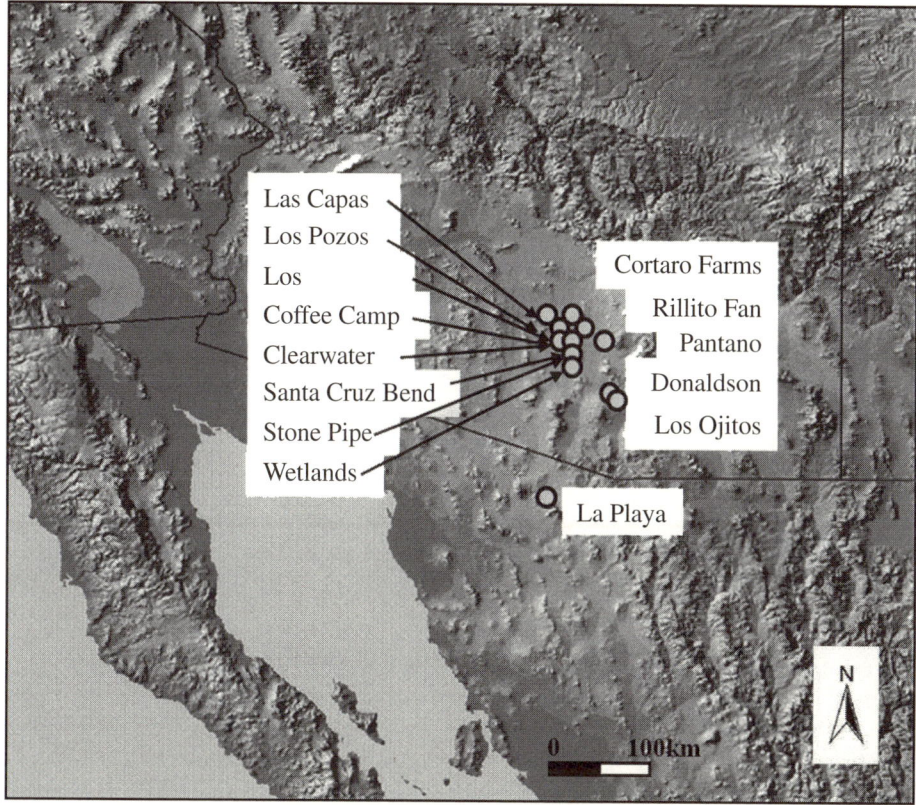

Figure 6-5. *Dispersion of Early Agricultural period sites in the Sonoran Desert.*

skeletal sample from the site of La Playa in northern Sonora to assess the potential for phenotypic differences between sexes and by archaeological phase (Watson 2008a). I identified significant differences in the mean values of buccolingual tooth dimensions of the first and second mandibular molars between sexes, with males exhibiting larger buccolingual dimensions in both teeth suggesting that some phenotypic differences exist between males and females within the sample. These were suggested to be the result of sexual dimorphism in the population (Watson 2008a). Significant differences in the buccolingual dimensions of the third mandibular molars were also documented between the San Pedro phase and the Cienega phase, indicating a slight decrease in average size over time. Although the observed reduction in size could be an artifact of the tremendous variability inherent in the third molar, it was postulated to reflect the impact of external forces such as a change in food processing techniques on the phenotypic expression of this dimension over the duration of the Early Agricultural period (Watson 2008a). The overall homogeneity observed in molar dimensions minimally indicates that the Early Agriculturalists at La Playa comprised a single related population.

A preliminary assessment of discrete dental traits from the same La Playa dental sample (Watson et al. 2006) also failed to identify any differences within the sample by sex or archaeological phase. This lends further support to the supposition that these burials represent a relatively homogenous biological population of early farmers in the region. The more pressing question however, is whether this homogenous population is one of immigrants or indigenous farmers. Comparing the dental nonmetric traits from the Early Agricultural period sample to those used in LeBlanc and colleagues (2008) identified a closer relationship to groups in the southern Southwest, northwest Mexico, and several Puebloan groups (Watson et al. 2006) and lends no support to the farmer migration hypothesis. These assessments are preliminary however, and they need to be tested with a larger sample of more closely contemporaneous groups.

Health

Early Agricultural period groups display low frequencies of skeletal lesions compared to later Formative populations in the Sonoran Desert (Fink and Merbs 1991; Merbs and Miller 1985; Van Gerven and Sheridan 1994). Lesions of infectious disease are relatively infrequent in the Early Agricultural period sample and the vast majority of evidence for infection (affecting 13 percent of individuals in the sample) are nonspecific bone lesions such as discrete periosteal lesions on long bones. This is in sharp contrast to Classic period Hohokam groups along the Salt and Gila rivers who exhibit nonspecific bone infection frequencies well above 50 percent (Benjamin and Merbs 1990; Fink and Merbs 1991). A handful of adult individuals ($n = 15$; 4 percent) do show signs of more serious systemic infections including osteomyelitis and reactive bone lesions on ribs, vertebrae, and hip bones. Evidence for anemic reactions are rare (<5 percent) compared to later Hohokam frequencies that exceed 80 percent (Abbott 2003; Benjamin and Merbs 1990). Evidence of functional stress is relatively limited although several males exhibit significant vertebral lipping on the lumbar vertebrae, indicative of a heavy stress on the lower back. Overall, individuals in the sample exhibit little functional stress as they aged, indicating a life relatively free of degenerative disease (McClelland 2005; Minturn and Lincoln-Babb 1995; Watson et al. 2006).

Two researchers considered cross-sectional dimensions of long bones from Early Agricultural period skeletons to investigate patterns related to subsistence economy (McClelland 2005; Ogilvie 2005). Both studies demonstrated that males display more elongated femoral cross sections than females display, which indicates more long-distance travel over rough terrain, a pattern more often characteristic of dedicated foraging populations. The female cross-sectional measurements are more characteristic of fully sedentary agricultural groups. These studies offer a glimpse into the nature of gender roles in this mixed-subsistence economy, in which males likely made long-distance gathering forays while females stayed close to the village and tended the fields. These data may also suggest that this pattern of long-bone morphology would not be expected, nor is it consistent with, populations of dedicated agriculturalists who had migrated into the area.

Diet

Information on oral health is available for a large number of teeth (*n* = 2,548) from Early Agricultural burials. Carious lesions are relatively abundant, affecting almost 60 percent of the individuals in the sample, resulting in an overall caries frequency of 11.7 percent (McClelland 2005; Minturn and Lincoln-Babb 1995; Watson 2008b). Caries frequency did not vary significantly between the San Pedro phase and the Cienega phase indicating that plant resources being consumed remained relatively consistent for the duration of the Early Agricultural period (Watson 2008b). Most prehistoric agricultural populations from the North American Desert West exhibit caries frequencies ranging between 15 percent and 50 percent (Berry 1985; Fink and Merbs 1991). Inhumations recovered from the Hohokam Classic period (800–500 B.P.) sites of Grand Canal, Casa Buena, and La Ciudad in the Phoenix basin demonstrate a combined caries rate of 16.3 percent (Fink and Merbs 1991). Although the overall caries frequency of the Early Agricultural period burials is below that of most later, fully agricultural groups, it is considered relatively high when compared to other populations who practiced a mixed subsistence economy (Larsen 1995; Turner 1979). These data indicate that the diet was relatively reliant on cariogenic carbohydrates (Watson 2005, 2008b), a pattern traditionally attributed to agricultural dependence around the world (Larsen 1997). However, in the Sonoran Desert there are a number of edible local plant species that are equally cariogenic, such as cactus (pads and fruit), mesquite beans, and agave. Researchers have recently begun to recognize the impact these local resources had on the oral health of early farmers and to support the proposition that they practiced a mixed subsistence economy (McClelland 2005; Watson 2005, 2008b). Very few studies have considered the impact of naturally occurring carbohydrates on the prevalence of dental disease in prehistory (Armelagos and Rose 1972; Hartnady and Rose 1991; Tayles et al. 2000). Tayles and colleagues (2000) documented that caries, attrition, and antemortem tooth loss (AMTL) all decreased dramatically from the Neolithic to the Iron Age (4000–1700 B.P.) in skeletal samples from three localities in Southeast Asia. They found that oral health actually improved with the transition to rice agriculture. In the Sonoran Desert, the consumption of wild cariogenic species likely resulted in poorer oral health during the Archaic period and functioned to effectively "mask" the transition to agriculture in the region (Watson 2008b).

AMTL affected 45 percent of Early Agricultural period individuals, resulting in an overall frequency of 18.3 percent (McClelland 2005; Minturn and Lincoln-Babb 1995; Watson 2008b). The frequency of tooth loss did not differ significantly between the San Pedro and Cienega phases. This dynamic disease process can result from a number of factors including caries, periodontal infection, severe attrition, and trauma. This frequency is low compared to 26 percent AMTL observed in sites from the Hohokam Classic period (Fink and Merbs 1991). However, as with caries, this represents a relatively elevated frequency when compared to foraging or mixed-subsistence economies (Larsen 1995; Turner 1979). Again, in the case of the Sonoran Desert it is likely to be related to the consumption of wild species rich in sticky carbohydrates (McClelland 2005; Watson 2005, 2008b).

Dental attrition during the Early Agricultural period was relatively heavy but also extremely variable. In a recent quantitative assessment of attrition rates at the site of La Playa, I identified that rates did not vary by archaeological phase but instead discovered that the angle of wear on molar occlusal surfaces did vary over time (Watson 2008a). In Huckell's 1995 study of the Cienega Creek skeletal sample, Minturn and Lincoln-Babb (1995) identified steep molar wear angles. Steep wear angles have been shown to be consistent with consumption of a highly processed agricultural diet (Smith 1984), therefore Minturn, Lincoln-Babb and Huckell used this evidence to conclude that the Cienega Creek population was fully dependent on agriculture. I repeated the same procedure with a larger skeletal sample from the La Playa site that included individuals dating to the earlier San Pedro phase (Watson 2008a). Examining molar wear angles in samples encompassing the full temporal range of the Early Agricultural period demonstrated that although wear rates did not change, there was a significant increase in slope from the San Pedro to the Cienega phase (Watson 2008a). I used these data to suggest that, although the overall contributions of plant resources to their diet did not change much for almost 2,000 years, these early farmers may have increased the mechanical processing of these resources during the Cienega phase and effectively softened the diet, altering wear planes.

There is a clear difference in the patterns observed in skeletal health among samples from the Early Agricultural period versus those recovered from Hohokam sites in southern Arizona. Despite the fact that cremation burial was the most common mortuary practice among the Hohokam beginning in the Pioneer period and offers little in the way of dental information, La Ciudad (Benjamin and Merbs 1990), Las Colinas ruins group (Harrington 1981), Pueblo Grande (Abbott 2003), and Snaketown (Haury 1976; Sayles 1938) all produced several large well-studied skeletal samples to which Early Agricultural groups can be compared. In general, a definite trend of decreasing health—through the phases of the Hohokam cultural sequence until poor health and mortality reach maximums during the Classic period (Abbott 2003)—is observed. This decline was followed by social collapse during the post–Classic period and an effective "abandonment" of the region (Cordell 1997).

The majority of individuals recovered from Hohokam sites display some skeletal lesions of infection, nutritional deficiencies, and/or bodily stress; porotic hyperostosis was especially prevalent, exceeding 80 percent in some samples (Benjamin and Merbs 1990; Abbott 2003). Caries and AMTL rates exceed 50 percent, and abscesses and rates of enamel hypoplasias exceed 30 percent for most settlements (Berry 1985; Fink and Merbs 1991). Overall, the skeletal evidence indicates that early farmers of the Sonoran Desert were very healthy compared with their Formative period descendants (Barnes 2002; McClelland 2005; Watson et al. 2006). The limited presence of infectious disease and nutritional deficiencies likely suggests that population densities were moderate and food security was relatively good. I have previously argued (Watson 2005) that this evidence suggests that Early Agricultural period populations in the Sonoran Desert were indeed undergoing the transition from foraging to agriculture and that pathology rates are more similar to those observed in foraging or mixed-subsistence economies (Larsen 1995). The patterns observed in the skeletal biology of these early populations do not support the migration hypothesis.

Discussion

Evidence from the archaeological record, language models, molecular studies, and skeletal biology of the earliest farmers in the Sonoran Desert does not support the acceptance or rejection of the hypothesis that agriculture was introduced into the area by migrating farmers from Mesoamerica. The same arguments are being debated in several regions of the globe but the Sonoran Desert has several unique issues that prevent definitive statements regarding the introduction of agriculture. This review of our current understanding and the issues involved also functions to highlight future directions to better test the migration hypothesis in the Sonoran Desert, and perhaps in other parts of the world.

The archaeological evidence for human utilization and occupation of the Sonoran Desert from the Middle Archaic to the end of the Early Agricultural period is problematic. Although sites and some dates exist for the Middle Archaic, there is a definite gap in the archaeological and radiocarbon record throughout much of the U.S. Southwest–Northwest Mexico. Many archaeologists use this gap to claim that humans completely abandoned the area during the hot, dry Altithermal climatic episode (Antevs 1955), thereby leaving an open niche for migrating agriculturalists in the Late Archaic when climate began to ameliorate. However, the simple presence of maize at several Late Archaic sites and its ubiquitous presence during the Early Agricultural period are hardly damning evidence for an influx of migrant farmers. Instead, the material culture associated with the post–Altithermal period is more suggestive of small semimobile bands that gradually incorporated maize into an existing local foraging strategy. It is at the end of the San Pedro phase, when canals are observed, and during the Cienega phase, when larger villages, structures, and more complex artifact assemblages are present, that we observe a potential increase in agricultural investment. Regardless of whether or not the Sonoran Desert was abandoned during the Middle Archaic, Early Agricultural period populations represent the earliest settlement of farmers in the region and constitute the biological foundation for the subsequent native occupation of the region. What are needed to better address the specifics of continuity/discontinuity in the Sonoran Desert are the discovery and investigation of more Middle and Late Archaic sites and skeletal samples from these periods.

Much of the linguistic data and theory supports a migration hypothesis, yet here I was able to use these same arguments to reinterpret the data to suggest that Proto–Uto-Aztecan groups could have been roaming throughout western North America during the Altithermal period, creating a large cultural and linguistic continuum in which independent groups utilized distributed strategies. It is possible that maize cultivation and agricultural technology, developed in Mesoamerica prompted by the improving conditions of the early Late Holocene, could have spread throughout this cultural continuum by a combination of diffusion and population movement, including demic diffusion and "leapfrogging" of river systems along the western Mexican coast as suggested by Hill (2001). Once in the Sonoran Desert, people, along with their language and technology, could have spread throughout the U.S. Southwest–Northwest Mexico and begun to settle in and lay claim to local resources and restrict interaction zones, favoring localist strategies as sedentism increased and home ranges decreased. Socially open cultural systems will register archaeologically as uniform artifact assemblages and therefore would be difficult to distinguish from stable populations with similar signatures. The artifact uniformity observed from Early Agricultural occupations in the Sonoran Desert may be the result of distributed strategies in which mobility and interaction were not restricted but led to the eventual settlement of sites by diverse populations who shared similar technologies. I believe that studies that would attempt to tie specific evidence for prehistoric group identity or "ethnicity" in the archaeological record to patterns in lexical data or language distribution would do more for advancing the broader application of these theories.

The evidence from several molecular studies, both of modern and ancient distributions of DNA markers, appears to indicate that there was minimal migration involved in the movement of maize and agricultural technology into the Sonoran Desert (Kemp 2006; Kemp et al. 2010; Malhi et al. 2002, 2003). Although limited in this region, studies of ancient DNA suggest that there is little evidence for a farmer migration (LeBlanc et al. 2007); however, they also present data that are suggestive of episodes of microevolution in the ancient Southwest. Clear distinctions can be observed in haplogroup frequencies between the northern and southern Southwest. Data presented from several studies of modern mtDNA haplogroup distributions indicate that much of the prehistoric Southwest had high frequencies of mtDNA haplogroup B, especially those belonging to the Uto-Aztecan language family (Kemp 2006; Malhi et al. 2003). This supports the supposition that after the introduction of agriculture and the settlement of the region there was a great deal of genetic homogeneity across distinctive Formative cultures. Subsequent episodes of prehistoric migration, epidemics, colonialism, and social reorganization functioned largely to strengthen or maintain the biological foundation of the region and perhaps reduce overall genetic variability. There is a great deal of biological homogeneity and continuity in the Southwest, which supports the idea that Early Agricultural period populations form the foundation of biological variation. The best possible scenario to further test the farmer-migration hypothesis would be to extract ancient DNA from Early Agricultural and Archaic period skeletal samples. The need for appropriate skeletal samples from earlier periods is again underscored. Yet, recent attempts to sample DNA from Early Agricultural period skeletal material has failed due to poor preservation.

The wide variety of evidence from patterns in skeletal biology of the Early Agricultural period samples suggests that an overarching biological and cultural continuity exists to connect the people living at relatively distant locations. Limited evidence for infections and nutritional deficiencies compared to later Hohokam groups (Berry 1985; Fink and Merbs 1991) indicates that populations were relatively healthy during the Early Agricultural period. This reflects a pattern more indicative of forager or mixed-subsistence groups than of fully dedicated agriculturalists. A contrary pattern is observed with relatively poor oral health but in fact reflects a diet heavily based on the consumption of local wild resources that are very cariogenic (Watson 2008b). Wild resources were an integral part of the subsistence strategy, a somewhat contrary notion to the idea that dedicated agriculturalists, with an existing subsistence-economy structure, migrated into the area. Changes observed in tooth wear angles indicate that, although the overall dietary base may have remained stable as a mixed-subsistence economy for much of the Early Agricultural period, there could have been an increase in mechanical processing during the Cienega phase (Watson 2008a). An increase in mechanical processing appears to mirror changes observed in sites and artifacts from the San Pedro to the Cienega phase and could be interpreted as increased investment associated with agricultural production. Future studies in skeletal biology addressing the migration hypothesis will remain somewhat limited until earlier burial samples can be recovered and studied. Yet, there is still much to be done to understand the dynamic nature of the Early Agricultural period and its lifestyle effects on the human body.

Our understanding of the Early Agricultural period is swiftly changing, but it continues to provide information about the earliest farming populations in the Sonoran Desert: groups who maintained a dynamic adaptation to their environment for nearly 2,000 years. Although the information reviewed here cannot be used to explicitly reject or support the farmer-migration hypothesis, it does suggest that cultural and biological interaction was far more complex prior to and during the Formative transition than previously proposed and further underscores the importance of deciphering the biological foundation of the native populations of the U.S. Southwest–Northwest Mexico.

Several researchers have proposed that the later Formative period cultures of the Hohokam, Mogollon, and Ancestral Puebloans developed from a common ancestral root (Berry and Berry 1986; Ciolek-Torrello 1998; Matson 1991). Indeed, the presence of maize throughout the U.S. Southwest–Northwest Mexico argues for sufficient interaction between these different ecological zones until approximately 2000 B.P. Maize was incorporated into the local foraging subsistence strategies on the Colorado Plateau, in the Mogollon highlands, and in the Sonoran Desert lowlands by 4000 B.P. (Huckell 2006); however, each area began intensively cultivating maize on different schedules. Lowland groups, already heavily dependent on river floodplain resources, began farming and relying more heavily on cultigens by 3600 B.P., far earlier than groups to the north and east. Following Matson's (1991) contention that maize first needed to be adapted to dry farming, the shorter growing season, and the cooler climate of the northern Southwest, maize species were likely better adapted to the warmer and wetter climate of the Sonoran Desert. Once cold-adapted varieties were developed, maize was incorporated by Basketmaker groups throughout the Colorado Plateau and by early Mogollon groups in southern New Mexico. After this time agricultural dependence followed somewhat similar trajectories, overall interaction decreased, and cultural differences developed between the respective areas of the Desert West.

Most models of agricultural adoption by local foraging populations postulate a slow and steady economic transformation in which cultural traits and technology change gradually over time. Rowley-Conwy (2004) acknowledges that, although these models are almost universally accepted by archaeologists of all theoretical paradigms, they do not account for some of the transformation dynamics visible within the archaeological record. He applies a three-phase model of agricultural adoption, developed by Zvelebil and Rowley-Conwy (1986), to the transition from the Mesolithic to the Neolithic in Britain, Ireland, and southern Scandinavia to conclude that the adoption of agriculture in these areas was a "rapid and massive socio-economic wave of disruption" (Rowley-Conwy 2004:S97). These conclusions provide an attractive alternative to explain agricultural adoption in the archaeological record of the Sonoran Desert, in which the transition could be argued as having been a rapid process.

The model of agricultural transition proposed by Zvelebil and Rowley-Conwy (1986) identifies three phases: Phase 1—availability: agriculture is available to foragers but plays little or no role in their economy; Phase 2—substitution: agriculture provides between 5 percent and 50 percent of the diet and represents the actual transition; and Phase 3—consolidation: agriculture provides the majority of the diet (over 50 percent). Rowley-Conwy (2004) further examines the concept of this transition phase by comparing the percentage of agricultural dependence across an ethnographic sample of societies divided by subsistence economies with data originally reported by Hunn and Williams (1982). The societies were distributed bimodally, whereby they either rely more heavily on hunted, herded, and gathered foods or on agricultural products; very few societies fell within 5 percent and 45 percent agricultural dependence. He interprets this to mean that the substitution phase is an "unstable intermediate area through which transitional societies are likely to move rapidly" (Rowley-Conwy 2004:S97). Patterns observed in the archaeological and skeletal biological records of the Sonoran Desert, from the Middle and Late Archaic through the course of the Early Agricultural period, fit into this scheme.

Conclusions

Evidence for the migration of agriculturalists into the Sonoran Desert appears to be limited, yet we still lack crucial data that would provide conclusive evidence for diffusion. Linguistic models and a few molecular studies support the migration of agricultural groups into the region, whereas most molecular studies and inferences from skeletal biology of the earliest farmers support the idea that domesticates and agricultural technology diffused into the region over a broad cultural continuum of biologically diverse groups from Mesoamerica. The archaeological record can be interpreted to potentially support either hypothesis. Each of these lines of evidence lacks some crucial ingredient to make its arguments convincing. In numerous places around the world, plants and technology have been demonstrated to move and to modify culture across dynamic language families, and vice versa. We have yet to be able to extract ancient DNA from skeletal material dating to the Early Agricultural period and there is a conspicuous lack of skeletal material dating from periods prior to this in the Sonoran Desert.

What can be gleaned from the information presented here is that the introduction of agriculture into the U.S. Southwest–Northwest Mexico provided the foundation for biological variation for the next several centuries, as large-scale Formative societies developed in situ from the earliest groups of local farmers in the region. Late Formative developments see numerous small- and large-scale migrations within the region as populations reconfigure both socially and biologically in response to social and environmental pressures. These relationships are again reconfigured with the arrival of nonnatives and subsequent colonial policies that largely dictated the historical constructions of native groups in the U.S. Southwest–Northwest Mexico. However, evidence from the Sonoran Desert indicates that biological variation does indeed coincide with the variation observed in the archaeological record and can be traced back to the origins of agriculture in the region.

Acknowledgments

I would like to thank Benjamin Auerbach, Bruce Huckell, John McClelland, and an anonymous reviewer for their valuable comments on earlier versions of this manuscript. I would also like to thank my long-term collaborators—John Carpenter, Elisa Villapando, Ethne Barnes, and Art Rohn—for their support and continued shared interest in this work. Lastly, I wish to thank my wife and son for their patience with and understanding of my passion for science and my effort to contribute to a better understanding of the human past.

This research was sponsored by grants from Instituto Nacional de Antropología e Historia and CONACyT, México, and the National Science Foundation (Grant No. BCS-0433986).

References

Abbott, David R.
    2003   *Centuries of Decline During the Hohokam Classic Period at Pueblo Grande*. University of Arizona Press, Tucson.
Adler, Michael A., Todd Van Pool, and Robert D. Leonard
    1996   Ancestral Pueblo Population Aggregation and Abandonment in the North American Southwest. *Journal of World Prehistory* 10:375–438.
Ammerman, Albert J., and Luigi L. Cavalli-Sforza
    1984   *The Neolithic Transition and the Genetics of Human Populations*. Princeton University Press, Princeton, New Jersey.
Antevs, Ernst
    1955   Geologic-Climatic Dating in the West. *American Antiquity* 20:317–355.
Armelagos, George J., and Jerome C. Rose
    1972   Factors Affecting Tooth Loss in Prehistoric Nubian Populations. *American Journal of Physical Anthropology* 37:428.
Barnes, Ethne
    2002   La Playa Burial Analysis. In *Rescate Arqueológico La Playa. Informe Técnico al Consejo de Arqueología del Instituto Nacional de Arqueología e Historia, Temporada 2001,*

edited by John P. Carpenter, Guadalupe Sánchez, and M. Elisa Villalplando. Archivo Técnico del INAH, México.

Bellwood, Peter
1997   Prehistoric Cultural Explanations for Widespread Linguistic Families. In *Archaeology and Linguistics: Aboriginal Australia in Global Perspective*, edited by Patrick McConvell and Nicholas Evans, pp. 123–134. Oxford University Press, Melbourne.

Benjamin, Oslynn, and Charles F. Merbs
1990   Cremation Burials from Los Hornos. In *Archaeological Investigations at La Ciudad de Los Hornos: Lassen Substation Parcel*, edited by Richard W. Effland, pp. 159–171. Arizona Archaeologist No. 24, Arizona Archaeological Society. Phoenix.

Bennett, Kenneth A.
1973   *The Indians of Point of Pines, Arizona: A Comparative Study of their Physical Characteristics*. Anthropological Papers of the University of Arizona No. 23. University of Arizona Press, Tucson.

Berry, Claudia F., and Michael S. Berry
1986   Chronological and Conceptual Models of the Southwest Archaic. In *Anthropology of the Desert West: Essays in Honor of Jesse D. Jennings*, edited by Carol J. Condie and Don W. Fowler, pp. 253–327. Anthropological Papers No. 110. University of Utah Press, Salt Lake City.

Berry, David R.
1985   Dental Paleopathology of Grasshopper Pueblo, Arizona. In *Health and Disease in the Prehistoric Southwest*, edited by Charles F. Merbs and Richard J. Miller, pp. 43–64. Arizona State University Anthropological Research Papers No. 34, Tempe.

Braidwood, Robert J.
1967   *Prehistoric Men*. 7th ed. Scott Foresman, Glenview, Illinois.

Cabana, Graciela S., Keith Hunley, and Frederika A. Kaestle
2008   Population Continuity or Replacement? A Novel Computer Simulation Approach and Its Application to the Numic Expansion (Western Great Basin, USA). *American Journal of Physical Anthropology* 135:438–447.

Carlyle, Shawn W., Ryan L. Parr, M. Geoffrey Hayes, and Dennis H. O'Rourke
2000   Context of Maternal Lineages in the Greater Southwest. *American Journal of Physical Anthropology* 113:85–101.

Carpenter, John P., Jonathan B. Mabry, and Guadalupe Sánchez
1999   Arqueología de los Grupos Yutoaztecas Tempranos. In *Avances y Balances de Lenguas Yutoaztecas, Homenaje a Wick R. Miller*, edited by José L. Moctezuma Zamarrón and Jane H. Hill. Edición Especial del Noroeste. Centro INAH Sonora, Conaculta, INAH, Sonora.

Carpenter, John P., Guadalupe Sánchez, and M. Elisa Villalpando
1997   *Prehistory of the Borderlands: Recent Archaeological Research in Northern Mexico and the Southern Southwest*. Arizona State Museum Archaeological Series 186, University of Arizona, Tucson.
2002   Of Maize and Migration: Mode and Tempo in the Diffusion of *Zea mays* in Northwest Mexico and the American Southwest. In *Traditions, Transitions, and Technologies: Themes in Southwestern Archaeology*, edited by Sarah Schlanger, pp. 245–258. University of Colorado Press, Boulder.
2005   The Late Archaic/Early Agricultural Period in Sonora, Mexico. In *New Perspectives on the Late Archaic Across the Borderlands*, edited by Bradley J. Vierra, pp. 13–40. University of Texas Press, Austin.

Cavalli-Sforza, Luigi L., and Anthony W. F. Edwards
    1967   Phylogenetic Analysis Models and Estimation Procedures. *American Journal of Human Genetics* 19:233–257.
Childe, V. Gordon
    1925   *The Dawn of European Civilization*. Kegan Paul, London.
    1965   *Man Makes Himself*. Watts, London.
Ciolek-Torrello, Richard (editor)
    1998   *Early Farmers of the Sonoran Desert. Archaeological Investigations at the Houghton Road Site, Tucson, Arizona*. Technical Series 72, Statistical Research, Tucson.
Clark, Jeffery J.
    2004   Tracking Cultural Affiliation: Enculturation and Ethnicity. In *Identity, Feasting and the Archaeology of the Greater Southwest*, edited by Barbara J. Mills, pp. 42–73. University Press of Colorado, Boulder.
Cohen, Mark N., and George J. Armelagos (editors)
    1984   *Paleopathology at the Origins of Agriculture*. Academic Press, New York.
Cordell, Linda S.
    1997   *Prehistory of the Southwest*. Academic Press, New York.
Corruccini, Robert S.
    1972   The Biological Relationship of Some Prehistoric and Historic Pueblo Populations. *American Journal of Physical Anthropology* 37:373–388.
Dart, Alan
    1986   *Archaeological Investigations at La Paloma: Archaic and Hohokam Occupations at Three Sites in the Northeastern Tucson Basin, Arizona*. Anthropological Papers No. 4. Institute for American Research, Tucson.
Dean, Jeffery S., William H. Doelle, and Janet D. Orcutt
    1994   Adaptive Stress: Environment and Demography. In *Themes in Southwest Prehistory*, edited by George J. Gumerman, pp. 53–86. University of Washington Press, Seattle.
Diamond, Jared, and Peter Bellwood
    2003   Farmers and Their Languages: The First Expansions. *Science* 300:597–603.
Diehl, Michael W.
    1997   *Archaeological Investigations of the Early Agriculture Period Settlement at the Base of A-Mountain, Tucson, Arizona*. Technical Report No. 96-21. Center for Desert Archaeology, Tucson.
    2005   *Subsistence and Resource Use Strategies of Early Agricultural Communities in Southern Arizona*. Anthropological Papers No. 34. Center for Desert Archaeology, Tucson.
Dobyns, Henry F.
    1983   *Their Number Become Thinned*. University of Tennessee Press, Knoxville.
    1989   Native Historic Epidemiology in the Greater Southwest. *American Anthropologist* 91:171–174.
    1993   Disease Transfer at Contact. *Annual Review of Anthropology* 22:273–291.
Doelle, William H., and Paul R. Fish (editors)
    1988   *Recent Research on Tucson Basin Prehistory: Proceedings of the Second Tucson Basin Conference*. Institute for American Research Anthropological Papers No. 10, Tucson.
Doolittle, William E., and Jonathan B. Mabry
    2006   Environmental Mosaics, Agricultural Diversity, and the Evolutionary Adoption of Maize in the American Southwest. In *Histories of Maize: Multidisciplinary Approaches to the Prehistory, Linguistics, Biogeography, Domestication and Evolution of*

Maize, edited by John Staller, Robert Tykot, and Bruce Benz, pp. 109–121. Elsevier, Burlington, Massachusetts.

Dongoske, Kurt
    1993   Burial Population and Mortuary Practices. In *Archaic Occupation on the Santa Cruz Flats: The Tator Hills Archaeological Project*, edited by Carl D. Halbirt and T. Kathleen Henderson, pp. 173–181. Northland Research, Inc., Flagstaff.

El-Najjar, Mahmoud Y.
    1978   Southwestern Physical Anthropology: Do the Cultural and Biological Parameters Correspond? *American Journal of Physical Anthropology* 48:151–158.

Ezzo, Joseph A., and William L. Deaver
    1998   *Data Recovery at the Costello-King Site (AZ AA:12:503[ASM]), a Late Archaic Site in the Northern Tucson Basin*. Technical Series No. 68. Statistical Research, Tucson.

Fink, Timothy M., and Charles F. Merbs
    1991   Paleonutrition and Palopathology of the Salt River Hohokam: A Search for Correlations. *Kiva* 56:293–317.

Fish, Paul R., Susanne K. Fish, Austin Long, and Charles H. Miksicek
    1986   Early Corn Remains from Tumamoc Hill, Southern Arizona. *American Antiquity* 51:563–572.

Fix, Alan
    2005   *Migration and Colonization in Human Microevolution*. Cambridge Studies in Biological and Evolutionary Anthropology 24. Cambridge University Press, Cambridge.

Fowler, Catherine
    1983   Lexical Clues to Uto-Aztecan Prehistory. *International Journal of American Linguistics* 49:224–257.
    1994   Corn, Beans, and Squash: Some Linguistic Perspectives from Uto-Aztecan. In *Corn and Culture in the Prehistoric New World*, edited by Sissel Johannessen and Christine A. Hastorf, pp. 445–467. Westview Press, Boulder.

Freeman, Andrea K. (editor)
    1998   *Archaeological Investigations at the Wetlands Site, AZ AA:12:90 (ASM)*. Technical Report No. 97-5. Center for Desert Archaeology, Tucson.

Gregory, David A. (editor)
    1999   *Excavations in the Santa Cruz River Floodplain: The Middle Archaic Component at Los Pozos*. Anthropological Papers No. 20. Center for Desert Archaeology, Tucson.
    2001   *Excavations in the Santa Cruz River Floodplain: The Early Agricultural Period Component at Los Pozos*. Anthropological Papers No. 21. Center for Desert Archaeology, Tucson.

Gregory, David A., and David R. Wilcox
    2007   *Zuni Origins. Toward a New Synthesis of Southwestern Archaeology*. The University of Arizona Press, Tucson.

Gutherie, Elaine, and Lorrie Lincoln-Babb
    1998   Human Remains from the Wetlands Site. In *Archaeological Investigations at the Wetlands Site, AZ AA:12:90 (ASM)*, edited by Andrea K. L. Freeman, pp. 129–145. Technical Report No. 97-5. Center for Desert Archaeology, Tucson.

Hackbarth, Mark R.
    1992   *Prehistoric and Historic Occupation of the Lower Verde Valley: The State Route 87 Verde Bridge Project*. Prepared for Arizona Department of Transportation Contract No. 89-28. Northland Research Inc., Flagstaff.

Hale, Kenneth L.
    1958   Internal Diversity in Uto-Aztecan I. *International Journal of American Linguistics* 24:101–107.

1959   Internal Diversity in Uto-Aztecan II. *International Journal of American Linguistics* 25:114–121.

Halverson, Melissa S., and Deborah A. Bolnick
2008   An Ancient DNA Test of a Founder Effect in Native American ABO Blood Group Frequencies. *American Journal of Physical Anthropology* 137:342–347.

Hard, Robert J.
1986   *Ecological Relationships Affecting the Rise of Farming Economies: A Test from the American Southwest*. Ph.D. dissertation, University of New Mexico, Albuquerque. University Microfilms, Ann Arbor.

Harrington, Richard J.
1981   Analysis of the Human Skeletal Remains from Las Colinas. In *The 1968 Excavations at Mound 8 Las Colinas Ruins Group, Phoenix, Arizona*, edited by Laurence C. Hammack and Alan P. Sullivan, pp. 251–256. Archaeological Series 154, Arizona State Museum, University of Arizona, Tucson.

Hartnady, Peter, and Jerome C. Rose
1991   Abnormal Tooth-Loss Patterns Among Archaic-Period Inhabitants of the Lower Pecos Region, Texas. In *Advances in Dental Anthropology*, edited by Mark A. Kelley and Clark S. Larsen, pp. 267–278. Wiley-Liss, New York.

Haury, Emil W.
1958   Evidence at Point of Pines for a Prehistoric Migration from Northern Arizona. In *Migrations in New World Culture History*, edited by Raymond H. Thompson, pp. 1–6. University of Arizona Bulletin Vol. 29(2). Social Science Bulletin No. 27. University of Arizona Press, Tucson.
1962   The Greater American Southwest. In *Courses Toward Urban Life: Some Archaeological Considerations of Cultural Alternatives*, edited by Robert J. Braidwood and Gordon R. Willey, pp. 106–131. Viking Fund Publications in Anthropology No. 32. Aldine, Chicago.
1976   *The Hohokam: Desert Farmers & Craftsmen. Excavation of Snaketown, 1964-1965*. University of Arizona Press, Tucson.

Heidke, James M.
1999   Cienega Phase Incipient Plainware from Southeastern Arizona. *Kiva* 64:311–338.

Hill, J. Brett, Jeffery J. Clark, William H. Doelle, and Patrick D. Lyons
2004   Prehistoric Demography in the Southwest: Migration, Coalescence, and Hohokam Population Decline. *American Antiquity* 69:689–716.

Hill, Jane H.
1999   Linguistics. *Archaeology Southwest* 13(1):8.
2000   Dating the Break-Up of Southern Uto-Aztecan. In *Avances y Balances de Lenguas Yutoaztecas, Homenaje a Wick R. Miller*, edited by Jose L. Moctezuma Zamarron and Jane H. Hill. Noroeste de México (Special Edition CD-ROM), Centro INAH Sonora, Hermosillo, Sonora.
2001   Proto-Uto-Aztecan: A Community of Cultivators in Central Mexico? *American Anthropologist* 103:913–934.

Hooton, Earnest A.
1930   *The Indians of Pecos Pueblo: A Study of Their Skeletal Remains*. Papers of the Southwest Expedition, No. 4, Phillips Academy, Andover, Massachusetts. Yale University Press, New Haven.

Huckell, Bruce B.
1990   Late Preceramic Farmer-Foragers in Southeastern Arizona: A Cultural and Ecological Consideration of the Spread of Agriculture into the Arid Southwestern

United States. Ph.D. dissertation, Arid Lands Resource Sciences, University of Arizona, Tucson.

1995   *Of Marshes and Maize: Preceramic Agricultural Settlements in the Cienega Valley, Southeastern Arizona*. University of Arizona Press, Tucson.

Huckell, Bruce B., and Lisa W. Huckell

1999   McEuen Cave. *Archaeology Southwest* 13(1):12.

Huckell, Bruce B., Lisa W. Huckell, and Suzanne K. Fish

1995   *Investigations at Milagro, A Late Preceramic Site in the Eastern Tucson Basin*. Technical Report No. 94-5. Center for Desert Archaeology, Tucson.

Huckell, Bruce B., and Barbara J. Roth

1992   Cortaro Points and the Archaic of Southern Arizona. *Kiva* 57:353–369.

Huckell, Lisa W.

2006   Ancient Maize from the American Southwest: What Does It Look Like and What Can It Tell Us? In *Histories of Maize: Multidisciplinary Approaches to the Prehistory, Linguistics, Biogeography, Domestication and Evolution of Maize*, edited by John Staller, Robert Tykot, and Bruce Benz, pp. 97–108. Elsevier, Burlington, Massachusetts.

Hunn, Eugene S., and Nancy M. Williams

1982   *Resource Managers: North American and Australian Hunter-Gatherers*. Westview Press, Boulder.

Kaestle, Frederika A., and David Glenn Smith

2001   Ancient Mitochondrial DNA Evidence for Prehistoric Population Movement: The Numic Expansion. *American Journal of Physical Anthropology* 115:1–12.

Kemp, Brian M.

2006   Mesoamerica and Southwest Prehistory, and the Entrance of Humans into the Americas: Mitochondrial DNA Evidence. Ph.D. dissertation, Department of Anthropology, University of California, Davis.

Kemp, Brian M., Angélica González-Oliver, Ripan S. Malhi, Cara Monroe, Kari Britt Schroeder, John McDonough, Gillian Rhett, Andres Resendéz, Rosenda I. Peñaloza-Espinosa, Leonor Buentello-Malo, Clara Gorodesky, and David Glenn Smith

2010   Evaluating the Farming/Language Dispersal Hypothesis with Genetic Variation Exhibited by Populations in the Southwest and Mesoamerica. *Proceedings of the National Academy of Sciences of the United States of America* 107:6759–6764.

Kohler, Timothy A., Matt Pier Glaude, Jean-Pierre Bocquet-Appel, and Brian M. Kemp

2008   The Neolithic Demographic Transition in the US Southwest. *American Antiquity* 77:645–669

Lamb, Sydney M.

1958   Linguistic Prehistory in the Great Basin. *International Journal of American Linguistics* 24:95–100.

Larsen, Clark S.

1995   Biological Changes in Human Populations with Agriculture. *Annual Review of Anthropology* 24:185–213.

1997   *Bioarchaeology: Interpreting Behavior from the Human Skeleton*. Cambridge University Press, New York.

LeBlanc, Stephen A., Lori Kreisman, Brian M. Kemp, Shawn W. Carlyle, Anne Dhody, Francis Smiley, and Thomas Benjamin

2007   Quids and Aprons: Ancient DNA from Artifacts from the American Southwest. *Journal of Field Archaeology* 32:161–175.

LeBlanc, Stephen A., Christy G. Turner II, and Michele E. Morgan

2008   Genetic Relationships Based on Discrete Dental Traits: Basketmaker II and Mimbres. *International Journal of Osteoarchaeology* 18:109–130.

Lindsay, Alexander J., Jr.

1983   Anasazi Population Movements to Southeastern Arizona. *American Archaeology* 6:190–198.

Longacre, William A.

1973   Population Dynamics at Grasshopper Pueblo. In *Demographic Anthropology Quantitative Approaches*, edited by Ezra B. W. Zubrow, pp. 169–184. School of American Research, Santa Fe, and University of New Mexico Press, Albuquerque.

Lorenz, Joseph G., and David Glenn Smith

1996   Distribution of Four Founding mtDNA Haplogroups among Native North Americans. *American Journal of Physical Anthropology* 101:307–323.

Lyons, Patrick D.

2003   *Ancestral Hopi Migrations*. Anthropological Papers of the University of Arizona No. 68. University of Arizona Press, Tucson.

Mabry, Jonathan B.

1997   *Archaeological Investigations of Early Village Sites in the Middle Santa Cruz Valley: Descriptions of the Santa Cruz Bend, Square Hearth, Stone Pipe, and Canal Sites*. Anthropological Papers No. 18. Center for Desert Archaeology, Tucson.

1998a   Archaic Complexes of the Late Holocene. In *Paleoindian and Archaic Sites in Arizona*, edited by Jonathan B. Mabry, pp. 73–87, Technical Report No. 97-7. Center for Desert Archaeology, Tucson.

1998b   Mortuary Patterns. In *Archaeological Investigations of Early Village Sites in the Middle Santa Cruz Valley: Analysis and Synthesis*, edited by Jonathan B. Mabry, pp. 697–738. Anthropological Papers No. 19. Center for Desert Archaeology, Tucson.

1999   Las Capas and Early Irrigation Farming. *Archaeology Southwest* 13(1):14.

2005   Changing Knowledge and Ideas about the First Farmers in Southeastern Arizona. In *New Perspectives on the Late Archaic Across the Borderlands*, edited by Bradley J. Vierra, pp. 41–83. University of Texas Press, Austin.

2007   *Las Capas: Early Irrigation and Sedentism in a Southwestern Floodplain*. Anthropological Papers 28, Center for Desert Archaeology, Tucson.

McClelland, John A.

2005   Bioarchaeological Analysis of Early Agricultural Period Human Skeletal Remains from Southern Arizona. In *Subsistence and Resource Use Strategies of Early Agricultural Communities in Southern Arizona*, edited by Michael W. Diehl, pp. 153–168. Anthropological Papers No. 34. Center for Desert Archaeology, Tucson.

Malhi, Ripan S., Jason A. Eshleman, Jonathan A. Greenberg, Deborah A. Weiss, Beth A. Schultz Shook, Frederika A. Kaestle, Joseph G. Lorenz, Brian M. Kemp John R. Johnson, and David Glenn Smith

2002   The Structure and Diversity Within New World Mitochondrial DNA Haplogroups: Implications for the Prehistory of North America. *American Journal of Human Genetics* 70:905–919.

Malhi, Ripan S., Holly M. Mortensen, Jason A. Eshleman, Brian M. Kemp, Joseph G. Lorenz, Frederika A. Kaestle, John R. Johnson, Clara Gorodezky, and David Glenn Smith

2003   Native American mtDNA Prehistory in the American Southwest. *American Journal of Physical Anthropology* 120:108–124.

Matson, Richard G.

1991   *The Origins of Southwestern Agriculture*. University of Arizona Press, Tucson.

Merbs, Charles F., and Robert J. Miller (editors)

1985   *Health and Disease in the Prehistoric Southwest*. Arizona State University, Tempe.

Miller, Wick R.
  1984   The Classification of Uto-Aztecan Languages Based on Lexical Evidence. *International Journal of American Linguistics* 50:1–24.
  1986   Numic Languages. In *Great Basin*, edited by William L. d'Azevedo, pp. 98–107. Handbook of North American Indians, Vol. 11, William C. Sturtevant, general editor, Smithsonian Institution, Washington, D.C.
Minturn, Penny Dufoe, and Lorrie Lincoln-Babb
  1995   Bioarchaeology of the Donaldson Site and Los Ojitos. In *Of Marshes and Maize: Preceramic Agricultural Settlements in the Cienega Valley, Southeastern Arizona,* edited by Bruce B. Huckell, pp. 106–116. University of Arizona Press, Tucson.
  2001   Appendix C. Human Osteological Remains. In *Excavations in the Santa Cruz River Floodplain: The Early Agricultural Period Component at Los Pozos*, edited by David A. Gregory. Anthropological Papers No. 21. Center for Desert Archaeology, Tucson.
Minturn, Penny Dufoe, Lorrie Lincoln-Babb, and Jonathan B. Mabry
  1998   Human Osteology. In *Archaeological Investigations of Early Village Sites in the Middle Santa Cruz Valley: Analyses and Synthesis*, edited by Jonathan B. Mabry, pp. 739–755. Anthropological Papers No. 19. Center for Desert Archaeology, Tucson.
Nichols, Johanna
  1997   The Epicentre of the Indo-European Linguistic Spread. In *Archaeology and Language I: Theoretical and Methodological Orientations*, edited by Roger Blench and Matthew Spriggs, pp. 122–148. Routledge, London.
Nichols, Michael J. P.
  1981   Old California Uto-Aztecan. In *Reports of California and Other Indian Languages*, Report I, edited by Alicelice Schlicher, Wallace L. Chafe, and Leanne Hinton, pp. 5–41. The Survey of California and Other Indian Languages, Berkeley.
Ochoa, Sarahi
  2004   La Industria Lítica de Bifaciales y Puntas de Proyectil en el Sitio de La Playa, Sonora. Unpublished Licenciatura thesis, Department of Anthropology, Universidad de las Américas, Puebla.
Ogilvie, Marsha D.
  2005   A Biological Reconstruction of Mobility Patterns in Late Archaic Populations. In *The Late Archaic Across the Borderlands: From Foraging to Farming*, edited by Bradley J. Vierra, pp. 84–112. University of Texas Press, Austin.
O'Rourke, Dennis H., Shawn W. Carlyle, and Ryan L. Parr
  1996   Ancient DNA: Methods, Progress, and Perspectives. *American Journal of Human Biology* 8:557–571.
O'Rourke, Dennis H., M. Geoffrey Hayes, and Shawn W. Carlyle
  2000   Spatial and Temporal Stability of mtDNA Haplogroup Frequencies in Native North Americans. *Human Biology* 72:15–34.
Parr, Ryan L., Shawn W. Carlyle, and Dennis H. O'Rourke
  1996   Ancient DNA Analysis of Fremont Amerindians of the Great Salt Lake Wetlands. *American Journal of Physical Anthropology* 99:507–518.
Ramenofsky, Ann
  1987   *Vectors of Death*. University of New Mexico Press, Albuquerque.
Reid, J. Jefferson, and Stephanie Whittlesey
  1997   *The Archaeology of Ancient Arizona*. University of Arizona Press, Tucson.
Renfrew, Colin
  1987   *Archaeology and Language: The Puzzle of Indo–European Origins*. Cambridge University Press, Cambridge.

Romney, Antone Kimball
   1957   The Genetic Model and Uto-Aztecan Time Perspective. *Davidson Journal of Anthropology* 3:35–41.
Roth, Barbara J.
   1989   Late Archaic Settlement and Subsistence in the Tucson Basin. Ph.D. dissertation, Department of Anthropology, University of Arizona, Tucson.
   1996   Regional Land Use in the Late Archaic of the Tucson Basin. *Early Formative Adaptations in the Southern Southwest*, edited by Barbara J. Roth, pp. 37–48. Monographs in World Archaeology No. 25. Prehistory Press, Madison.
Roth, Barbara J., and Keith Wellman
   2001   New Insights into the Early Agricultural Period in the Tucson Basin: Excavations at the Valley Farms Site (AZ AA:12:736). *Kiva* 67:59–79.
Rowley-Conwy, Peter
   2004   How the West Was Lost: A Reconsideration of Agricultural Origins in Britain, Ireland, and Southern Scandinavia. *Current Anthropology* 45(S4):83–113.
Sayles, Edwin B.
   1938   Cremations. In *Excavations at Snaketown I. Material Culture*, edited by Harold S. Gladwin, Emil W. Haury, Edwin B. Sayles, and Nora Gladwin. Medallion Papers, No. 25, Gila Pueblo, Globe, Arizona.
Schillaci, Michael A., and Christopher M. Stojanowski
   2005   Craniometric Variation and Population History of the Prehistoric Tewa. *American Journal of Physical Anthropology* 126:404–412
Schurr, Mark R., and David A. Gregory
   2002   Fluoride Dating of Faunal Materials by Ion-Selective Electrode: High Resolution Relative Dating at an Early Agricultural Period Site in the Tucson Basin. *American Antiquity* 67:281–299.
Sliva, Jane R.
   2001   Flaked Stone Artifacts. In *Excavations in the Santa Cruz River Floodplain: The Early Agricultural Period Component at Los Pozos*, edited by David A. Gregory, pp. 91–106. Anthropological Papers No. 21, Center for Desert Archaeology, Tucson.
Smith, B. Holly
   1984   Patterns of Molar Wear in Hunter-Gatherers and Agriculturalists. *American Journal of Physical Anthropology* 63:39–56.
Smith, Bruce D.
   1998   *The Emergence of Agriculture.* Scientific American Library, New York.
Stannard, David E.
   1992   *American Holocaust.* New York: Oxford University Press.
Tayles, Nancy, Kate Domett, and K. Nelson
   2000   Agriculture and Dental Caries? The Case of Rice in Prehistoric Southeast Asia. *World Archaeology* 32:68–83.
Thomas, David H. (editor)
   1989   *Columbian Consequences: Archaeological and Historical Perspectives on the Spanish Borderlands West*, Vol. 1. Smithsonian Institution, Washington, D.C.
Torroni, Antonio, Theodore G. Schurr, Margaret F. Cabel, Michael D. Brown,
James V. Neel, Merethe Larsen, David G. Smith, Carlos M. Vullo, and Douglas C. Wallace
   1993   Asian Affinities and Continental Radiation of the Four Founding Native American mtDNAs. *American Journal of Human Genetics* 53:563–590.
Turner, Christy G., II
   1979   Dental Anthropological Indications of Agriculture among the Jamon People of Central Japan. *American Journal of Physical Anthropology* 51:619–635.

1986   The First Americans: The Dental Evidence. *National Geographic Research* 2:39–46.

Van Gerven, Dennis P., and Susan G. Sheridan

1994   Life and Death at Pueblo Grande: The Demographic Context. In *The Bioeth-nography of a Classic Period Hohokam Population*, edited by Dennis P. Van Gerven and Susan G. Sheridan, pp. 5–24. The Pueblo Grande Project, Vol. 6. Soil Systems Publications in Archaeology, No. 20, Phoenix.

Villalpando, M. Elisa, and John P. Carpenter

2001   *Proyecto Salvamento Arqueológico La Playa SON:F:10:3, Tercer Informe, Temporadas 1998-1999 y 2000*. Archivo Técnico Centro INAH Sonora.

2004   *Proyecto Arqueológico La Playa, VI Informe, Informe de la Temporada 2003, Análisis de los Materiales Arqueológicos, Propuesta para la temporada 2004*. Archivo Técnico Centro INAH, Sonora.

Wallace, Henry D.

1995   Mortuary Remains. In *Archaeological Investigations at Los Morteros, a Prehistoric Settlement in the Northern Tucson Basin*, by Henry D. Wallace, pp. 721–762. Anthropological Papers No. 17. Center for Desert Archaeology, Tucson.

Waters, Michael R., and C. Vance Haynes

2001   Late Quaternary Arroyo Formation and Climate Change in the American Southwest. *Geology* 29:399–402.

Watson, James T.

2005   Cavities on the Cob: Dental Health and the Agricultural Transition in Sonora, Mexico. Unpublished Ph.D. dissertation, Department of Anthropology and Ethnic Studies, University of Nevada, Las Vegas.

2008a   Changes in Food Processing and Occlusal Dental Wear During the Early Agricultural Period in Northwest Mexico. *American Journal of Physical Anthropology* 135:92–99.

2008b   Prehistoric Dental Disease and the Dietary Shift from Cactus to Cultigens in Northwest Mexico. *International Journal of Osteoarchaeology* 18:202–212.

Watson, James T., Ethne Barnes, and Art Rohn

2006   *Demography, Disease, and Diet of the Human Skeletal Sample from La Playa*. Paper presented at 71st Annual Meeting of the Society for American Archaeology, San Juan.

Watson, James T., and Jessica Cerezo-Román

2010   *The Performative Transition of Mortuary Ritual in the Southern Southwest*. Paper presented at 75th Annual Meeting of the Society for American Archaeology, St. Louis.

Wellman, Keith D. (editor)

2000   *Farming Through the Ages: 3,400 Years of Agriculture at the Valley Farms Site in the Northern Tucson Basin*. SWCA Cultural Resource Report No. 98-226. SWCA Environmental Consultants, Tucson.

Wilcox, David R., David A. Gregory, and J. Brett Hill

2008   Zuni in the Puebloan and Southwestern Worlds. In *Zuni Origins: Toward a New Synthesis of Southwestern Archaeology*, edited by David A. Gregory and David R. Wilcox. University of Arizona Press, Tucson.

Wills, Wirt H.

1988   *Early Prehistoric Agriculture in the American Southwest*. University of Washington Press, Seattle.

1990   Cultivating Ideas: The Changing Intellectual History of the Introduction of Agriculture in the American Southwest. In *Perspectives on Southwestern Prehistory*, edited by Paul E. Minnis and Charles L. Redman, pp. 319–331. Westview Press, Boulder.

1992   Plant Cultivation and the Evolution of Risk Prone Economies in the Prehistoric American Southwest. In *Transitions to Agriculture in Prehistory*, edited by Anne B. Gebauer and T. Douglas Price, pp. 153–176. Monographs in World Archaeology No. 4. Prehistory Press, Madison.

1995   Archaic Foraging and the Beginning of Food Production in the American Southwest. In *Last Hunters-First Farmers: New Perspectives on the Prehistoric Transition to Agriculture*, edited by T. Douglas Price and Anne B. Gebauer, pp. 215–242. School of American Research Press, Santa Fe.

Wöcherl, Helga (editor)

2003   *Archaeological Investigations at the El Taller, AZ AA:12:92 (ASM), and Rillito Fan, AZ AA:12:788 (ASM), Sites along Eastbound I-10 between Sunset and Ruthrauff Roads, Tucson, Pima County, Arizona.* Technical Report No. 2003-08. Center for Desert Archaeology, Tucson.

Wöcherl, Helga, and Jeffery J. Clark

1997   The Square Hearth Site, AZ AA:12:745 (ASM). In *Archaeological Investigations of Early Village Sites in the Middle Santa Cruz Valley: Descriptions of the Santa Cruz Bend, Square Hearth, Stone Pipe, and Canal Sites*, edited by Jonathan B. Mabry, pp. 229–280. Anthropological Papers No. 18, Center for Desert Archaeology, Tucson.

Zvelebil, Marek, and Peter Rowley-Conwy

1986   Foragers and Farmers in Atlantic Europe. In *Hunters in Transition*, edited by Marek Zvelebil, pp. 67–93. Cambridge University Press, Cambridge.

7. Giants among Us? Morphological Variation and Migration on the Great Plains

Benjamin M. Auerbach

Summary Statement: Over the last millennium, the Great Plains of North America have been inhabited by people of multiple linguistic heritages and archaeological traditions. In the Central and Northern Plains regions (encompassing modern Kansas, Nebraska, and the Dakotas), for example, Siouan-speaking groups (e.g., Mandan) and Caddoan-speaking groups (e.g., Pawnee and Arikara) shared similar subsistence strategies despite their (purportedly) uncommon origins. Such discontinuities between culture and biology have occasionally led to ambiguity in interpreting affinities and origins of groups on the Plains. The Caddoan-speaking groups are one example of this uncertainty.

This study uses skeletal remains from archaeological sites located throughout the Plains, as well as from neighboring regions, in an attempt to elucidate some of these relationships while documenting morphological variation in the last millennium. Osteometric measurements are used to calculate five morphologies: stature, body mass, body breadth, and limb proportions (brachial and crural indices). These are compared among the skeletal samples, including groups possibly related or ancestral to the Pawnee and Arikara (such as members of the Itskari phase). Results provide evidence for continuity as well as migration and gene flow from multiple neighboring regions in shaping the variation of humans on the Great Plains over the last 1,000 years. Additional implications for the study of morphological variation in humans in the Americas are discussed.

Human Variation in the Americas: The Integration of Archaeology and Biological Anthropology, edited by Benjamin M. Auerbach. Center for Archaeological Investigations, Occasional Paper No. 38. © 2010 by the Board of Trustees, Southern Illinois University. All rights reserved. ISBN 978-0-88104-095-1.

Introduction

The Great Plains of North America bridge the distinct landscapes of the sub-boreal forest, Western Plateau, Great Basin, Southwest, Gulf Coast, Southeast, Prairie, and Great Lakes regions. Given this geography and temporal variation in climates (Park 1996), the cultural and, potentially, biological diversity of the prehistoric indigenous peoples reflects the heterogeneity of the Plains and the surrounding regions. Archaeologists have spent more than a century documenting the record of human existence on the Plains (Hill et al. 1996). As the resolution of these studies change with increasing data, researchers have constantly shifted taxonomies in order to better understand ethnogenesis, ecology, and intergroup dynamics (Krause 1998). Debate, then, continues about cultural history and group relationships among the occupants of the Plains (Wood 1998).

Over the last five decades, human morphological variation among the indigenous precontact groups occupying the Great Plains has contributed to the archaeological corpus (Bass 1961, 1981). Because of differences in preservation, burial techniques, and excavation history, skeletal variation research has centered on the Central and Northern Plains, regions roughly bounded between the Arkansas River and the northernmost reaches of the Missouri River in modern North Dakota and separated from each other by the Niobrara River (Wood 1998). Moreover, the majority of this research has focused on cranial size and shape variation (Byrd and Jantz 1994; Cole and Cole 1994; Jantz 1973; Key 1983, 1994; Key and Jantz 1981; McKeown 2000; Owsley et al. 1981), though some researchers have examined postcranial differences among samples from throughout the Plains (Cole 1994; Crumbley 1986; Jantz 1997; Puskarich 1984; Ruff 1994a; Wescott 2001; Wescott and Cunningham 2006; Zobeck 1983). Most of these researchers either investigated variance within and between groups or examined temporal changes within groups (e.g., cranial shape change, limb bone cross-sectional properties). The former type of studies have sought to discern genetic relationships among populations represented by skeletal samples, and the latter generally have had a goal of understanding changes relating to behavior, environmental stress, and microevolutionary processes.

Collectively, this research has indicated a dynamic population history and helped to clarify some relationships among Central and Northern Plains groups (Jantz and Owsley 1994), though ambiguity persists (Owsley 1997). Much of this uncertainty surrounds the transition from Late Woodland cultures to the Plains Village traditions (ca. 1000 B.P.) and, thus, the origin of historic nations such as the Mandan, Pawnee, and Arikara. For example, Patrick Key's cranial studies (1983, 1994) verified previous skeletal biologists' conclusions that the ancestral Arikara component of the Initial Coalescent (ca. 700 B.P.) most likely resulted from a northern migration of peoples identified with Central Plains traditions. (See the following section for a more detailed discussion of the archaeology of these regions.) Archaeology, however, does not provide clear support for this pattern of movement (Steinacher and Carlson 1998; cf. Blakeslee 1994) and may, therefore, call into question the mechanisms by which the temporal sequence in cranial shape change occurred. The archaeological evidence does suggest that significant changes took place at the transition to the Coalescent tradition, when it is apparent that village fortification and increased violence occurred (Johnson 1998; Mitchell 2007), most likely between Siouan speakers (ancestral Mandan) and Caddoan speakers (ancestral Arikara). The process by which these changes took place—by demic diffusion and displacement, a demographic shift, or a combination of these—remains unresolved (Bamforth and Nepstad-Thornberry 2007; Johnson 2003).

This paper seeks to examine patterns of morphological variation in body size, shape, and proportions among archaeological Arikara groups dating to the latest pre- and protohistoric periods (the Initial and Extended Coalescent) and compare these with groups that may relate to their ancestors. Previous studies have shown that some of these morphologies—namely limb proportions and body breadth—take many centuries to change, most likely in response to climatic factors (Auerbach 2007; Holliday 1997a, 1999; Ruff 1994b, 2002; Temple et al. 2008). If the ancestors of the Arikara were recent arrivals to the Northern Plains, then the pre- and protohistoric Arikara should demonstrate significantly different body shapes and proportions from groups that had occupied the region for longer periods. By comparing morphologies that are arguably stable over time, migration patterns may be discernible and population affinities proposed.

It should be noted that the correlation of biological variation with cultural identity, let alone with history, is problematic. Indeed, addressing their connection was a focus of the conference that gave rise to this volume (see the introductory chapter of this volume). Some authors, such as Blakeslee (1994:9–11), have argued that archaeologists and biological anthropologists trend toward oversimplification of group relationships, with a tendency (1) to draw direct connections between antecedent and subsequent cultures or phases and (2) to assume a dichotomy of either continuity *or* replacement (i.e., migration) when making these connections. Part of this results from difficulties distinguishing the trade of material culture from local production, either by migrants or endemic people. Moreover, even when relationships among groups may be drawn from craniometric analysis, relating these to linguistic or cultural heritage risks oversimplification (Blakeslee 1981), especially as language and biology, while unquestionably linked, are not synonymous (Bolnick et al. 2004; Campbell 1997; cf. Nichols and Peterson 1998).

Skeletal biologists, though, often reveal some of the complexities of past population histories through osteometric analysis. In Plains research, for example, studies have indicated that some cemeteries (but not necessarily the habitations near them) consist of groups expressing different amounts of morphological variability (e.g., Key and Jantz 1990), present notable substructure relating to within-group identity and breeding patterns (e.g., Byrd and Jantz 1994), or represent peoples of more than one ancestry (e.g., Owsley and Jantz 1978). In addition, entire regions may contain skeletal samples that did not have discernible ancestry or evidence of descendants (Key 1994). In the case of the latter study, the skeletal analyses introduced interesting implications for indeterminate archaeological evidence for Late Woodland group affiliations.

Before attempting to illuminate group relationships using the morphological data collected for this study, additional background would be useful. To that end, an overview of the patterns of population movement and the development of cultural groups as understood by archaeological investigations is presented first. Figure 7-1 provides a reference for the regions and some of the archaeological distributions discussed. Following this summary, a synopsis of the significance of body shape, size, and proportions among humans is presented.

Archaeological Context: From Plains Woodland to Plains Village

The history of Great Plains indigenous population movements and affinities continues to undergo debate and revision (Krause 1998). There is insufficient space here to do justice to the multiple sources of evidence, analyses, and interpretations presented by researchers. Instead, this section provides a sketch of archaeological evidence leading up to the development of the Coalescent tradition, the focus of this paper. This summary includes a discussion of ideas about contributory cultures from the Central and Northern Plains. Readers who wish to obtain a more detailed accounting of Great Plains archaeology should read publications by Lehmer (1971), Grange (1979), Jantz and Owsley (1994), Frison and Mainfort (1996), Hofman (ed. 1996), Wood (ed. 1998), DeMallie (2001), and Ahler and Kay (2007), and the works contained therein.

There is abundant evidence that the Great Plains have been occupied for an extensive period, dating back at least 11,500 years B.P. (Green et al. 1998; Hofman and Graham 1998), if not before (Hofman 1996). The majority of human existence on the Plains consisted of a broad-spectrum foraging subsistence with widely dispersed, temporary habitations prior to 2500 B.P. (Hofman and Graham 1998). It is important to note that the dynamic Plains environment throughout this 9,000-year period (Johnson and Park 1996) produced varied cultures and multiple subsistence strategies (Frison 1998; Kay 1998); the Paleo-Indian and Archaic periods were not uniform. Skeletal remains of humans from this extensive period are rare, with few exceptions (e.g., Carlson et al. 1999; Millar 1978), and their study is beyond the scope of this paper's analyses.

In the eastern margins of the Plains, changes in settlement patterns toward more permanent occupations of riparian environments, a shift in broad-spectrum subsistence strategies that gave rise to the growing of cultivars, and population increases marked the beginnings of the Plains Woodland cultures approximately 2500 years B.P. (Adair 1996; Johnson 2001; Johnson and Johnson 1998). This transition was gradual and did not occur universally or simultaneously across the Plains (Adair 1996; Perttula and McGuff 1985). Hunting and gathering remained important aspects of the Plains Woodland as well, including long-range procurement of bison (Johnson and Johnson 1998). Debate about the origins and diffusion of the Woodland settlement and subsistence patterns continues, namely in relation to the influence of Hopewellian culture (Johnson 2001). The Hopewell culture predominated the Prairie to the east of the Mississippi River at the beginning of the Early Plains Woodland, but at least one phase (Kansas City) also occurred within a portion the lower Missouri River valley. These phases may be

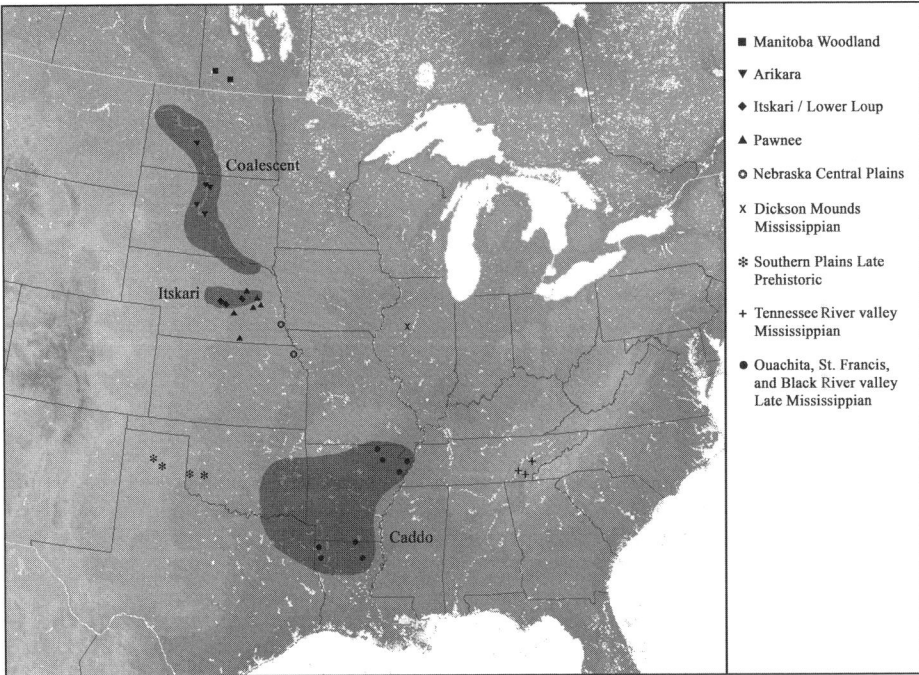

Figure 7-1. *Sites and some archaeological tradition regions sampled. See Table 7-1 for site names, affiliations, and antiquity information.*

associated with the rise of Woodland cultures in the Plains (Johnson and Johnson 1998). Regardless of ultimate influences, the spread of changes associated with the Woodland focus follow a general east-and-south to west-and-north temporal pattern across the Plains and, in conjunction with the Kansas City phase, have been interpreted to reflect an exchange of ideas and either gene flow or population replacement with peoples to the east (Blakeslee 1994).

The developments of the Early Woodland (ca. 2500–2000 B.P.) intensified and proliferated throughout the Central and Northern Plains during the Middle Woodland (ca. 2000–1500 B.P.) and into the Late Woodland (ca. 1500–1000 B.P.). This cultural shift is associated with larger, semipermanent villages where inhabitants increasingly domesticated local plants and began a limited incorporation of exotics, namely maize and squash (Adair 1996), while continuing to hunt both smaller game and large ungulates. Most of the relationships among sites have focused on pottery, which in turn has been used to define broad cultural traditions and phases within these regions.[1]

There is emerging evidence that during the Middle Plains Woodland, populations related to ancestors of the Caddo migrated from the region between the Arkansas and Red rivers and into the Central Plains. The relationship of any Woodland Plains groups and ancestral Caddoans is proposed to explain the shared language family found among historic Caddoans, Wichita, Pawnee, Arikara, and others; these groups exhibit linguistic distinctions that follow a geographic cline and temporal pattern of differentiation (Goddard 2001; Parks 1979). Linguistic analyses, for instance, argue for at least 2,000 years of divergence between Wichita dialects (of the Southern Plains) and the languages of the Pawnee and Arikara (Parks 2001). That Caddoan-speaking peoples migrated onto the Plains is not under debate; a general agreement exists that Caddoan languages exhibit a south-to-north history of divergence most parsimoniously explained by the movement of groups of the language family's speakers, initially from the Arkansas River valley. Matching the separation of these languages to archaeological evidence, however, awaits more analyses. This is further complicated by problems in understanding continuity versus replacement of peoples between the Woodland and subsequent cultures, specifically the Central Plains tradition and Coalescent tradition (and even between these). For example, Schlesier (1994) makes an argument that some of the variants of the Middle Woodland in modern northeastern Nebraska—namely the Valley variant—show technological similarities to earlier archaeological Woodland sites from the Arkansas River valley, though his argument is not proffered by other researchers (e.g., Johnson and Johnson 1998). It is also uncertain if these similarities indicate trade, cultural borrowing, or colonization events and, if the latter, whether peoples maintained breeding isolation or intermingled with other Plains Woodland groups.

Available data support the presence of Caddoan-speakers in at least the Southern Plains when two broad cultural horizons emerged in the Central and Northern Plains around 1200–1000 B.P., demarcating the end of the Woodland; these were the Central Plains and the Middle Missouri traditions. The shift to these traditions is marked by an intensification of the subsistence and habitation trends that developed throughout the Woodland (Wedel 2001; Wood 2001): Populations settled into large seasonally occupied villages wherein horticulture dependence grew alongside continued use of wild plants (Johnson 1998). Seasonal hunting expeditions for bison continued, but this more sedentary lifestyle, termed the Plains Village culture, became predominant with few exceptions (e.g., the Pomona variant; Roper 2007). The Plains Village culture would persist in the Central and Northern Plains through contact with Europeans.

Linking Late Woodland variants with either the Central Plains or Middle Missouri traditions continues to be under intense scrutiny and discussion (Roper 2007; Tiffany 2007). At the center of the debate is whether the changes in subsistence and village structure, neither of which was monolithic throughout the Central or Northern Plains, were the result of endemic processes, cultural borrowing, or colonization with or without replacement. This, in turn, has a significant impact on whether the Arikara and Pawnee—ostensibly descendants of some of the Central Plains traditions—were recent (i.e., post-Woodland) migrants to the Central and Northern Plains. Different levels of archaeological visibility offered by mobile foragers versus more sedentary horticulturalist-hunters, as well as difficulties discerning genetic from trade relationships (if indeed these are exclusive), contribute to the ambiguity of these connections (Johnson 1992). Like the emergence of the Woodland culture in the Central Plains region, the development of the Central Plains (Plains Village) tradition appears to have occurred in a mosaic pattern, with some signs of gradual transformation and continuity between Woodland and Central Plains variants (Roper 2007) contrasted with other areas that exhibit "leapfrog" migrations, occupation gaps, or sudden changes (Johnson 2001; Johnson 2007; Krause 1995). Middle Missouri traditions, contrastingly, present continuity with Late Woodland phases (i.e., Great Oasis), and the shift to Plains Village lifeways likely resulted from acculturation of new subsistence patterns from neighbors to the southeast, in what is now modern Minnesota (Winham and Calabrese 1998). The southeastern groups may have influenced endemic peoples and then migrated to the Missouri River valley (Tiffany 2007). These population movements, in turn, have been argued as evidence for the migration of Siouan-speaking ancestors to the Hidatsa and Mandan into the Northern Plains (Toom 1996; Wood 1967), though the relationships of Middle Missouri phases to these historic nations is complex and remains unresolved (Johnson 2003; Winham and Calabrese 1998). In fact, some studies have suggested a direct Mississippian influence on the formation of the Initial Middle Missouri tradition (Lehmer 1954; Tiffany 2007), citing the Steed-Kisker phase of the Central Plains tradition as evidence supporting this connection. This is tenuous, as researchers have not reached a consensus about the relationships between central Mississippians and either far northern Mississippi River or central Missouri River valley groups (Henning 2005).

Between 700 and 600 B.P., aspects of the Central Plains tradition transitioned into the Initial Coalescent variant (Krause 2001), an outgrowth of the general Plains Village subsistence. Steinacher and Carlson (1998) cite this change as emerging from two of the Central Plains variants (defined by pottery characteristics [Logan 1996a])—the Lower Loup and the St. Helena phases—an argument partially supported by craniometric comparisons (Key 1983; cf. Owsley 1997). However, other studies have suggested that St. Helena is a *derivative* of the Initial Coalescent (Ludwickson et al. 1993) and that alternative phases of the Central Plains tradition gave rise to the Coalescent (Johnson 1998). Regardless of which phases directly contributed, current consensus argues that unfavorable environmental conditions in the Loup River valley and surroundings caused Central Plains tradition populations to migrate north into the Missouri River valley of modern South Dakota, which displaced the Middle Missouri tradition peoples already living in the region (Johnson 1998). Archaeological remnants of habitations demonstrate that many of the Middle Missouri River valley villages that formed during this period were fortified with moats and palisades (Blakeslee 1994). This phenomenon was possibly coupled with violent interactions among the Middle Missouri groups and with the displacement of Initial Coalescent peoples (Krause 2001; Willey and Emerson 1993; cf. Zimmerman 1985). The development of fortified villages could also relate to drastic changes occurring in the Mississippian cultural sphere at this time (Pauketat 2004), but the evidence is unsubstantiated.

Whether groups that are historically identified as Mandan, Hidatsa, Arikara, or Pawnee existed as distinct populations prior to the last 700 years is ambiguous. There is archaeological evidence that links the Terminal Middle Missouri tradition (ca. 700 B.P.) with proto-Mandan cultures (Wood 2001); it is possible that population pressures arising from movements of the Central Plains peoples at the Initial Coalescent led to the shift of the Extended Middle Missouri tradition into ancestral Mandan culture. It is not clear if these Initial Coalescent peoples were ancestors to both the Arikara and the Pawnee or to the Arikara alone; debate continues about the relationship of the Lower Loup (also known as the Itskari) phase of the Central Plains tradition to these two groups, and strong arguments have linked this group to the Pawnee (Grange 1979; Jantz 1977; Key 1983; Logan 1996b; Schlesier 1994) or cited an ambiguous relationship or no direct relationship (Johnson 1998; Roper 1993; Steinacher and Carlson 1998). Linguistic evidence supports a recent divergence between the Pawnee and the Arikara and also clearly distinguishes the Siouan-family speakers (Mandan and Hidatsa) from these Caddoan-family speakers (Goddard 2001; Parks 2001). This, coupled with archaeological evidence, suggests that the Arikara did not arise from a combination of peoples of the Central Plains and Terminal Middle Missouri traditions but instead developed directly from peoples practicing Central Plains–Initial Coalescent traditions. In turn, Initial Coalescent peoples either were related to but separate from the Lower Loup/Itskari peoples—who in time became the Pawnee—or the ancestral Pawnee separated from the ancestral Arikara to give rise to or replace the Lower Loup phase. Even without linguistic evidence, gene flow from the Siouan speakers, however, cannot be excluded from any of these groups (Owsley 1997), especially after European contact (Jantz 1973; Key 1983; Key and Jantz 1981; Key 1983; McKeown 2000; Owsley et al. 1981).

By 500 B.P., these historic nations were established and variants of the Plains Village lifeway extended throughout the Missouri River valley, shared among the Mandan, Hidatsa, Arikara, and Pawnee (Lehmer 1971). The Initial Coalescent evolved into the Extended and then Postcontact Coalescent; during the latter cultural horizon, European trade items became common at both Arikara and Mandan sites. European contact influenced the movement of the Sioux into the Northern Plains, displacing the Mandan and Arikara, populations that were also experiencing demographic decimation from European diseases (Owsley and Rose 1997). It is incidentally interesting that, despite these stresses, all of these Northern and Central Plains groups historically exhibited tall statures, which is likely convergence related to their shared lifeways, though, again, genetic relatedness cannot be barred as a contributory factor (Steckel and Prince 2001).

In summary, then, some ambiguity persists about the relationships of prehistoric peoples in the Central and Northern Plains with the historical Caddoan speakers of these regions. There is competing evidence for multiple sources of the Plains Woodland peoples, including in situ changes to late Archaic populations, Caddoan-speaking groups migrating from the Arkansas River and Red River valleys, and Hopewellian populations that colonized the lower Missouri River valley (near modern-day Kansas City). In fact, it is likely that all of these populations moved through the Plains and contributed to the Woodland peoples, who in turn helped give rise to the Plains Village traditions.[2] When the Caddoan speakers arrived in the Central Plains is not certain; they may have been part of the discontinuity observed between the Late Woodland and initial Central Plains traditions. There is no question that Caddoan groups contributed to the transition between the Central Plains traditions and the Initial Coalescent. Whether the Itskari phase peoples were direct ancestors of the Pawnee or the Arikara still is unresolved, though skeletal and linguistic evidence make a strong case for a direct relationship with the Pawnee, at least with those of the Skiri-speaking dialect (Key 1983; Schlesier 1994). Finally, craniometric evidence also argues for genetic isolation between the Siouan speakers and the Arikara until population decimations around the time of European contact forced these groups together.

These conclusions, then, yield specific questions that this paper seeks to address. Are the Arikara as varied in postcranial morphology as they are in cranial morphology, and does this follow a temporal trend (as found by Jantz [1973] and by Key [1983])? Is it possible to better resolve the relationship of the Itskari with the Pawnee and with the Arikara? Given the likely recent separation of the Arikara and Pawnee, this may not be discernible. Have the Pawnee and Arikara maintained the morphologies of the Caddo, or has sufficient time passed for these groups to significantly differ? In addition, are there morphological similarities among the Arikara and neighboring Woodland and Mississippian groups? Ideally, the similarities and differences of these groups with the Siouan speakers of the Northern Plains would also be assessed, but, unfortunately, no skeletal samples for the Mandan or Hidatsa were available for this study. Comparisons of body shape, size, and proportions reconstructed from skeletal remains, however, may be used to assess variation among the Arikara, Pawnee, Itskari, Mississippians, and some Late Woodland peoples of the Plains.

Body Shape, Size, and Proportions

Researchers have demonstrated that aspects of human morphological variation covary with environmental factors and so may be applied in models of human population ecology, affinities, and movement (see King, this volume). The influence of climate and subsistence are two such sets of environmental variables. Most observations of human variation in relation to these factors fit broader models that correspond body shape, size, and proportions to the efficient maintenance of body temperature (Holliday 1997a; Mayr 1956; Ruff 1994b; Trinkaus 1981), to the effects of subsistence and stress on growth and development (Danforth 1999a, 1999b; Johnston et al. 1976; Malina et al. 2004), and to secular change in these factors (Genoves 1966; Jantz and Jantz 1999; Katzmarzyk and Leonard 1998), among other influences (e.g., Alexander 1984). In this study, analyses will focus on four morphologies: stature, body mass, bi-iliac (pelvic) breadth, and intralimb proportions (brachial and crural indices).

Climate and subsistence have independent and combinatory effects on various aspects of morphology (Auerbach 2007). Humans, like all homeothermic organisms, need to maintain core body temperatures efficiently. The proportion of surface area to body mass helps regulate heat dissipation (Beals 1972; Newman 1956; Roberts 1953), among other factors (Schreider 1964). This principle follows from the ecogeographic patterns observed in body mass among species with wide geographic ranges, where larger (more massive) individuals generally occur at higher, and therefore colder, latitudes (Bergmann 1847; Mayr 1956; Roberts 1978). Modeling the human torso as a cylinder, Ruff (1991, 1994b) demonstrated that body breadth (bi-iliac breadth) determines surface area-to-mass ratios; individuals with different statures but identical body breadths mathematically have the same ratio. Subsequent analyses of past groups from Europe and Africa have indicated that variation in bi-iliac breadth among humans over wide geographic areas follows a pattern matching predictions based on the cylindrical model (Holliday 1997a; Ruff 2002; Ruff et al. 1997). Further research has correlated this variation with climate, namely temperature (Auerbach 2007; King, this volume), though additional factors, including obstetrics and energetics, attenuate variation in pelvic breadth (Kurki 2007; Ruff 1991; Ruff et al. 1997; Tague 2005; Weaver and Steudel-Numbers 2005). Statures, under the cylindrical model, would not be expected to covary with climate, yet some studies indicate a relationship between stature and humidity or precipitation (Ruff 1994b; Stinson 1990), though it is likely that this is due to a third common factor, namely diet.[3] Indeed, the relationship between stature and subsistence forms the basis for an extensive and ongoing body of research (Bogin 2002; Bogin and Keep 1999; Walker et al. 2006). The relationships among mass, stature, and these environmental factors, however, are not independent of genetic variation (Little and Malina 1986) and are complicated by biological variation in humans arising from allometry (Holliday and Ruff 2001; Jantz and Jantz 1999; Sylvester et al. 2008). In fact, these studies have shown that taller individuals tend to have disproportionately longer legs (tibiae) relative to thighs (femora); this trend complicates some interpretations of variation in crural indices (tibial length relative to femoral length), which, along with brachial indices (forearm length relative to arm length), have been shown to strongly covary with climatic variables (Holliday 1997b; Ruff 1991; Trinkaus 1981).

Despite the complexity of correlating these morphologies with specific environmental factors, comparisons of body shape, size, and proportions among past human groups have been used to assess population movement and affinities. For example, studies of Pleistocene humans in Europe (Holliday 1995, 1997a, 1997b, 1999; Ruff 2002; Trinkaus 1981) has shown that the first modern humans to occupy Europe had significantly different shapes and proportions from Neanderthals, which generally possessed more "cold adapted" morphologies. The general pattern that has emerged is that intralimb proportions take many generations to change in relation to climatic expectations, though total limb lengths shortened more rapidly over time in these comparisons. It should be noted that Holliday (1997b) argued that, because there is considerable range in intralimb indices within prehistoric groups, the use of brachial and crural indices may be imprecise indicators of migration; he instead favored the relative length of the torso to the lower limbs, as this had less variance within his samples, though this drastically reduces the number of archaeological skeletons that can be used in analyses. Even so, other studies have been able to discern group migration based on limb proportions (Temple et al. 2008; cf. Kurki et al. 2008). Body breadth may be the most conservative of any of these morphologies in the rate of change over time as tropically proportioned humans colonized Europe (Ruff 1994b), a finding reflected when examining the "cold-filtered" wide bodies common to colonizers of the Americas (Auerbach 2009). In contrast, multiple studies have demonstrated that stature is plastic and changes rapidly between generations under different nutritional and stress levels (Malina et al. 1983; Takamura et al. 1988).

A basic assumption of this study, then, is that, with the exception of stature, these morphological dimensions would not have had enough time to undergo significant change among the groups studied. Based on this tenet and the questions posed at the end of the previous section, the following hypotheses are proposed:

1. The Arikara and Pawnee are descendants of recent migrants to the Northern and Central Plains and will exhibit higher intralimb indices and narrower body breadths than other groups occupying these (higher latitude) regions.

2. The body proportions, shape, and size of the Arikara and Pawnee will not significantly differ from the Caddoan-speaking Mississippians (St. Francis, Black, and Ouachita river valleys), as these groups share a common ancestral population and insufficient time has elapsed for changes in their morphologies.

3. The Pawnee will not significantly differ from the Itskari peoples in body morphology, as these groups demonstrate evidence for shared ancestry.

4. There will be significant differences in morphology between the Arikara, Pawnee, and Itskari groups and both the Late Prehistoric Southern Plains and Tennessee River watershed Mississippians. As stature is under both genetic and environmental influence, it is not expected to show any pattern among these groups, whereas body mass, body breadth, and limb proportions should differ, as these have been indicated as changing less rapidly as revealed in previous studies of population movement.

Materials and Methods

A total of 522 adult human skeletons (283 males, 239 females) were measured and used in this study's analyses. These represent 33 sites (or site aggregates) from six regions, and span the last millennium; sites are listed in Table 7-1 with temporal, geographic, and affinity information. Figure 7-1 provides the geographic distribution of these sites. Many archaeological Plains skeleton samples are now unavailable or have restricted access, and Dr. Daniel Wescott very generously shared his data for some of these. Sites sampled by him are so designated in Table 7-1. The author measured skeletons from all other sites. Interobserver errors between the author and Wescott, based on comparisons of Arikara skeletons independently measured by both, are all below 1 percent (Auerbach 2007).

Measurements obtained for this study are listed in Table 7-2, and the morphological dimensions calculated from them are provided in Table 7-3. Metric measures were only taken from adults, as determined by complete epiphyseal closure of limb bones and vertebrae. The sexes of skeletons were assigned using characteristics of the pelvis and of the cranium (Bruzek 2002; Krogman and Işcan 1986; Phenice 1969). Though the author took more measurements by which to reconstruct additional morphological dimensions (e.g., length of the torso relative to the lower limbs), these measurements were not available for a significant percentage of the skeletons, and Wescott's data set does not contain them. For this reason, the dimensions have been selected to maximize the available sample. Bi-iliac breadth is the only exception to this restriction; this dimension was not

Table 7-1. *Sites Sampled*

Site[a]	n (♂/♀)	Region	Affiliation/archaeological taxon[b]	Average date (B.P.)[c]	Location[d]	Source[e]
Snowflake	10 (8/2)	Northern Plains	Middle Woodland	1000	Manitoba	CMC
Antler Plain/ Souris River	15 (11/4)	Northern Plains	Late Woodland	800	Manitoba	CMC
Dickson Mounds	53 (26/27)	Prairie	Middle Mississippian	650	Illinois	ISM
St. Francis & Black River valley burial mounds[f]	28 (14/14)	Southeast	Caddoan/Late Mississippian	500	Arkansas	NMNH
Ouachita River valley burial mounds[f]	31 (15/16)	Southeast	Caddoan/Late Mississippian	350	Louisiana	NMNH
Hiwassee Island	40 (20/20)	Southeast	Middle Mississippian	500	Tennessee	FHMM
Ledford Island	41 (19/22)	Southeast	Late Mississippian	400	Tennessee	FHMM
Toqua	37 (18/19)	Southeast	Late Mississippian	300	Tennessee	FHMM
Larson	32 (16/16)	Northern Plains	Arikara/Postcontact Coalescent	250	South Dakota	UTKDA
Cheyenne River	26 (15/11)	Northern Plains	Arikara/Postcontact Coalescent	200	South Dakota	UTKDA
Mobridge A	41 (25/16)	Northern Plains	Arikara?/Extended Coalescent	300	South Dakota	NMNH
Sully A & D	20 (12/8)	Northern Plains	Arikara/Extended Coalescent	300	South Dakota	NMNH

Table 7-1.—*Continued*

Site[a]	n (♂/♀)	Region	Affiliation/archaeological taxon[b]	Average date (B.P.)[c]	Location[d]	Source[e]
Greenshield	11 (6/5)	Northern Plains	Arikara/Historic	140	North Dakota	Wescott 2001
Dinsmore Mound[*]	7 (2/5)	Central Plains	Nebraska variant of Central Plains	800?	Kansas	Wescott 2001
Kelly Ossuary[*]	7 (5/2)	Central Plains	Nebraska variant of Central Plains	725	Nebraska	Wescott 2001
O'Hanlon[*]	5 (4/1)	Central Plains	Nebraska variant of Central Plains	775	Nebraska	Wescott 2001
Linwood[‡]	9 (8/1)	Central Plains	Pawnee/Coalescent	150	Nebraska	Wescott 2001
Barcal[‡]	8 (3/5)	Central Plains	Pawnee/Coalescent	200	Nebraska	Wescott 2001
25MK14[‡]	2 (1/1)	Central Plains	Pawnee/Coalescent	130	Nebraska	Wescott 2001
Burkett[‡]	6 (3/3)	Central Plains	Pawnee/Coalescent	200	Nebraska	Wescott 2001
Wright[‡]	6 (4/2)	Central Plains	Pawnee/Coalescent	180	Nebraska	Wescott 2001
Genoa[‡]	13 (5/8)	Central Plains	Pawnee/Coalescent	100	Nebraska	Wescott 2001
Clarks[‡]	10 (8/2)	Central Plains	Pawnee/Coalescent	120	Nebraska	Wescott 2001
Pike Village[‡]	7 (7/0)	Central Plains	Pawnee/Coalescent	150	Nebraska	Wescott 2001
Sondergaard[**]	8 (2/6)	Central Plains	Itskari/Lower Loup	600	Nebraska	Wescott 2001
Wozney[**]	2 (2/0)	Central Plains	Itskari/Lower Loup	700	Nebraska	Wescott 2001

Christensen[**]	7 (5/2)	Central Plains	Itskari/Lower Loup	600	Nebraska	Wescott 2001
Sandstone Creek[†]	12 (8/4)	Southern Plains	Late Prehistoric	700?	Oklahoma	Wescott 2001
McLemore[†]	16 (4/12)	Southern Plains	Late Prehistoric	650	Oklahoma	Wescott 2001
Antelope Creek[†]	5 (5/0)	Southern Plains	Late Prehistoric	600	Texas	Wescott 2001
Courson B[†]	4 (1/3)	Southern Plains	Late Prehistoric	700	Texas	Wescott 2001
Alibates Ruin[†]	3 (1/2)	Southern Plains	Late Prehistoric	600	Texas	Wescott 2001

[*]Sites combined into "Nebraska" sample.
[‡]Sites combined into "Pawnee" sample.

[**]Sites combined into "Itskari" sample.
[†]Sites combined into "Southern Plains Late Prehistoric" sample.

[a]Some sites have been aggregated when they are temporally and geographically proximate and represent the same cultural tradition.
[b]Note that these are general traditions and variants based on documentation available with sites and may not reflect the most specific taxonomic level determined from the archaeological evidence.
[c]Temporal depths provide the average date for sites based on available dating techniques presented in published literature.
[d]This represents the modern state or province in which sites were located. Although more specific locations were documented and are available in the data set from which these are taken (Auerbach 2007), specific longitudes and latitudes were not used in this analysis. Locations, therefore, are provided only for geographic reference.
[e]Measurements shared by Dr. Daniel J. Wescott are designated as "Wescott 2001," and the locations of their curation at the time of measurement may be obtained from his dissertation (Wescott 2001). If I measured the skeletons from a site, the collection holding the remains at the time of measurement is designated. Collection abbreviations: **CMC**: Canadian Museum of Civilisation, Gatineau, Quebec; **FHMM**: Frank H. McClung Museum, Knoxville, Tennessee; **ISM**: Illinois State Museum, Springfield, Illinois; **NMNH**: National Museum of Natural History (Smithsonian), Washington, D.C.; **UTKDA**: University of Tennessee Department of Anthropology, Knoxville, Tennessee.
[f]See appendix 1 in Auerbach (2007) for detailed descriptions of the sites combined to yield these Manitoba Woodland aggregate samples.

Table 7-2. *Osteometric Measurements Used in Analyses*

Element[a]	Measurement	Abbreviation	Reference
Humerus	Maximum length	HML	Martin (1928), Humerus #1
Radius	Maximum length	RML	Martin (1928), Radius #1
Femur	Bicondylar length	FBL	Martin (1928), Femur #2
	Femoral head anteroposterior diameter	FH	Martin (1928), Femur #19
Tibia	Maximum length	TML	Martin (1928), Tibia #2
Os coxae	Bi-iliac breadth	BIB	Martin (1928), Pelvis #2

Note: See Auerbach (2007) for measurement errors and detailed descriptions of measurement techniques.
[a]All limb measurements are taken bilaterally when possible and averaged to minimize the effects of directional bilateral asymmetry (Auerbach and Ruff 2006).

Table 7-3. *Morphologies Compared in Analyses*

Dimension	Calculation[a]	Reference
Stature (non-Plains samples)	Males: $.160 \times FBL + .126 \times TML + 47.11$ Females: $.176 \times FBL + .117 \times TML + 41.75$	Auerbach and Ruff (2010)
Stature (Plains samples)	Males: $.188 \times FBL + .076 \times TML + 54.13$ Females: $.168 \times FBL + .104 \times TML + 50.55$	Auerbach and Ruff (2010)
Body mass	$2.268 \times FH - 36.5$	Grine et al. (1995)
Brachial index	$RML \div HML \times 100$	Holliday (1999)
Crural index	$TML \div FBL \times 100$	Davenport (1933)
Body breadth	BIB	Ruff (1991)

[a]All measurements are in millimeters.

obtained by Wescott, and so only a portion of the skeletons measured by the author are available for comparison in this dimension. However, it should be noted that in the case of two samples not significantly differing in stature, significant differences in body mass are indicative of significant differences in body breadth (Auerbach and Ruff 2004; Ruff et al. 1997; Ruff et al. 2005).

The methods for estimating stature and body mass are based on previous work by the author. Stature may be reconstructed using either "anatomical" or "mathematical" methods, as outlined elsewhere (Lundy 1985; Raxter et al. 2006). As most of the elements necessary for reconstructing statures using the anatomical method were not available in the skeletal samples, mathematical methods were used in this paper's analyses. Using anatomical stature reconstructions, Auerbach and Ruff (2010) devised new mathematical stature estimation equations specific to archaeological groups in North America. These were developed to be regionally specific, based on lower limb proportions, and were demonstrated to be more precise and accurate than other available formulas. Using the guidelines of Auerbach and Ruff (2010), the "Temperate" and "Great Plains" formulas were applied to the skeletal samples used in this study. Similar to stature estimation techniques, two methods exist for body mass estimation: "mechanical" and "morphometric" (Auerbach and Ruff 2004). The practical application of these methods was investigated at length by Auerbach and Ruff (2004). As mechanical methods, specifically the measurement of the femoral head and subsequent estimation of body mass, maximize the available skeletal sample, that method was selected for this analysis. Comparisons made elsewhere (Auerbach 2007, 2009) between the femoral head body mass estimations and stature/bi-iliac breadth (morphometric) body mass estimations showed the femoral head estimation equations created by Grine and associates (1995) produced the closest correspondence—with the least systematic bias—for archaeological skeletal samples from the Americas. For this reason, the Grine and associates (1995) equations were used instead of an average of all femoral head equations (as prescribed by Auerbach and Ruff 2004).

Comparisons of morphologies among groups were conducted to assess the hypotheses outlined above. As they are ratio data, and therefore cannot be used reliably in parametric analyses depending on variance (Albrecht 1978; Atchley et al. 1976; cf. Hills 1978), especially without arcsine transformation (Sokal and Rohlf 1995), the intralimb (brachial and crural) indices were compared among groups using geometric mean-scale length measurements of the four component bones. This follows the methods developed by Darroch and Mosimann (1985), in which the long bone length data are scaled for isometric size effects (Albrecht et al. 1995). The scaled data were then subjected to a principal components analysis to demonstrate relative group differences;[4] Temple and colleagues (2008) recently conducted a similar analysis on limb elements, following after Holliday (1999). Multivariate analyses of variance (MANOVAs) were used to compare stature, mass, and body breadth (separately, as fewer samples present this dimension) across time and among sites. The Pawnee, Itskari phase, Nebraska phase, and Late Prehistoric Southern Plains sites were combined into these four groups for all analyses, as designated in Table 7-1. This was done because the sample sizes for individual sites were not large enough for adequate statistical power, even though this method obscures variation among the samples within each group. Sites were also combined to provide more comparable sample sizes among these groups and the Arikara and Mississippian samples. Furthermore, as the goal of this study is to examine broad group differences, intercemetery differences, while of interest, are beyond the scope of analyses (with the notable exception of temporal differences among Arikara sites).

All analyses were conducted using Microsoft Excel 2008 and Stata 10.1 for Macintosh.

Results

Descriptive statistics for the five morphologies under examination are presented in Table 7-4. Mean statures calculated for the Arikara are very similar to those reported by Prince (1998) based on the Boas anthropometric data, so the stature estimates are deemed reasonably accurate; no anthropometric measurements are available to corroborate the body mass estimations. It is noteworthy that the higher statures, larger body masses, and wider body breadths tend to be found in the Caddoan-speaking groups for both sexes. Across the entire sample (not shown in Table 7-4), males tend to have greater ranges and variance than females in body mass (male interquartile range = 7.05 kg; female interquartile range = 5.98 kg) and stature (male interquartile range = 7.77 cm; female interquartile range = 6.44 cm). Females, however, have greater variance and ranges in bi-iliac breadth (male interquartile range = 19.38 mm; female interquartile range = 21.00 mm). There is no difference between males and females in ranges of brachial or crural indices.

Overall *t*-tests indicate that males and females are significantly different in stature, body mass, and bi-iliac breadth ($p < .01$). A comparison of brachial and crural indices using Mann-Whitney *U*-tests likewise shows significant differences between the sexes in brachial indices ($p < .05$) but not in crural indices ($p = .65$). Owing to the significant overall sexual dimorphism in these morphologies, all statistical analyses are conducted by sex unless otherwise indicated.

Percent sexual dimorphism, calculated within groups using mean dimensions and reported in Table 7-4, is not similar among groups. Plains groups, with the exception of Ledford Island, have a higher percent sexual dimorphism than Mississippian groups. Greater sexual dimorphism in body breadth occurs at higher latitudes, with the exception of the Souris River sample. This pattern, unexpectedly, does not occur among body masses, though both the highest (18.49 percent for Southern Plains Late Prehistoric) and lowest (Cheyenne River at 12.22 percent) sexual dimorphism in body mass may be found in the Great Plains.

As noted in the introduction, previous studies have demonstrated temporal changes in cranial shape among the protohistoric and historic Arikara (Jantz 1977; Key 1983). A one-way analysis of variance (ANOVA) comparing stature, body mass, and body breadth among the Coalescent Arikara sample reveals no significant differences in any of these dimensions among males or among females. Box plots (Figure 7-2) show the differences among these groups in these three dimensions. Despite nonsignificant differences among the sites, there are some temporal trends: both stature and body mass demonstrate decreasing values among males. Both the box plots and an examination of the percent sexual dimorphisms reported in Table 7-4 also indicate that sexual dimorphism in these dimensions decreased from the Extended Coalescent into the Postcontact Coalescent. If Greenshield, a historic (though smaller) Arikara sample from North Dakota is also examined, the trend appears to continue in mean male stature but not for body mass. Given this temporal trend, but lack of significant differences among sites, the Extended Coalescent Arikara (Mobridge and Sully) and

Table 7-4. *Descriptive Statistics for Morphological Dimensions by Site and Sex*

Morphology	Group/Site[a]	Sex (n)	Mean	Std. Dev.	% Sexual Dimorphism[b]
Stature (cm)	Snowflake	M (4)	167.92	4.74	4.45
		F (2)	160.45	2.10	
	Antler Plain/Souris River	M (9)	170.07	6.13	8.02
		F (4)	156.43	3.08	
	Dickson Mounds	M (25)	168.96	5.14	6.01
		F (26)	158.80	5.39	
	St. Francis & Black Rivers	M (13)	169.74	5.82	6.82
		F (14)	158.17	2.86	
	Ouachita River	M (11)	165.72	4.45	5.42
		F (9)	156.73	6.65	
	Hiwassee Island	M (19)	165.11	7.08	5.80
		F (18)	155.53	4.77	
	Ledford Island	M (18)	166.18	4.38	7.40
		F (21)	153.88	3.28	
	Toqua	M (18)	163.83	5.06	4.71
		F (18)	156.12	6.33	
	Larson	M (16)	166.23	4.46	6.40
		F (16)	155.59	5.29	
	Cheyenne River	M (15)	166.68	3.66	6.26
		F (11)	156.25	4.65	
	Mobridge A	M (22)	168.30	4.66	8.21
		F (16)	154.49	4.37	
	Sully A & D	M (12)	167.07	3.70	8.43
		F (7)	152.99	1.61	
	Greenshield	M (6)	165.61	5.40	7.92
		F (4)	152.50	.67	
	Nebraska	M (7)	166.00	2.59	6.60
		F (3)	155.04	0.81	
	Pawnee	M (31)	166.28	4.41	6.08
		F (23)	156.17	5.16	
	Itskari	M (12)	165.97	6.01	5.78
		F (14)	156.38	3.98	
	Southern Plains Late Prehistoric	M (17)	166.36	3.89	7.11
		F (14)	154.54	4.82	
Body mass (kg)	Snowflake	M (3)	68.60	3.85	12.32
		F (2)	60.15	4.34	
	Antler Plain/Souris River	M (10)	72.17	3.63	13.61
		F (4)	62.35	4.84	
	Dickson Mounds	M (25)	67.86	4.71	14.88
		F (27)	57.76	4.73	

Table 7-4.—*Continued*

Morphology	Group/Site[a]	Sex (n)	Mean	Std. Dev.	% Sexual Dimorphism[b]
	St. Francis & Black Rivers	M (14)	69.13	6.84	17.14
		F (14)	57.28	4.33	
	Ouachita River	M (11)	68.73	4.19	17.77
		F (10)	56.52	6.10	
	Hiwassee Island	M (19)	66.40	5.35	17.61
		F (19)	54.71	3.99	
	Ledford Island	M (18)	64.63	4.14	17.78
		F (21)	53.14	4.10	
	Toqua	M (18)	64.57	6.42	14.11
		F (19)	55.46	4.84	
	Larson	M (16)	66.47	4.63	13.07
		F (16)	57.78	4.49	
	Cheyenne River	M (15)	66.55	4.98	12.22
		F (11)	58.42	4.98	
	Mobridge A	M (24)	70.09	5.15	18.36
		F (16)	57.22	6.37	
	Sully A & D	M (12)	68.57	3.42	14.48
		F (8)	58.64	2.79	
	Greenshield	M (6)	71.99	7.09	13.97
		F (5)	61.93	3.04	
	Nebraska	M (6)	70.48	7.37	15.56
		F (3)	59.51	1.30	
	Pawnee	M (28)	71.88	4.99	15.50
		F (24)	60.74	4.74	
	Itskari	M (8)	65.84	5.61	15.67
		F (7)	55.52	5.21	
	Southern Plains Late Prehistoric	M (21)	67.83	4.12	18.49
		F (19)	55.29	3.15	
BIB (mm)	Antler Plain/Souris River	M (9)	282.89	16.67	1.91
		F (3)	277.50	9.73	
	Dickson Mounds	M (21)	283.00	14.82	5.23
		F (23)	268.20	12.51	
	St. Francis & Black Rivers	M (13)	282.04	18.47	4.50
		F (11)	269.36	11.61	
	Ouachita River	M (8)	273.69	18.61	2.56
		F (8)	266.69	9.92	
	Hiwassee Island	M (15)	271.90	13.78	2.60
		F (14)	264.82	13.56	
	Ledford Island	M (12)	269.88	13.51	3.91
		F (11)	259.32	14.90	
	Toqua	M (11)	271.91	12.94	3.81
		F (13)	261.54	16.22	

Table 7-4.—*Continued*

Morphology	Group/Site[a]	Sex (n)	Mean	Std. Dev.	% Sexual Dimorphism[b]
	Larson	M (13)	281.23	12.54	4.08
		F (16)	269.75	13.58	
	Cheyenne River	M (15)	277.60	10.12	5.84
		F (8)	261.38	9.21	
	Mobridge A	M (20)	283.23	15.52	4.51
		F (13)	270.46	19.36	
	Sully A & D	M (12)	279.46	11.10	7.39
		F (8)	258.81	15.05	
Brachial index	Snowflake	M (4)	79.35	3.28	
		F (2)	76.99	.45	
	Antler Plain/Souris River	M (6)	79.82	2.93	
		F (3)	77.21	1.85	
	Dickson Mounds	M (24)	78.05	2.07	
		F (27)	76.56	2.20	
	Saint Francis & Black Rivers	M (14)	77.65	1.94	
		F (14)	76.90	2.52	
	Ouachita River	M (10)	78.40	1.90	
		F (11)	76.28	1.64	
	Hiwassee Island	M (20)	77.90	1.77	
		F (20)	76.08	2.51	
	Ledford Island	M (18)	77.91	2.76	
		F (20)	76.28	1.66	
	Toqua	M (17)	76.06	2.16	
		F (19)	75.93	1.95	
	Larson	M (16)	79.10	2.88	
		F (16)	78.25	2.28	
	Cheyenne River	M (14)	79.67	2.11	
		F (8)	77.85	1.12	
	Mobridge A	M (23)	78.92	2.53	
		F (14)	78.72	2.27	
	Sully A & D	M (12)	79.28	1.51	
		F (7)	78.84	1.05	
	Greenshield	M (3)	77.35	3.42	
		F (5)	76.62	2.15	
	Nebraska	M (2)	77.27	.16	
		F (3)	76.42	2.98	
	Pawnee	M (28)	80.00	3.11	
		F (11)	76.87	2.69	
	Itskari	M (8)	80.16	3.63	
		F (7)	78.79	2.14	
	Southern Plains Late Prehistoric	M (14)	78.67	2.46	
		F (17)	79.30	2.43	

Table 7-4.—*Continued*

Morphology	Group/Site[a]	Sex (n)	Mean	Std. Dev.	% Sexual Dimorphism[b]
Crural index	Snowflake	M (4)	85.07	3.68	
		F (2)	82.47	2.81	
	Antler Plain/Souris River	M (9)	85.10	1.66	
		F (4)	84.41	.94	
	Dickson Mounds	M (25)	84.43	2.34	
		F (25)	84.33	1.89	
	St. Francis & Black Rivers	M (13)	84.67	2.05	
		F (14)	83.67	2.68	
	Ouachita River	M (11)	83.73	2.81	
		F (9)	84.37	1.13	
	Hiwassee Island	M (19)	83.88	2.52	
		F (18)	83.01	2.27	
	Ledford Island	M (18)	84.25	1.83	
		F (21)	83.93	1.39	
	Toqua	M (18)	83.61	1.28	
		F (18)	83.02	2.42	
	Larson	M (16)	86.74	1.76	
		F (16)	85.92	1.95	
	Cheyenne River	M (15)	86.03	1.55	
		F (11)	84.63	2.86	
	Mobridge A	M (22)	85.66	1.64	
		F (16)	86.11	2.39	
	Sully A & D	M (12)	86.92	2.24	
		F (7)	87.00	2.15	
	Greenshield	M (6)	85.73	2.06	
		F (4)	83.19	3.03	
	Nebraska	M (7)	83.98	1.12	
		F (3)	82.91	2.24	
	Pawnee	M (31)	84.31	2.75	
		F (23)	82.75	2.11	
	Itskari	M (12)	84.58	1.93	
		F (14)	85.97	2.62	
	Southern Plains Late Prehistoric	M (17)	85.24	1.93	
		F (14)	84.20	2.17	

[a]As noted in Materials and Methods, some sites were combined to permit statistical analyses. See Table 7-1 for individual site details.
[b]Sexual dimorphism is calculated as [(male mean – female mean)/(male mean)] × 100. Note that sexual dimorphism is not calculated for intralimb indices.

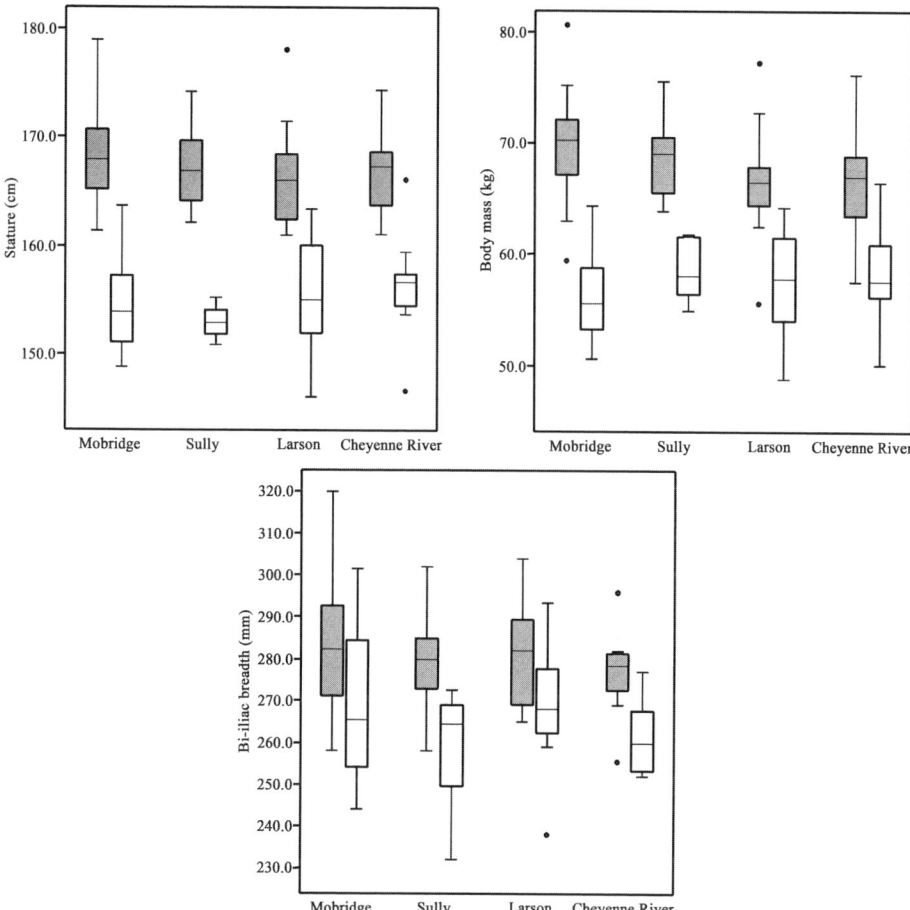

Figure 7-2. *Box plots of stature, body mass, and bi-iliac breadth among Extended Coalescent (Mobridge and Sully) and Protohistoric (Larson and Cheyenne River) Arikara. Males are shaded boxes, and females are white boxes. Dots represent individual outliers.*

the Postcontact Coalescent Arikara (Larson and Cheyenne River) were combined in subsequent analyses.

MANOVAs show that there are significant differences in stature and body mass among samples but not across the average temporal depth of sites. Bi-iliac breadth, interestingly, does not significantly differ over time or across groups. Hochberg's GT2 post hoc tests, which assume unequal samples sizes but equal variance, specify the significant differences among groups; the resulting homogeneous subsets for stature and body mass are presented in Tables 7-5 and 7-6. From these subsets, some trends are worth highlighting. Among males, the Mississippian groups from the Southeast (Toqua, Hiwassee Island, and Ledford Island) tend to be significantly less massive and shorter than many of the Central and Northern Plains groups. This trend is generally found among females in

Table 7-5. *Hochberg's GT2 Post-hoc Test Homogeneous Subsets (p < .05), Stature Group Means*

a. Males

Sample	Subset 1	Subset 2
Toqua	163.8	
Hiwassee Island	165.1	165.1
Greenshield	165.6	165.6
Ouachita River	165.7	165.7
Pawnee	165.8	165.8
Itskari	166.0	166.0
Nebraska	166.0	166.0
Ledford Island	166.2	166.2
Southern Plains LP	166.2	166.2
Protohistoric Coalescent	166.4	166.4
Extended Coalescent		167.9
Snowflake		167.9
Dickson Mounds		169.0
St. Francis & Black Rivers		169.7
Antler Plain/Souris River		170.1

b. Females

Sample	Subset 1	Subset 2
Greenshield	152.5	
Ledford Island	153.9	
Extended Coalescent	154.0	154.0
Southern Plains LP	154.5	154.5
Nebraska	155.0	155.0
Pawnee	155.4	155.4
Hiwassee Island	155.5	155.5
Protohistoric Coalescent	155.9	155.9
Toqua	156.1	156.1
Itskari	156.4	156.4
Antler Plain/Souris River	156.4	156.4
Ouachita River	156.7	156.7
St. Francis & Black Rivers		158.2
Dickson Mounds		158.8
Snowflake		160.5

Table 7-6. *Hochberg's GT2 Post-hoc Test Homogeneous Subsets (p < .05), Body Mass Group Means*

a. Males

Sample	Subset 1	Subset 2
Toqua	64.6	
Ledford Island	64.6	
Itskari	65.8	65.8
Hiwassee Island	66.4	66.4
Protohistoric Caddo	66.5	66.5
Southern Plains LP	67.7	67.7
Dickson Mounds	67.9	67.9
Snowflake	68.6	68.6
Ouachita River	68.7	68.7
St. Francis & Black Rivers	69.1	69.1
Extended Coalescent	69.6	69.6
Nebraska	70.5	70.5
Pawnee		71.1
Greenshield		72.0
Antler Plain/Souris River		72.2

b. Females

Sample	Subset 1	Subset 2
Ledford Island	53.1	
Hiwassee Island	54.7	
Southern Plains LP	55.3	55.3
Toqua	55.5	55.5
Itskari	55.5	55.5
Ouachita River	56.5	56.5
St. Francis & Black Rivers	57.3	57.3
Extended Coalescent	57.7	57.7
Dickson Mounds	57.8	57.8
Protohistoric Coalescent	58.0	58.0
Nebraska	59.5	59.5
Snowflake		60.2
Pawnee		60.4
Greenshield		61.9
Antler Plain/Souris River		62.3

mass but not in stature. There are no significant differences between the samples from the Ouachita or St. Francis and Black River valleys and any of the Central and Northern Plains Caddoan speakers in stature or body mass, and it is noteworthy that the St. Francis and Black River samples are, on average, among the tallest and most massive, especially in comparison with other southeastern samples. Also, the small samples from southern Manitoba (Souris River and Snowflake) are among the tallest and most massive of any of these samples. Indeed, the samples with the greatest statures and masses are located at the extreme northern and southern ends of the central axis of the Great Plains. Neither the southern Caddoans nor Manitoba groups, however, are significantly different from the Arikara, Pawnee, Itskari, or Southern Plains Late Prehistoric samples. Even though it is not a significant pattern, and it spans a subsample of the total data set, these trends also tend to be found in comparisons of bi-iliac breadths among these groups.

Does this pattern also occur among intralimb indices? A comparison of the mean brachial indices among the groups (Table 7-4) indicates that, although there is no clear clinal pattern, males from the Central and Northern Plains tend to have *higher* brachial indices than those from the southeastern United States, including the Caddoan-speaking groups from Arkansas and Louisiana. Females also reflect this trend, and in both sexes, the small historic Arikara sample from the Greenshield site has lower brachial indices, unlike the high indices observed among the Coalescent and historic Arikara. In the crural indices, male averages demonstrate a difference between the southeastern samples and the Northern Plains samples, where the Plains groups again exhibit higher indices. Unlike the pattern observed among male brachial indices, the crural indices among males from the St. Francis and Black River sites are higher and more similar to those of the Itskari, Arikara (both Coalescent and historic), and the Southern Plains Late Prehistoric samples. A similar, but not identical, pattern is not found among the females, except that Coalescent samples all have the highest crural indices and the southeastern U.S. sites tend to have the lowest values, except for females from the Ouachita River sites.

These patterns are more fully borne out in the results from the principal components analysis of the limb lengths scaled using the geometric mean. The sexes were combined in this analysis, as the scaled data do not have the significant sex differences found in brachial indices (Auerbach 2007; Holliday and Ruff 2001) and as the general patterns of variation among groups for these limbs are similar between males and females. Two components with eigenvalues greater than 1.0 were yielded by the analysis. Component loading eigenvectors are given in Table 7-7. Principal component 1 (PC1), which explains 41.42 percent of the total variance, reflects the relative lengths of the distal limb segments to proximal segments; individuals with lower values on this axis have higher intralimb indices (i.e., longer radii and tibiae relative to humeri and femora, respectively). The second component (PC2; 38.29 percent of the total variance) contrasts the length of the upper and lower limbs. Individuals with lower values on the axis of component 2 have longer upper limbs relative to lower limbs, a pattern that is driven more by humeral and tibial lengths.

Table 7-7. *Principal Component Loadings for Geometric Mean-Standardized Limb Lengths*

Principal components	HML	RML	FBL	TML	% variance
Eigenvectors of PC1	.551	−.790	.743	−.420	41.42
Eigenvectors of PC2	−.698	−.475	.454	.783	38.29

Figure 7-3 exhibits the mean group eigenvalues for these two components. Reflecting the examination of group mean values for the intralimb indices, Great Plains groups, with the exceptions of the Middle Woodland Snowflake (Manitoba) and Northern Plains Greenshield (historic Arikara) samples, have lower values on PC1; they have higher intralimb indices than the samples from south and east of the Plains. This division is less clear on PC2, though it is notable that all of the Coalescent site Arikara, the Pawnee, and the Itskari have the longest lower limbs relative to upper limbs. As these also tend to be among the tallest samples, there may be a slight allometric effect on this pattern. However, individuals from the Souris River Late Woodland sample have the tallest statures, yet they plot considerably lower on PC2, and Ledford Island Mississippians—despite higher loadings on PC2—were both considerably shorter than the Plains groups. In fact, a Pearson's correlation analysis of the scaled limb lengths with the geometric mean among individuals shows no significant correlations; this indicates no allometric effect of body size on limb length (Holliday 1999). (However, compare this result with Jantz and Jantz [1999] and Sylvester and colleagues [2008].)

Discussion

General Patterns

Comparisons of the five morphological dimensions examined reveal patterns in body shape, size, and proportions among Great Plains samples from the last millennium. Both taller and more massive groups—especially males—lived in the Plains than in sites located along the Tennessee River watershed. This corroborates previous findings (Prince 1998; Steckel and Prince 2001); Central and Northern Plains groups were, effectively, "giants" among groups bordering them to the southeast and southwest (Auerbach 2007). The Itskari sample, however, had notably smaller mean statures and masses. Interestingly, taller statures and greater body masses are associated with the Dickson Mounds sample and with Caddoan-speaking groups sampled from the St. Francis, Black, and Ouachita river valleys, all of whom were Mississippians like the groups from the Tennessee River. These patterns are reflected as well in comparisons of bi-iliac breadth, as the widest body breadths are associated with the Mississippian Caddoan-speaking groups and the samples from the Northern Plains. Differences among these groups are apparent only in limb dimensions, where the Great Plains groups (except for

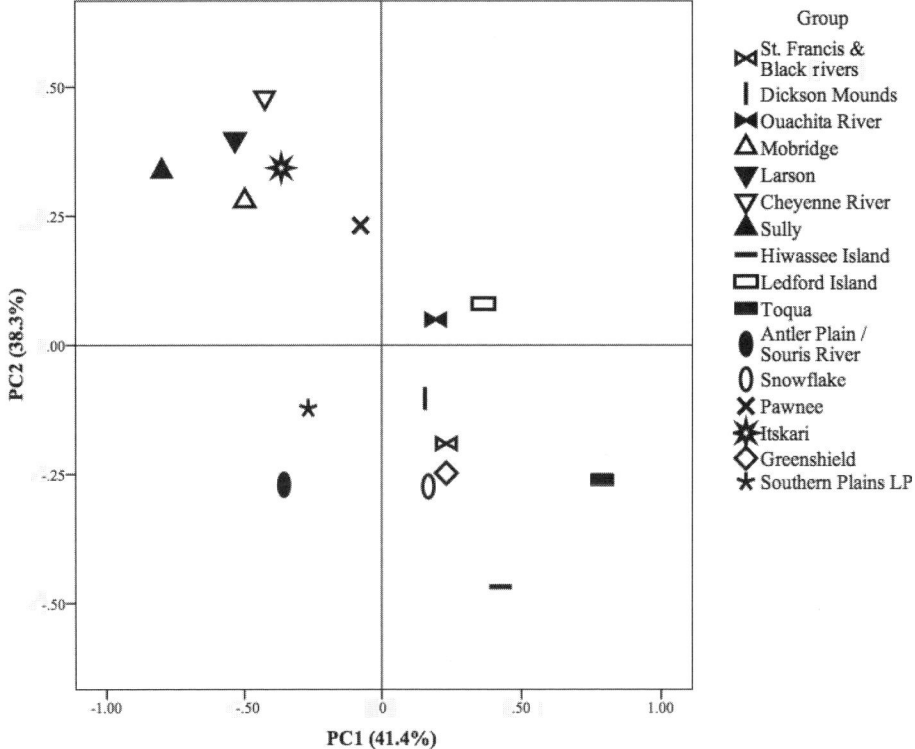

Figure 7-3. *Mean group values for component 1 and component 2, taken from principal components analysis results of limb lengths scaled by the geometric mean, males and females combined. See Table 7-7 for component eigenvalues and loadings.*

the Middle Woodland Manitoba and historic Arikara) have the highest brachial and crural indices, whereas the Mississippian Caddoan-speaking samples have indices falling between the Plains samples and the lower-proportioned Tennessee River watershed Mississippians. These latter samples also present longer upper limb lengths relative to lower limb lengths; the Pawnee, Coalescent Arikara, and Itskari had the longest relative lower limb lengths.

Together, these results are suggestive of two clines. First, from north to south, groups generally became less massive and had shorter lower limb lengths relative to upper limb lengths. Also, from northwest to southeast, groups exhibit shorter statures, narrower body breadths, and lower intralimb indices. There are important exceptions, however, such as mean statures and weights of the Itskari or the tall statures among St. Francis and Black River valley samples.

The relationships of these morphological relationships will be considered further below. Before discussing these details, it should be noted that these patterns of morphological variation are, admittedly, limited. Some groups and time periods have few—if any—samples, while others undoubtedly have internal variation or substructure that has not been documented in this study but surely affects any conclusions that may be drawn about the population from which the samples were taken. However, the general trends that may be concluded from the comparisons provide information about population change, affinities, and, possibly, movement on the Great Plains.

Secular Trends in the Coalescent

As noted in the introduction, studies of craniometric change among Coalescent sites have revealed a secular trend in neurocranial height and dolichocephaly from the Initial Coalescent through the historic period (Jantz 1977; Key 1983; McKeown 2000). Comparisons among the four Coalescent sites sampled here—Mobridge, Sully, Larson, and Cheyenne River—which in turn represent two general time periods (Extended Coalescent and Postcontact Coalescent), reflect the presence of secular change among the Arikara. As shown in Figure 7-2, both male body mass and stature exhibit slight decreases between the Extended Coalescent and Postcontact Coalescent periods, whereas females show increases in these dimensions. Consequently, sexual dimorphism in stature and body mass decreased over time among these samples. It is curious that no clear pattern is present in bi-iliac breadth. This may be due to sampling, but given the trends observed in the other morphologies using the same samples, it is evident that decreases in body mass through time may be due, in part, to the decreases in stature. If extended to include the small historic sample from the Greenshield site, the decreases in mean male stature continue, though both males and females from this site had the highest mean body masses of any of the Arikara.

These results indicate that the Arikara experienced higher stress and/or admixture with shorter groups within a brief period between 300 and 200 years ago. The presence of European trade goods and intensification in the growth of maize (mostly for trade) over other crops and to the exclusion of other plants during this time period argue for significant changes in subsistence economies (Johnson 1998). This change may have directly contributed to the decreases in mean stature and, in most cases, body mass among males, in addition to less sexual dimorphism in these morphologies (Stini 1974). Additional population-level stresses would undoubtedly have resulted from the recent incursion of the Sioux into the Northern Plains from the Northeastern Plains (Owsley and Bruweldheide 1997), a movement that reduced Arikara mobility and gathering. Whether interbreeding between the individuals at these Extended and Postcontact Coalescent sites and neighboring groups (e.g., the Mandan, Hidatsa, or Sioux) would have contributed to these subtle secular changes cannot be addressed, though these groups likely maintained clear cultural separations until population decimations during the late eighteenth century pushed them toward consolidation (Jantz 1997; O'Shea 1984).

Higher body masses in the Greenshield sample, however, solicit further examination. It is notable that all the Northern Plains groups have higher body masses than most of the comparative samples employed here. Indeed, in comparison with a wider indigenous sample from North American archaeological sites, these Plains samples exhibit the highest body masses of any subarctic group living during the last millennium (Auerbach 2007). The Greenshield masses, though, are among the highest of these, a finding that makes them more similar to the very massive Souris River and Mobridge samples, which both predate the more temporally and geographically proximate Cheyenne River sample. A high body mass within the Greenshield sample may simply result from a nonrepresentative sample, given the small number of skeletons for this site. However, given the high correlation between bi-iliac breadth and body mass ($r = .60–.69$ for both sexes in the Plains sample)—a relationship previously explained (Ruff 1991)—and the lower mean statures in this sample in comparison with the Postcontact Coalescent, one may conclude that these North Dakota skeletons had higher bi-iliac breadths than the Coalescent samples for which bi-iliac breadths were measured. This would fit with the general north-to-south cline in body breadth exhibited overall in the sample and whether it is indicative of gene flow, a founder effect, or subtle adaptive change in body breadth over time among the Arikara awaits further study.

Group Variation, Affinities, and Movement

The patterns of variation among the groups represented in this study indicate an interesting morphological mosaic. As demonstrated in Figure 7-3 in and the comparisons in Table 7-6, the Coalescent Arikara and Pawnee present high intralimb indices while also exhibiting among the widest body breadths and highest body masses of any sample. This is a paradoxical relationship based on the thermoregulatory expectations from previous studies (Holliday 1995, 1997b; Ruff 1994b), which would argue for groups to share narrower bi-iliac breadths and higher brachial and crural indices, or vice versa. Holliday (1999) showed that such expectations might be oversimplified and argued that the earliest Upper Pleistocene colonizers of Europe developed shorter overall limb lengths across generations while retaining high intralimb indices when exposed to colder average climates. This tendency, however, was not the case for the groups explored in this study; the Northern and Central Plains samples have higher relative lower and upper limb lengths—that is, longer limbs relative to stature—than the Tennessee River valley samples (Auerbach 2007). Only the Pawnee and Itskari, interestingly, depart from this trend and are more similar to the southeastern Mississippians in relative limb lengths (but not intralimb indices). Moreover, the St. Francis and Black River samples significantly differ from the other southern and eastern samples in having high body masses, wide body breadths, and long relative limb lengths.

If these proportions are indeed relatively stable over temporal periods exceeding 1,000 years, a tendency strongly argued by Holliday (1999) and Temple and colleagues (2008), then these morphologies provide some tantalizing clues for group relationships. The preponderance of the evidence supports morphological affinities among all the Plains groups. All exhibited higher body masses, taller statures, and higher intralimb indices than groups to the southeast and, as demonstrated elsewhere (Auerbach 2007, 2009), than groups in the U.S. Southwest and in the Eastern Woodlands. Two peripheral samples also demonstrated morphological characteristics, with the notable exception of the higher intralimb indices or longer relative lower limb lengths: the Dickson Mounds Mississippians and the St. Francis and Black River valley Caddoan speakers. This is surprising, as the Caddoan speakers lived in what is modern-day Arkansas and so would be expected to exhibit the narrower body breadths of other southeastern groups. Southern Plains groups, though, also had relatively shorter lower limbs (both in comparison to stature and to the upper limb length) when compared with most Northern and Central Plains groups, except the Pawnee and Itskari, who again also had shorter lower limb lengths relative to stature. Higher masses and statures also were found among the Northern Plains samples from the Middle Woodland in Manitoba (Snowflake) and Greenshield, but their lower intralimb indices and relative limb lengths separate them from most other Plains samples.

Combining these results, the morphological affinities may relate to gene flow or shared ancestry. As explicated in the introduction, there is no controversy about a shared linguistic origin for the Arikara, Pawnee, and Southern Caddo. The morphological variations in part support this relationship on a biological level; the St. Francis and Black River Caddoan-speaking samples share many morphologies with the Pawnee, Itskari, Arikara, and Southern Plains samples. With the exception of intralimb indices, these Caddoan ancestors were distinguished from other Mississippians in the Southeast and were more similar to the Plains samples than the other Caddoan-speaking sample from farther south (the Ouachita River sites) was. One might conclude, then, that the Caddoan St. Francis and Black River valley samples, Southern Plains Late Prehistoric, Arikara, Itskari, and Pawnee share a common ancestry and/or later gene flow. Given the variety of environments that these groups inhabited, and the short time period separating them, this would be the most likely resolution for their morphological similarities.

This relationship is not conclusive. Dissimilarities in limb proportions among these groups—if limb proportions are as stable as body breadth—separates all Plains groups from the Arkansas and Louisiana Caddoan-speaking samples. Likewise, the Greenshield site Arikara were different in limb morphologies while also apparently sharing the wider body breadths of the Coalescent groups, yet their direct relationship is more supported by archaeology than any other relationship is. As the heritability of limb proportions versus pelvic morphology has not been assessed, it also is possible that the former are more phenotypically plastic than the latter. Similar morphologies among the Dickson Mounds and Plains samples likewise argue for similar adaptations, shared ancestry, or gene flow between these groups; the last is unprecedented, as no direct link between the Plains samples and the Mississippians in the western Prairie has ever been proposed (with the notable possible exception of the Steed-Kisker phase).

In addition, as noted above, the tendency for higher body masses and wider bi-iliac breadths among the Arikara and Pawnee is unexpected. Were these groups recent migrants, as archaeological data and this study would support, then they would be expected to have narrower body breadths and lower body masses. Their higher intralimb indices, however, do match the expectation that they descend from ancestors who migrated recently, most likely from the south (see the introduction). The southeastern Mississippians, in fact, were the narrowest and least massive—both expected based on thermoregulatory morphological patterns—but also had among the lowest brachial and crural indices. An examination of the origins of this trend is beyond the scope of this study, as the sample used lacks the temporal depth necessary to determine whether the trend is the result of recent migration of populations, founder effects, or the endemic violation of morphological patterns associated with thermoregulation. For example, it is possible that the Mississippian samples from Toqua, Hiwassee Island, and Ledford Island were obtained from populations whose ancestors recently migrated or experienced gene flow from colder climates. Their narrower body breadths, however, argue against this. Likewise, the higher indices of the Arikara and Pawnee support a more recent southern origin, but their high masses and wide body breadths do not, except in light of the presence of similarly high values for these traits among the St. Francis and Black River valley samples.

Relationships among the Coalescent Arikara, Pawnee, and Itskari likewise remain somewhat ambiguous. It has been established that all three groups share most morphologies, which in turn supports a very recent, common ancestry for them. The Itskari were not as massive as either of the later groups, though their statures were not significantly different, arguing for slightly narrower bi-iliac breadths. Also, both the Itskari and Pawnee differed from the Arikara in their relative limb lengths (to stature). An important caveat, though, is that the Greenshield historic Arikara had different limb proportions from all of these groups, but as explored above, the ultimate origins of this cannot be assessed. As the Itskari predate both the Pawnee and Coalescent Arikara, it may be concluded that the Itskari were ancestral to both groups or descended from a common ancestor. Based on the morphological comparisons, they differ less from the Pawnee than from the Coalescent Arikara and are more like the Extended Coalescent Arikara than Postcontact Arikara. These are slight differences, and these groups are overall more like each other than they are like any other groups. Any argument for discontinuity between the Itskari and either later group, therefore, is not supported, and it is provisionally likely that the Itskari contributed to the ancestry of the Pawnee and the Arikara.

Conclusions and Future Directions

This study set out to employ morphological analyses to address some of the archaeological ambiguities concerning relationships among Great Plains groups following the development of the Plains Village tradition. Although these relationships have not been conclusively resolved, the analysis has addressed the questions and hypotheses set out in the introduction:

- Similar morphologies, in addition to previous craniometric, archaeological, and linguistic evidence, strongly support a recent common ancestry for the Arikara and Pawnee. They likely represent recent migrants to the Central and Northern Plains. High intralimb indices—which are dissimilar from the lower intralimb indices observed among Middle Woodland samples from southern Manitoba—are strong evidence of this. However, these groups present high body masses and wide body breadths, which either resulted from gene flow with other higher latitude groups or were retained from more southern Caddoan-speaking ancestors. As these morphologies are not significantly different from the (non-Caddoan) Late Woodland Souris River sample from southern Manitoba, the Mississippians from Dickson Mounds, or from the Caddoan St. Francis and Black River valley samples from Arkansas, the origin of this morphology remains inclusive.

- The Caddoan speakers' samples from Arkansas and Louisiana differ from the Pawnee and Coalescent Arikara in limb indices (both relative limb lengths and intralimb indices), presenting slightly lower brachial and crural indices despite living in much warmer and more southern climes. This argues for differences in plasticity in limb proportions versus body breadth, differential gene flow for one or both sets of groups, or other contributory factors not assessed in this study. Nutrition and stress do not appear to be factors, as no significant differences in stature or body mass are indicated. Despite this, linguistic evidence and some similar morphological characteristics (high body masses, wide body breadths, and tall statures) argue for common ancestry for all the Caddoan-speaking groups.

- Neither the Pawnee nor Coalescent Arikara differed in most morphological traits from the Itskari sample, though the Pawnee shared similarities in relative limb length with the Itskari that they did not with the Coalescent Arikara. All three groups are likely directly related, arguing against discontinuity between the Central Pains Itskari/Lower Loup phase and the Arikara or Pawnee. Again, similarities in intralimb indices and body mass with samples from the Late Prehistoric Southern Plains as well as southern Manitoba Souris River Late Woodland make relationships among the Central Plains and southern Northern Plains groups with their neighbors ambiguous.

- The Arikara exhibit signs of secular change in body mass and stature from the Extended Coalescent to the historic period. Generally, male statures and body masses decreased, and sexual dimorphism also decreased. This likely resulted from a number of factors, including changes in subsistence and increased population stress due to European contact and displacement and the resulting colonization of the Sioux in the Northern Plains. Differences in limb proportions between the protohistoric and historic periods may indicate additional gene flow into the Arikara.

A few broader conclusions should be emphasized. The relationships of body mass, bi-iliac breadth, stature, and limb proportions are not resolved, but this study clearly indicates that these morphologies independently vary to different degrees. The presence of high intralimb indices and relatively long limbs among the Arikara and Pawnee does not match morphological expectations based on thermoregulation or account for the large body breadths of the St. Francis and Back River valley samples. Indeed, the variation in limb proportions appears to be decoupled from variation in body breadth and, in turn, body mass. These argue that the ecogeographic patterns observed in Europe and Africa were not occurring in North America during the last millennium before European colonization or that gene flow and/or population migration occurred frequently or at a faster pace than morphological adaptation. Given the patterns described in this study and elsewhere by the author (Auerbach 2007), the last situation appears to be more likely, though the first possibility cannot be excluded (see King, this volume; Jantz et al., this volume). These possibilities await additional assessment.

In summary, it is evident that the population history of the Great Plains over the last millennium remains complex. Based on the archaeological evidence, previous craniometric analyses, and the morphological data presented here, one may conclude that a combination of population migration and gene flow yielded the populations preceding the arrival of and encountered by Europeans on the Central and Northern Plains. Some of this gene flow and/or migration unquestionably emerged from the southern and eastern border of the Plains (in modern Arkansas), though contributions from the north (southern Manitoba) and east (Illinois and Mississippi river valleys) cannot be rejected. Future archaeological analyses and additional studies of morphologies from skeletal remains, including genetic analyses, will help to resolve some of these uncertainties.

Acknowledgments

A number of people made the research for this paper possible. First, I give many thanks to Dr. Danny Wescott for his continuing generosity in sharing data, as well as his thoughts and interpretations about the indigenous peoples of the Great Plains. In addition, I send my gratitude to the curators and staff at various skeletal collections in North America for continuing to permit access to their invaluable resources, especially the Smithsonian Institution, the Illinois State Museum, the Frank H. McClung Museum, and the Canadian Museum of Civilisation. I thank the participants of the Twenty-Fifth Annual Visiting Scholar Conference and the anonymous reviewers who gave beneficial feedback about the structure of this chapter, the archaeology of the Plains, and interpretations of the morphological data. Thanks as well to Dr. Chris Ruff, without whom this research never would have occurred. Funding from the National Science Foundation (a Graduate Research Fellowship and a Doctoral Dissertation Improvement Grant, #0550673) supported data collection for this chapter; none of the opinions or research conducted herein reflect their policies.

Notes

1. Some researchers have made a strong argument that, as pottery often is produced as a gender-specific technology, classifications of archaeological sites based on ceramics may not present a complete picture of intergroup relationships (Steinacher and Carlson 1998). This is made more important by the evi-

dence that male out-migration (female residence) is argued from comparisons of dental trace element analyses (Schneider and Blakeslee 1990).

2. Note that Key (1994) makes a case for the Hopewell as genetically isolated from other Woodland peoples of the Central Plains.

3. Indeed, as the author has shown elsewhere (Auerbach 2007), disentangling climate and subsistence variables is highly difficult given the dependence of one (subsistence) on the other (climate).

4. This method has acknowledged drawbacks, and the appropriate use of the geometric mean to scale measurements continues to be a topic of discussion and debate (Albrecht et al. 1993; Corruccini 1987; Jungers et al. 1995; Smith 2005).

References

Adair, Mary J.
   1996   Woodland Complexes in the Central Great Plains. In *Archaeology and Paleoecology of the Central Great Plains*, edited by Jack L. Hofman, pp. 101–122. Arkansas Archeological Survey, Fayetteville.
Ahler, Stanley A., and Marvin Kay
   2007   *Plains Village Archaeology: Bison-Hunting Farmers in the Central and Northern Plains*. University of Utah Press, Salt Lake City.
Albrecht, Gene H.
   1978   Some Comments on the Use of Ratios. *Systematic Zoology* 27:67–71.
Albrecht, Gene H., Bruce R. Gelvin, and Steve E. Hartman
   1993   Ratios as a Size Adjustment in Morphometrics. *American Journal of Physical Anthropology* 91:441–468.
   1995   Ratio Adjustments in Morphometrics: A Reply to Dr. Corruccini. *American Journal of Physical Anthropology* 96:193–197.
Alexander, R. McNeill
   1984   Stride Length and Speed for Adults, Children, and Fossil Hominids. *American Journal of Physical Anthropology* 63:23–27.
Atchley, William R., Charles T. Gaskins, and Dwane Anderson
   1976   Statistical Properties of Ratios. I. Empirical Results. *Systematic Zoology* 25:137–148.
Auerbach, Benjamin M.
   2007   Human Skeletal Variation in the New World During the Holocene: Effects of Climate and Subsistence Across Geography and Time. Unpublished Ph.D. dissertation, Center for Functional Anatomy and Evolution, Johns Hopkins University, Baltimore.
   2009   Body Mass, Stature, and Proportions of the Kennewick Early Holocene Skeleton. In *Kennewick Man*, edited by Douglas W. Owsley and Richard L. Jantz. Texas A&M Press, College Station, in press.
Auerbach, Benjamin M., and Christopher B. Ruff
   2004   Human Body Mass Estimation: A Comparison of "Morphometric" and "Mechanical" Methods. *American Journal of Physical Anthropology* 125:331–342.
   2006   Limb Bone Bilateral Asymmetry: Variability and Commonality among Modern Humans. *Journal of Human Evolution* 50:203–218.
   2010   Stature Estimation Formulae for Indigenous North American Populations. *American Journal of Physical Anthropology* 141:190–207.

Bamforth, Douglas B., and Curtis Nepstad-Thornberry
    2007   The Shifting Social Landscape of the Fifteenth-Century Middle Missouri Region. In *Plains Village Archaeology: Bison-Hunting Farmers in the Central and Northern Plains*, edited by Stanley A. Ahler and Marvin Kay, pp. 139–154. University of Utah Press, Salt Lake City.

Bass, William M.
    1961   Variation in the Physical Types of the Prehistoric Plains Indians. Unpublished Ph.D. dissertation, Department of Anthropology, University of Pennsylvania, Philadelphia.
    1981   Skeletal Biology of the United States Great Plains: A History and Personal Narrative. *Plains Anthropologist* 26:3–18, Part 2.

Beals, Kenneth L.
    1972   Head Form and Climatic Stress. *American Journal of Physical Anthropology* 37:85–92.

Bergmann, Carl
    1847   Ueber die Verhältnisse der wärmeökonomie der Thiere zu ihrer Grosse. *Göttinger Studien* 3:595–708.

Blakeslee, Donald J.
    1981   Toward a Cultural Understanding of Human Microevolution on the Great Plains. *Plains Anthropologist* 26:93–106.
    1994   The Archaeological Context of Human Skeletons in the Northern and Central Plains. In *Skeletal Biology in the Great Plains*, edited by Douglas W. Owsley and Richard L. Jantz, pp. 9–32. Smithsonian Institution, Washington, D.C.

Bogin, Barry
    2002   *Patterns of Human Growth.* 2nd ed. Cambridge Studies in Biological Anthropology No. 23. Cambridge University Press, New York.

Bogin, Barry, and Ryan Keep
    1999   Eight Thousand Years of Economic and Political History in Latin America Revealed by Anthropometry. *Annals of Human Biology* 26:333–351.

Bolnick, Deborah A., Beth A. Shook, Lyle Campbell, and Ives Goddard
    2004   Problematic Use of Greenberg's Linguistic Classification of the Americas in Studies of Native American Genetic Variation. *American Journal of Human Genetics* 75:519–522.

Bruzek, Jaroslav
    2002   A Method for Visual Determination of Sex, Using the Human Hip Bone. *American Journal of Physical Anthropology* 117:157–168.

Byrd, John E., and Richard L. Jantz
    1994   Osteological Evidence for Distinct Social Groups at the Leavenworth Site. In *Skeletal Biology in the Great Plains*, edited by Douglas W. Owsley and Richard L. Jantz, pp. 203–208. Smithsonian Institution, Washington, D.C.

Campbell, Lyle
    1997   *American Indian Languages: The Historical Linguistics of Native America.* Oxford Studies in Anthropological Linguistics. Oxford University Press, New York.

Carlson, Gayle F., John R. Bozell, Terry L. Steinacher, Marjorie Brooks Lavvorn, and George W. Gill
    1999   The Sidney Burial: A Middle Plains Archaic Mortuary Site from Western Nebraska. *Plains Anthropologist* 44:105–119.

Cole, Maria S., and Theodore M. Cole
    1994   Metric Variation in the Supraorbital Region in Northern Plains Indians. In *Skel-*

etal Biology in the Great Plains, edited by Douglas W. Owsley and Richard L. Jantz, pp. 209–218. Smithsonian Institution, Washington, D.C.

Cole, Theodore M.
1994   Size and Shape of the Femur and Tibia in Northern Plains Indians. In *Skeletal Biology in the Great Plains*, edited by Douglas W. Owsley and Richard L. Jantz, pp. 219–234. Smithsonian Institution, Washington, D.C.

Corruccini, Robert S.
1987   Shape in Morphometrics: Comparative Analyses. *American Journal of Physical Anthropology* 73:289–303.

Crumbley, William R.
1986   Variation in Tibial Morphology of Three Arikara Skeletal Populations. Unpublished Ph.D. dissertation, Department of Anthropology, The University of Tennessee, Knoxville.

Danforth, Marie E.
1999a   Coming Up Short: Stature and Nutrition among the Ancient Maya of the Southern Lowlands. In *Reconstructing Ancient Maya Diet*, edited by Christine White, pp. 103–117. University of Utah Press, Salt Lake City.
1999b   Nutrition and Politics in Prehistory. *Annual Review of Anthropology* 28:1–25.

Darroch, John N., and James E. Mosimann
1985   Canonical and Principal Components of Shape. *Biometrika* 72:241–252.

Davenport, Charles B.
1933   The Crural Index. *American Journal of Physical Anthropology* 17:333–353.

DeMallie, Raymond J. (editor)
2001   *Plains*. Handbook of North American Indians, Vol. 13, Pt. 1, William C. Sturtevant, general editor, Smithsonian Institution, Washington, D.C.

Frison, George C.
1998   The Northwestern and Northern Plains Archaic. In *Archaeology on the Great Plains*, edited by Raymond Wood, pp. 140–172. University Press of Kansas, Lawrence.

Frison, George C., and Robert C. Mainfort (editors)
1996   *Archaeology and Bioarchaeological Resources of the Northern Plains*. Arkansas Archeological Survey, Fayetteville.

Genoves, Santiago
1966   Some Comments on the "Secular Trend" of Stature in the Last Generations. *American Anthropologist* 68:499–504.

Goddard, Ives
2001   The Languages of the Plains: Introduction. In *Plains*, edited by Raymond J. DeMallie, pp. 61–70. Handbook of North American Indians Vol. 13, Pt. 1, William C. Sturtevant, general editor, Smithsonian Institution, Washington, D.C.

Grange, Roger T.
1979   An Archaeological View of Pawnee Origins. *Nebraska History* 60:134–160.

Green, Thomas J., Bruce Cochran, Todd W. Fenton, James C. Woods, Gene L. Titmus, Larry Tieszen, Mary Anne Davis, and Susanne J. Miller
1998   The Buhl Burial: A Paleoindian Woman from Southern Idaho. *American Antiquity* 63:437–456.

Grine, Frederick E., William L. Jungers, Paul V. Tobias, and Osbjorn M. Pearson
1995   Fossil *Homo* Femur from Berg Aukas, Northern Namibia. *American Journal of Physical Anthropology* 97:151–185.

Henning, Dale R.
2005   The Evolution of the Plains Village Tradition. In *North American Archaeology*,

edited by Timothy R. Pauketat and Diana DiPaolo Loren, pp. 161–186. Blackwell Publishing, New York.

Hill, Matthew E., Jack L. Hofman, and Karolyn Kinsey
  1996   A History of Archeological Research on the Central Plains. In *Archaeology and Paleoecology of the Central Great Plains*, edited by Jack L. Hofman, pp. 29–40. Arkansas Archeological Survey, Fayetteville.

Hills, Michael
  1978   On Ratios: A Response to Atchley, Gaskins, and Anderson. *Systematic Zoology* 27:61–62.

Hofman, Jack L.
  1996   Early Hunter-Gatherers of the Central Great Plains: Paleoindian and Mesoindian (Archaic) Cultures. In *Archaeology and Paleoecology of the Central Great Plains*, edited by Jack L. Hofman, pp. 41–100. Arkansas Archeological Survey, Fayetteville.

Hofman, Jack L. (editor)
  1996   *Archaeology and Paleoecology of the Central Great Plains*. Arkansas Archeological Survey, Fayetteville.

Hofman, Jack L., and Russell W. Graham
  1998   The Paleo-Indian Cultures of the Great Plains. In *Archaeology on the Great Plains*, edited by Raymond Wood, pp. 87–139. University Press of Kansas, Lawrence.

Holliday, Trenton W.
  1995   *Body Size and Proportions in the Late Pleistocene Western Old World and the Origins of Modern Humans*. Unpublished Ph.D. dissertation, Department of Anthropology, University of New Mexico, Albuquerque.
  1997a Body Proportions in Late Pleistocene Europe and Modern Human Origins. *Journal of Human Evolution* 32:423–448.
  1997b Postcranial Evidence of Cold Adaptation in European Neandertals. *American Journal of Physical Anthropology* 104:245–258.
  1999   Brachial and Crural Indices of European Late Upper Paleolithic and Mesolithic Humans. *Journal of Human Evolution* 36:549–566.

Holliday, Trenton W., and Christopher B. Ruff
  2001   Relative Variation in Human Proximal and Distal Limb Lengths. *American Journal of Physical Anthropology* 116:26–33.

Jantz, Lee M., and Richard L. Jantz
  1999   Secular Change in Long Bone Length and Proportion in the United States, 1800–1970. *American Journal of Physical Anthropology* 119:57–67.

Jantz, Richard L.
  1973   Microevolutionary Change in Arikara Crania: A Multivariate Analysis. *American Journal of Physical Anthropology* 38:15–26.
  1977   Craniometric Relationships of Plains Populations: Historical and Evolutionary Implications. In *Trends in Middle Missouri Prehistory: A Festschrift Honoring the Contributions of Donald J. Lehmer*, edited by Raymond Wood, *Plains Anthropologist Memoir* 13:162–176.
  1997   Cranial, Postcranial, and Discrete Trait Variation. In *Bioarcheology of the North Central United States*, edited by Douglas W. Owsley and Jerome C. Rose, pp. 240–247. Arkansas Archeological Survey, Fayetteville.

Jantz, Richard L., and Douglas W. Owsley (editors)
  1994   *Skeletal Biology in the Great Plains*. Smithsonian Institution, Washington, D.C.

Johnson, Alfred E.
  1992   Early Woodland in the Trans-Missouri West. *Plains Anthropologist* 37:129–136.

2001   Plains Woodland Tradition. In *Plains*, edited by Raymond J. DeMallie, pp. 159–172. Handbook of North American Indians Vol. 13, Part 1, William C. Sturtevant, general editor, Smithsonian Institution, Washington, D.C.

Johnson, Ann Mary, and Alfred E. Johnson
1998   The Plains Woodland. In *Archaeology on the Great Plains*, edited by Raymond Wood, pp. 201–234. University Press of Kansas, Lawrence.

Johnson, Craig M.
1998   The Coalescent Tradition. In *Archaeology on the Great Plains*, edited by Raymond Wood, pp. 308–344. University Press of Kansas, Lawrence.

2003   *A Chronology of Middle Missouri Plains Village Sites*. Publications in Anthropology. Smithsonian Institution, Washington, D.C.

2007   Jones Village: An Initial Middle Missouri Frontier Settlement. In *Plains Village Archaeology: Bison-Hunting Farmers in the Central and Northern Plains*, edited by Stanley A. Ahler and Marvin Kay, pp. 41–52. University of Utah Press, Salt Lake City.

Johnson, William C., and Kyeong Park
1996   Late Wisconsian and Holocene Environmental History. In *Archaeology and Paleoecology of the Central Great Plains*, edited by Jack L. Hofman, pp. 3–28. Arkansas Archeological Survey, Fayetteville.

Johnston, Francis E., Howard Wainer, David Thissen, and Robert McVean
1976   Hereditary and Environmental Determinants of Growth in Height in a Longitudinal Sample of Children and Youth of Guatemalan and European Ancestry. *American Journal of Physical Anthropology* 44:469–475.

Jungers, William L., Anthony B. Falsetti, and Christine E. Wall
1995   Shape, Relative Size, and Size-Adjustments in Morphometrics. *Yearbook of Physical Anthropology* 38:137–161.

Katzmarzyk, Peter T., and William R. Leonard
1998   Climatic Influences on Human Body Size and Proportions: Ecological Adaptations and Secular Trends. *American Journal of Physical Anthropology* 106:483–503.

Kay, Marvin
1998   The Central and Southern Plains Archaic. In *Archaeology on the Great Plains*, edited by Raymond Wood, pp. 173–200. University Press of Kansas, Lawrence.

Key, Patrick J.
1983   *Craniometric Relationships among Plains Indians*. Department of Anthropology Report of Investigations No. 34. The University of Tennessee, Knoxville.

1994   Relationships of the Woodland Period in the Northern and Central Plains: The Craniometric Evidence. In *Skeletal Biology in the Great Plains*, edited by Douglas W. Owsley and Richard L. Jantz, pp. 179–188. Smithsonian Institution, Washington, D.C.

Key, Patrick J., and Richard L. Jantz
1981   A Multivariate Analysis of Temporal Change in Arikara Craniometrics: A Methodological Approach. *American Journal of Physical Anthropology* 55:247–259.

1990   Statistical Assessment of Population Variability: A Methodological Approach. *American Journal of Physical Anthropology* 82:53–59.

Krause, Richard A.
1995   Attributes, Modes and 10th Century Potting Practices in Northcentral Kansas. *Plains Anthropologist* 40:307–352.

1998   A History of Great Plains Prehistory. In *Archaeology on the Great Plains*, edited by Raymond Wood, pp. 48–86. University Press of Kansas, Lawrence.

2001   Plains Village Tradition: Coalescent. In *Plains*, edited by Raymond J. DeMallie, pp. 196–206. Handbook of North American Indians Vol. 13, Pt. 1, William C. Sturtevant, general editor, Smithsonian Institution, Washington, D.C.

Krogman, Wilton, and M. Yasar Isçan
1986   *The Human Skeleton in Forensic Medicine*. Charles C. Thomas, Springfield, Illinois.

Kurki, Helen K.
2007   Protection of Obstetric Dimensions in a Small-Bodied Human Sample. *American Journal of Physical Anthropology* 133:1152–1165.

Kurki, Helen K., Jamie K. Ginter, Jay T. Stock, and Susan Pfeiffer
2008   Adult Proportionality in Small-Bodied Foragers: A Test of Ecogeographic Expectations. *American Journal of Physical Anthropology* 136:28–38.

Lehmer, Donald J.
1954   The Sedentary Horizon of the Northern Plains. *Southwestern Journal of Anthropology* 10:139–159.
1971   *Introduction to Middle Missouri Archaeology*. National Park Service, U.S. Department of the Interior, Washington, D.C.

Little, Bertis B., and Robert M. Malina
1986   Gene Flow and Variation in Stature and Craniofacial Dimensions among Indigenous Populations of Southern Mexico, Guatemala, and Honduras. *American Journal of Physical Anthropology* 70:505–512.

Logan, Brad
1996a   The Plains Village Period on the Central Plains. In *Archaeology and Paleoecology of the Central Great Plains*, edited by Jack L. Hofman, pp. 123–133. Arkansas Archeological Survey, Fayetteville.
1996b   The Protohistoric Period on the Central Plains. In *Archaeology and Paleoecology of the Central Great Plains*, edited by Jack L. Hofman, pp. 134–139. Arkansas Archeological Survey, Fayetteville.

Ludwickson, John, James N. Gudnerson, and Craig Johnson
1993   Select Exotic Artifacts from Cattle Oiler (39ST224): A Middle Missouri Tradition Site in Central South Dakota. In *Prehistory and Human Ecology of the Western Prairies and Northern Plains: Papers in Honor of Robert A. Alex (1941–1988)*, edited by Joseph A. Tiffany, pp. 151–168. *Memoir 27 Plains Anthropologist*, Lincoln, Nebraska.

Lundy, John K.
1985   The Mathematical Versus Anatomical Methods of Stature Estimate from Long Bones. *American Journal of Forensic Medicine and Pathology* 6:73–76.

McKeown, Ashley H.
2000   Investigating Variation among Arikara Crania Using Geometric Morphometry. Unpublished Ph.D. dissertation, Department of Anthropology, The University of Tennessee, Knoxville.

Malina, Robert M., Maria E. Peña Reyes, Swee Kheng Tan, Peter H. Buschang, Bertis B. Little, and Slawomir Koziel
2004   Secular Change in Height, Sitting Height and Leg Length in Rural Oaxaca, Southern Mexico: 1968–2000. *Annals of Human Biology* 31:615–633.

Malina, Robert M., Henry A. Selby, Peter H. Buschang, Wendy L. Aronson, and Richard G. Wilkinson
1983   Adult Stature and Age at Menarche in Zapotec-Speaking Communities in the Valley of Oaxaca, Mexico, in a Secular Perspective. *American Journal of Physical Anthropology* 60:437–449.

Mayr, Ernst
1956   Geographical Character Gradients and Climatic Adaptation. *Evolution* 10:105–108.

Millar, James F. V.
  1978   *The Gray Site: An Early Plains Burial Ground.* Parks Canada Manuscript Report 304 (2 vols.), Winnipeg.
Mitchell, Mark D.
  2007   Conflict and Cooperation in the Northern Middle Missouri, A.D. 1450–1650. In *Plains Village Archaeology: Bison-Hunting Farmers in the Central and Northern Plains*, edited by Stanley A. Ahler and Marvin Kay, pp. 155–169. University of Utah Press, Salt Lake City.
Newman, Marshall T.
  1956   Adaptation of Man to Cold Climates. *Evolution* 10:101–105.
Nichols, Johanna, and David A. Peterson
  1998   Personal Pronouns: A Reply to Campbell. *Language* 74:605–614.
O'Shea, John M.
  1984   *Mortuary Variability: An Archaeological Investigation.* Academic Press, Orlando.
Owsley, Douglas W.
  1997   Retrospective Analysis and Prospects for the Future. In *Bioarcheology of the North Central United States*, edited by Douglas W. Owsley and Jerome C. Rose, pp. 295–302. Arkansas Archeological Survey, Fayetteville.
Owsley, Douglas W., and Karin L. Bruweldheide
  1997   Bioarcheological Research in Northeastern Colorado, Northern Kansas, Nebraska, and South Dakota. In *Bioarcheology of the North Central United States*, edited by Douglas W. Owsley and Jerome C. Rose, pp. 7–56. Arkansas Archeological Survey, Fayetteville.
Owsley, Douglas W., and Richard L. Jantz
  1978   Intracemetery Morphological Variation in Arikara Crania from the Sully Site (39SL4), Sully County, South Dakota. *Plains Anthropologist* 23:139–147.
Owsley, Douglas W., and Jerome C. Rose
  1997   *Bioarcheology of the North Central United States.* Arkansas Archeological Survey, Fayetteville.
Owsley, Douglas W., Gale D. Slutzky, Mark F. Guagliardo, and Lynn M. Deitrick
  1981   Interpopulation Relationships of Four Post-contact Coalescent Sites from South Dakota: Four Bear (39DW2), Oahe Village (39HU2), Stony Point Village (39ST235), and Swan Creek (39WW7). *Plains Anthropologist* 26:31–42.
Park, Kyeong
  1996   Late Wisconsian and Holocene Environmental History. In *Archaeology and Paleoecology of the Central Great Plains*, edited by Jack L. Hofman, pp. 3–28. Arkansas Archeological Survey, Fayetteville.
Parks, Douglas R.
  1979   The Northern Caddoan Languages: Their Subgrouping and Time Depths. *Nebraska History* 60:197–213.
  2001   Caddoan Languages. In *Plains*, edited by Raymond J. DeMallie, pp. 80–93. Handbook of North American Indians Vol. 13, Part 1, William C. Sturtevant, general editor, Smithsonian Institution Press, Washington, D.C.
Pauketat, Timothy R.
  2004   *Ancient Cahokia and the Mississippians.* Cambridge University Press, New York.
Perttula, Timothy K., and Paul McGuff
  1985   Woodland and Caddoan Settlement in the McGee Creek Drainage, Southeast Oklahoma. *Plains Anthropologist* 30:219–235.
Phenice, Terrell W.
  1969   A Newly Developed Visual Method of Sexing the Os Pubis. *American Journal of Physical Anthropology* 30:297–301.

Prince, Joseph M.
 1998 The Plains Paradox: Secular Trends in Stature in 19th Century Nomadic Plains Equestrian Indians. Unpublished Ph.D. dissertation, Department of Anthropology, The University of Tennessee, Knoxville.

Puskarich, Cheryl L.
 1984 Metric Variation in the Arikara Pelvis. Unpublished Ph.D. dissertation, Department of Anthropology, The University of Tennessee, Knoxville.

Raxter, Michelle H., Benjamin M. Auerbach, and Christopher B. Ruff
 2006 Revision of the Fully Technique for Estimating Statures. *American Journal of Physical Anthropology* 130:374-384.

Roberts, Derek F.
 1953 Body Weight, Race, and Climate. *American Journal of Physical Anthropology* 11:533–558.
 1978 *Climate and Human Variability*. Cummings Publishing Company, Menlo Park, California.

Roper, Donna C.
 1993 A Culture-History of the Pawnee. Report to the Smithsonian Institution, Repatriation Office, Washington, D.C.
 2007 The Origins and Expansion of the Central Plains Tradition. In *Plains Village Archaeology: Bison-Hunting Farmers in the Central and Northern Plains*, edited by Stanley A. Ahler and Marvin Kay, pp. 53–63. University of Utah Press, Salt Lake City.

Ruff, Christopher B.
 1991 Climate and Body Shape in Hominid Evolution. *Journal of Human Evolution* 21:81–105.
 1994a Biomechanical Analysis of Northern and Southern Plains Femora: Behavioral Implications. In *Skeletal Biology in the Great Plains*, edited by Douglas W. Owsley and Richard L. Jantz, pp. 235–246. Smithsonian Institution, Washington, D.C.
 1994b Morphological Adaptation to Climate in Modern and Fossil Hominids. *Yearbook of Physical Anthropology* 37:65–107.
 2002 Variation in Human Size and Shape. *Annual Review of Anthropology* 31:211–232.

Ruff, Christopher B., Markku Niskanen, Juho-Antti Junno, and Paul Jamison
 2005 Body Mass Prediction from Stature and Bi-iliac Breadth in Two High Latitude Populations, with Application to Earlier Higher Latitude Humans. *Journal of Human Evolution* 48:381–392.

Ruff, Christopher B., Erik Trinkaus, and Trenton W. Holliday
 1997 Body Mass and Encephalization in Pleistocene *Homo. Nature* 387:173–176.

Schlesier, Karl H.
 1994 Commentary: A History of Ethnic Groups in the Great Plains, A.D. 150–1550. In *Plains Indians, A.D. 500–1500: The Archaeological Past of Historic Groups*, edited by Karl H. Schlesier, pp. 308–381. University of Oklahoma Press, Norman.

Schneider, Kim N., and Donald J. Blakeslee
 1990 Evaluating Residence Patterns among Prehistoric Populations: Clues from Dental Enamel Composition. *Human Biology* 62:71–83.

Schreider, Eugene
 1964 Ecological Rules, Body-Heat Regulation, and Human Evolution. *Evolution* 18:1–9.

Smith, Richard J.
 2005 Relative Size Versus Controlling for Size. *Current Anthropology* 46:249–273.

Sokal, Robert R., and F. James Rohlf
 1995 *Biometry: The Principles and Practices of Statistics in Biology Research*. 3rd ed. W. H. Freeman, New York.

Steckel, Richard H., and Joseph M. Prince
  2001   Tallest in the World: Native Americans of the Great Plains in the Nineteenth
          Century. *The American Economic Review* 91:287–294.
Steinacher, Terry L., and Gayle F. Carlson
  1998   The Central Plains Tradition. In *Archaeology on the Great Plains*, edited by Ray-
          mond Wood, pp. 235–268. University Press of Kansas, Lawrence.
Stini, William A.
  1974   Adaptive Strategies of Human Populations Under Nutritional Stress. In *Bio-
          social Interrelations in Population Adaptation*, edited by Elizabeth S. Watts, Francis E.
          Johnston, and Gabriel W. Lasker, pp. 19–41. Mouton Publishers, Paris.
Stinson, Sara
  1990   Variation in Body Size and Shape among South American Indians. *American
          Journal of Human Biology* 2:37–51.
Sylvester, Adam D., Patricia A. Kramer, and William L. Jungers
  2008   Modern Humans Are Not (Quite) Isometric. *American Journal of Physical An-
          thropology* 137:371–383.
Tague, Robert G.
  2005   Big-Bodied Males Help Us Recognize That Females Have Big Pelves. *American
          Journal of Physical Anthropology* 127:392–405.
Takamura, Kazuyuki, Shiro Ohyama, Teruki Yamada, and Noburu Ishinishi
  1988   Changes in Body Proportions of Japanese Medical Students Between 1961 and
          1986. *American Journal of Physical Anthropology* 77:17–22.
Temple, Daniel H., Benjamin M. Auerbach, Masato Nakatsukasa, Paul W. Sciulli, and
Clark S. Larsen
  2008   Variation in Limb Proportions Between Jomon Foragers and Yayoi Agricultur-
          alists from Prehistoric Japan. *American Journal of Physical Anthropology* 137:164–174.
Tiffany, Joseph A.
  2007   Examining the Origins of the Middle Missouri Tradition. In *Plains Village Ar-
          chaeology: Bison-Hunting Farmers in the Central and Northern Plains*, edited by Stanley
          A. Ahler and Marvin Kay, pp. 3–14. University of Utah Press, Salt Lake City.
Toom, Dennis L.
  1996   Archeology of the Middle Missouri. In *Archeological and Bioarcheological Re-
          sources of the Northern Plains*, edited by George C. Frison and Robert C. Mainfort,
          pp. 56–76. Arkansas Archeological Survey, Fayetteville.
Trinkaus, Erik
  1981   Neandertal Limb Proportions and Cold Adaptation. In *Aspects of Human Evolu-
          tion*, edited by Chris B. Stringer, pp. 187–224. Taylor and Francis, London.
Walker, Robert, Michael Gurven, Kim Hill, Andrea Migliano, Napoleon Chagnon,
Roberta de Souza, Gradimir Djurovic, Raymond Hames, A. Magdalena Hurtado,
Hillard Kaplan, Karen Kramer, William J. Oliver, Claudia Valeggia, and Taro Yamauchi
  2006   Growth Rates and Life Histories in Twenty-two Small-Scale Societies. *American
          Journal of Human Biology* 18:295–311.
Weaver, Timothy D., and Karen Steudel-Numbers
  2005   Does Climate or Mobility Explain the Differences in Body Proportions Be-
          tween Neandertals and Their Upper Paleolithic Successors? *Evolutionary Anthropol-
          ogy* 14:218–223.
Wedel, Waldo R.
  2001   Plains Village Tradition: Central. In *Plains*, edited by Raymond J. DeMallie, pp.
          173–185. Handbook of North American Indians Vol. 13, Pt. 1, William C. Sturtevant,
          general editor, Smithsonian Institution, Washington, D.C.

Wescott, Daniel J.
  2001   Structural Variation in the Humerus and Femur in the American Great Plains and Adjacent Regions: Differences in Subsistence Strategy and Physical Terrain. Unpublished Ph.D. dissertation, Department of Anthropology, The University of Tennessee, Knoxville.
Wescott, Daniel J., and Deborah L. Cunningham
  2006   Temporal Changes in Arikara Humeral and Femoral Cross-Sectional Geometry Associated with Horticultural Intensification. *Journal of Archaeological Science* 33:1022–1036.
Willey, Patrick S., and Thomas E. Emerson
  1993   The Osteology and Archaeology of the Crow Creek Massacre. In *Prehistory and Human Ecology of the Western Prairies and Northern Plains*, edited by Joseph A. Tiffany. *Plains Anthropologist Memoir* 38:227–269.
Winham, R. Peter, and Francis A. Calabrese
  1998   The Middle Missouri Tradition. In *Archaeology on the Great Plains*, edited by Raymond Wood, pp. 269–307. University Press of Kansas, Lawrence.
Wood, Raymond
  1967   *An Interpretation of Mandan Culture History*. River Basin Surveys Papers No. 39, Bureau of American Ethnology, Bulletin 198. Smithsonian Institution, Washington, D.C.
  1998   Introduction. In *Archaeology on the Great Plains*, edited by Raymond Wood, pp. 1–15. University Press of Kansas, Lawrence.
  2001   Plains Village Tradition: Middle Missouri. In *Plains*, edited by Raymond J. DeMallie, pp. 186–195. Handbook of North American Indians Vol. 13, Pt. 1, William C. Sturtevant, general editor, Smithsonian Institution, Washington, D.C.
Wood, Raymond (editor)
  1998   *Archaeology on the Great Plains*. University Press of Kansas, Lawrence.
Zimmerman, Larry J.
  1985   *Peoples in Prehistoric South Dakota*. University of Nebraska Press, Lincoln.
Zobeck, Terry S.
  1983   Postcraniometric Variation among the Arikara. Unpublished Ph.D. dissertation, Department of Anthropology, The University of Tennessee, Knoxville.

8. Skeletal Evidence of Cultural Variation: Mutilation Related to Warfare

Christopher W. Schmidt, Rachel Lockhart Sharkey,
Christopher Newman, Anna Serrano, Melissa Zolnierz,
Jeffrey A. Plunkett, Anne Bader

Summary Statement: At the heart of this chapter is the expression of human cultural variation: in particular, the manner in which individuals who lived between 6,000 and 800 years ago in Indiana were treated after being killed. Rather than being static throughout time and space, the manifestations of violence in prehistory are quite varied and evident when studied in detail. The archaeological remains that were analyzed herein come from along the Ohio River, which was an area of intense occupation throughout prehistory. Although this is the first study to document mutilation in Indiana populations, several sites from Kentucky have had similar mutilations reported. Thus, the Indiana sites appear to be part of a regionwide phenomenon, although they exhibit some idiosyncrasies (like tongue removal). Through time, violent acts were perpetrated in different ways, with later populations emphasizing killing greater numbers of people and earlier populations emphasizing removing heads and limbs as "trophies" using rather consistent procedures to do so. Overall, the implications of this paper are that ancient people changed how they viewed the proper treatment of warfare victims; more detail should be paid to cultural behaviors of the past, even well-documented ones like warfare, so that subtle but important aspects of human cultural diversity can be discovered.

Introduction

A fundamental aspect of anthropological and archaeological research is to document and explicate human variation, both cultural and biological.

Human Variation in the Americas: The Integration of Archaeology and Biological Anthropology, edited by Benjamin M. Auerbach. Center for Archaeological Investigations, Occasional Paper No. 38. © 2010 by the Board of Trustees, Southern Illinois University. All rights reserved. ISBN 978-0-88104-095-1.

Studying each of these is necessary if we are to elucidate the myriad relationships we see among societies. Groups that are closely linked biologically often have different cultural characteristics, that much is well-known. But, what factors lead to behavioral divergences? The current chapter explores the distribution of a particular cultural practice, mutilation, in the Eastern Woodlands of the United States in order to understand the factors that led to its various manifestations in the archaeological record.

Middle and Late Archaic skeletons with evidence of decapitation, glossectomy, dismemberment, and scalping have been discovered recently in southern Indiana. At least one victim was found in each of five cemeteries located along the Ohio River, the traumata most likely being perimortem. These particular cases from Indiana are similar in trauma types and patterning to contemporaneous examples from Kentucky and Tennessee and appear to be part of a regional phenomenon (e.g., Mensforth 2001; Smith 1993, 1995; Snow 1948). Although some forms of violence may have waned during the early part of the Woodland period (e.g., Milner 1995), by the late prehistoric mutilations—particularly scalping—became widespread throughout the Eastern Woodlands, as well as the rest of the continent (e.g., Berryman 1980; Holliman and Owsley 1994; Lambert 2002, 2008; Maschner and Reedy-Maschner 2007; Miller 1994; Milner 1995, 2007; Milner et al. 1991; Owsley and Berryman 1975; Smith 1997; Walker 2001; Willey 1990).

A factor that distinguishes the Middle/Late Archaic violence (which occurred 6000–3000 B.P.) from that of late prehistory (ca. 1000–500 B.P.) is the mortuary treatment of the mutilated bodies. During the Middle/Late Archaic, mutilation victims have similar cut mark patterns and are buried singly in manners stylistically similar to other burials in those cemeteries. By contrast, late prehistory mutilation victims have cut marks that are highly variable (Olsen and Shipman 1994), often have carnivore damage from being left on the surface after death (Milner et al. 1991), and are sometimes found in shared graves (those with more than one person) lacking the care given to those whose deaths did not result from violence. Thus, the developing picture is that rather than representing a monolithic pattern of warfare-related damage, the osteological record points to a shift in the circumstances of mutilation and in the treatment of mutilation victims.

Violence in small-scale societies of the Americas is well documented in the ethnographic record (Chacon 2007; Ember 1978; Ember and Ember 1997; Harner 1972). From the Jivaro of South America to the Indians of nineteenth-century California, head-hunting, as well as other means of taking human "trophies," was a rather common practice among numerous recent tribal-level societies (Chacon 2007; Ember and Ember 1997; Lambert 2007). Often, violence yielded a victim who was used, in whole or in part, to demonstrate physical or spiritual prowess or to ameliorate supernatural phenomena (Ember and Ember 1997: Lambert 2007).

Such behavior is not limited to the very recent; archaeological evidence exists for pre-Columbian violence in every region of North America including the Arctic (Maschner and Reedy-Maschner 2007) and Canada (Williamson 2007) as well as the American Northwest (Lovisek 2007), Southwest (LeBlanc 1999; Schaafsma 2007), West Coast (Andrushko et al. 2005; Lambert 1997, 2007), Plains (Owsley et al. 1977, 2007; Willey 1990), Southeast (Jacobi 2007), Northeast (Snow 2007), and Eastern Woodlands (Lockhart and Schmidt 2007, 2008; Mensforth 2001, 2007; Smith 1993, 1995). Although the motivations for the violence will likely remain poorly understood, the osteological evidence is unambiguous to the extent that violence explains much of the circumstances surrounding the deaths of those who were eventually mutilated. What is less well understood is how mutilation and trophy taking changed over the years. Clearly, by late prehistory the taking of human heads and scalps was allied to military action (e.g., Willey 1990). But is this explanation sufficient for victims of mutilation 4,000-5,000 years earlier? It may be that in order to more fully understand trophy taking in general, and in the Middle/Late Archaic in particular, analyses should not be limited to perimortem trauma, which is similar from place to place and from time to time. Refining our understanding of trophy taking also demands study of the trophy elements themselves as well as the eventual mortuary disposition of those who suffered mutilation.

This chapter describes five southern Indiana cemeteries along the Ohio River with evidence of interpersonal violence. Each site bears a common thread in the methodology of ritual violence; however, each site also has certain idiosyncrasies indicative of a local manifestation of what was probably a geographically broad phenomenon within the Ohio River valley (Figure 8-1). All the sites described here are either curated by or currently on loan for analysis at the University of Indianapolis.

Firehouse Site (12D563)

Located on a high bluff overlooking the Ohio River floodplain in southeastern Indiana, Firehouse (12D563) was excavated by Jeff Plunkett of Landmark Archaeology, Incorporated, in 2004. It produced over 100 features and thousands of lithic and osseous artifacts, including what may be the single greatest number of Riverton points ever described by professional archaeologists from a one locality (currently numbering in the hundreds). Other lithic tools include a small cache of hafted axes and limestone slabs that may have lined the site. Bone tools include combs, pins, and several atlatl fragments.

A group of five skeletons was located near the southern margin of the site although the burials were not clustered as indicative of a cemetery. It is unclear if they were deposited within domiciles or a plaza. All the interred were adults, and at least one was female. Burial 4 was a highly fragmented, tightly bundled burial for which age and sex could not be determined. Burial 1 was a tightly flexed old adult male who had a broken right tibia and fibula that had healed with lateral displacement of the distal aspect. Burial 3 was excavated by Plunkett and students from the University of Indianapolis under the direction of the lead author. This young adult male was loosely flexed, with his back extended and his heals pulled up toward his sacrum. His left arm was extended with his left hand resting just laterally to his left hip. His right arm was missing below the humerus and his head was absent (Figure 8-2).

Excavations revealed that five Riverton or Riverton-like projectile points were located around his thorax. A sixth point was found immediately adjacent to

Figure 8-1. *Locations of Middle/Late Archaic sites from west to east: Blue-grass (12W162), Meyer (12SP1082); 12HR6; 12FL73; Firehouse (12D563).*

his second lumbar vertebra, which had a healed crescentic notch missing from the proximal aspect of the spinous process. From anatomical right to left, the point had passed through the spinous process, obliterating part of it; it then struck the left transverse process and came to rest on the exterior aspect of the lamina. The damage to the lumbar vertebra was clearly antemortem and was not immediately associated with the activities that led to his death; it also makes clear that the violence he suffered at his death was not the only violence he encountered during his life. Unfortunately, none of the other points penetrated bone, so it is not obvious that he was killed by these points. But, they are not displayed in a single area like a cache and they are located near scapula and rib fragments that may have impact-related fractures.

Figure 8-2. *Adult male from Firehouse site (12D563) missing head and right forearm. (Photograph by Jeff Plunkett.)*

The first two cervical vertebrae are missing along with the entire skull. There are deep cut marks on cervical vertebrae 3 through 5 indicating that the skull was removed while soft tissues were still present. The cut marks are located on the left side of the vertebrae and indicate both chopping and incisive movements of the cutting instrument. The distal right humerus bears cut marks indicative of chopping and cutting, and these are found on the anterior, medial, and lateral surfaces with no marks on the posterior aspect. The cuts on the anterior surface suggest forceful slicing, while the lateral marks indicate chopping.

The skeleton has what look like stab marks on several of the ribs. Although there is some taphonomically related damage elsewhere on the ribs, the marks are clearly not taphonomic in origin. The stab marks located on the ventral and lateral portions of the bones penetrate just a few millimeters into the cortex. No cut or stab marks are found on the sternum. There are no other cut or stab marks on the skeleton and no other individuals at the site bear such evidence of violence.

12HR6

This site is located southwest of Firehouse on the Ohio River in Harrison County. The remains are extremely fragmentary, having been damaged by floods and looters who plundered the site for years. They were recovered in the 1990s by archaeologists from the Indiana Department of Natural Resources, Division of Historic Preservation and Archaeology, as well as by Plunkett. The majority of the remains are cranial fragments representing at least 20 people, mostly adults. How the bodies were initially interred is unclear, but the looting pits indicate close burial, suggesting some type of cemetery. Artifacts from the site place it in the Late Archaic, making the human remains about 3,000–4,000 years old.

Among the remains are two fragments of particular interest to the current study. One left temporal fragment has distinct cut marks above the external auditory meatus. When the piece is held in anatomical position, the roughly parallel lines are horizontally oriented and indicative of scalping (Figure 8-3). Unfortunately, no other elements that match this particular cranial fragment have been found, limiting the certainty of the scalping assertion. None of the remaining cranial fragments, of which there are several hundred, bear cut marks.

A distal right humerus fragment also has cut marks. The incisions are along the anterior surface as well as the medial and lateral margins and are strikingly similar to the humeral cut marks found at Firehouse in size and shape. Additionally, neither has any cuts on the posterior aspect. The size of the humerus is consistent with that of a male, although sex determination on such a fragment is tentative. Of the dozens of postcranial fragments, this is the only one that has cut marks.

At this time, determination of whether the cranial fragment and the humeral fragment represent a single individual or two separate people is not possible. The fragments are of similar color and robusticity, but a survey of all remains shows that many bone fragments are comparable, minimizing any confidence that gross observation alone will determine the actual relationship between these bones. Perhaps eventual chemical or biomolecular study can answer this question. Until then, it will be assumed that at least one person, and no more than two, suffered from perimortem trauma.

Bluegrass Site (12W162)

Located in Warrick County in an upland lacustrine plain near Bluegrass Creek, the Bluegrass site was excavated by Russell Stafford of Indiana State University in the 1980s (Stafford et al. 2000). Although near the Ohio River, it is located the farthest inland of all the sites discussed here. The site is somewhat large and includes habitation components. The material culture, especially the bone pins, is similar to what is seen at the Black Earth site in southern Illinois (Stafford et al. 2000). Although not as large as Black Earth, similarities include numerous features, dense middens, many lithic and osseous artifacts, and evidence of yearly use—albeit probably seasonal.

Bluegrass dates to the terminal Middle Archaic and produced 82 interments, nearly all of which were single burials. Skeletons were found in flexed, tightly

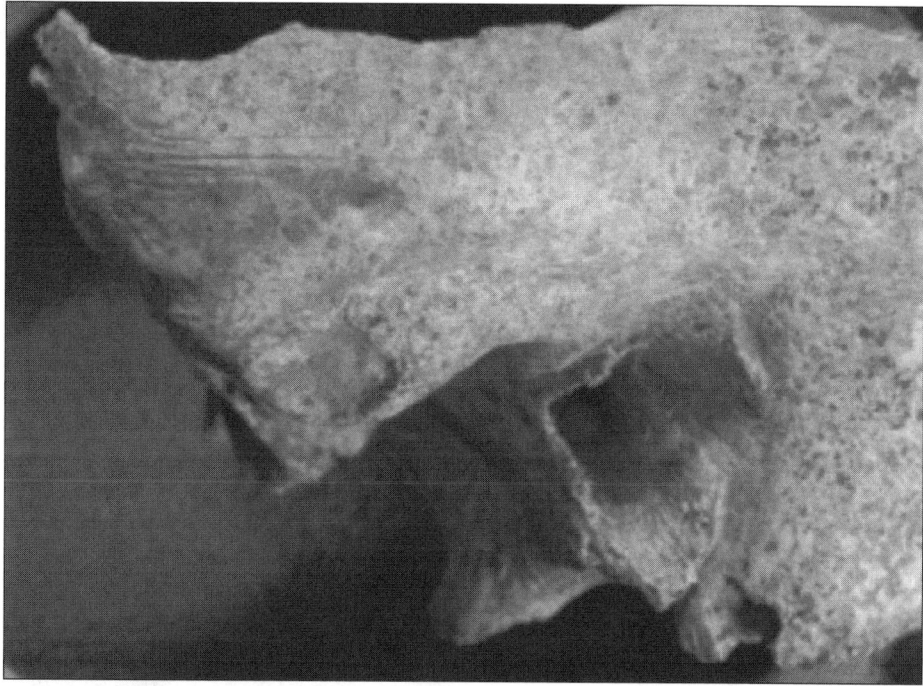

Figure 8-3. *Possible scalping cut marks on a temporal fragment from 12HR6. (Photograph by Rachel Lockhart Sharkey.)*

flexed, and extended positions (Mays 1997). Males and females were roughly equal in number and nearly 14 percent of the burial population were children under one year of age. Among the extended skeletons was a single young adult female missing her skull (including C1 and C2) and her left forearm (Figure 8-4).

Despite the fact that this site predates Firehouse and 12HR6 by as may as 2,000 years, the cut mark morphology and placement are nearly identical to those of the younger, more eastern sites. The forearm cuts are limited to the anterior medial and anterior lateral surfaces. No cut marks are found on the posterior side. The chopping and cutting marks on the cervical vertebrae are a bit heavier than what is seen at Firehouse, but overall the placement of the cut marks is strikingly similar (Figure 8-5). A second individual at this site is missing limbs and head, although it is not clear at this time if all were removed antemortem.

Meyer Site (12SP1082)

Archaeologist Anne Bader and a crew of volunteers, including the lead author and students from the University of Indianapolis, excavated the Meyer site in 2004. This cemetery, located on a backwater slough near the Ohio River in Spencer County, produced over 20 individuals. Dating this site to the terminal Middle Archaic comes from stylistic similarities of decorated bone pins that are almost identical to those from Bluegrass.

Figure 8-4. *Extended female from Blue-grass site (12W162) missing head and left forearm. (Photography by Russell Stafford.)*

Unfortunately, the site was discovered because a landowner was remodeling part of his home and several burials were damaged with augers and heavy machinery. Most of the archaeological excavation consisted of salvaging exposed burials; however, of the burials excavated in situ, all were flexed and on their side. The burials comprised adult males, females, and subadults and were clustered enough to imply burial within a cemetery. No distinct habitation area was located as the excavation plan was simply to locate and conserve the human remains.

One of the burials was a subadult interred on his right side, loosely flexed. Radiographs of his erupting dentition suggest he was 12-to-15 years old, while morphological and metric features of the cranium, mandible, and dentition indicate he was male (Buikstra and Ubelaker 1994; Ditch and Rose 1972; Owsley 1982; Scott and Parham 1979). His right arm was extended and in his right hand was his skull, C1, and C2 (Figure 8-6). Cut marks are present on the cervical vertebrae that remained with the rest of the body.

The mandible was in its anatomical position, indicating that soft tissues were in adherence when the head was removed. Cut marks are present on the distal aspect of the right ascending ramus, the inferior margin of the right corpus, and on the medial (lingual) aspect of the right corpus along the mylohyoid line. The ramus and inferior corpus cut marks include chops and incisions. These

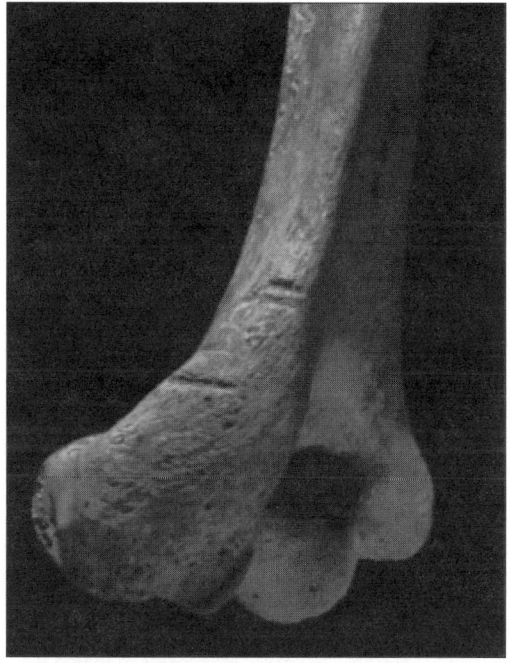

Figure 8-5. *Detail of lateral cut marks on distal left humerus from Bluegrass female. (Photograph by Rachel Lockhart Sharkey.)*

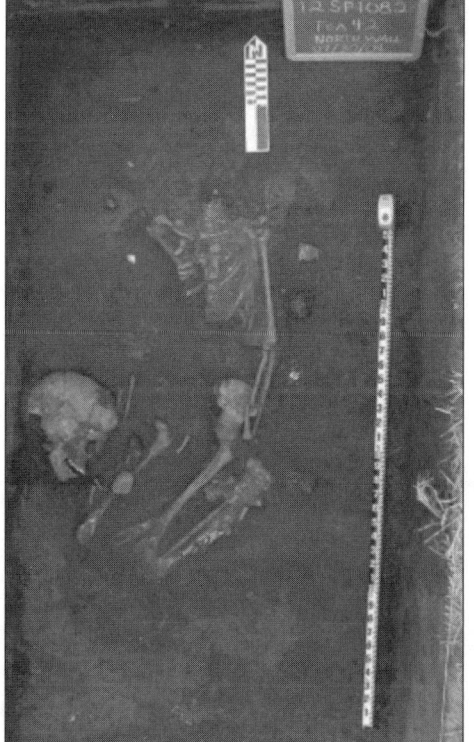

Figure 8-6. *Burial 42 from Meyer site (12SP1082). (Photograph by Anne Bader.)*

were not created as an artifact of head removal; detailed analysis of the cut marks themselves by forensic anthropologist Steve Symes of Mercyhurst College indicates that the directions of these cuts do not line up with the cervical cut marks (Lockhart et al. 2009).

Of the mandibular cut marks, perhaps the most intriguing are those found on the mylohyoid line. Like the other mandibular cuts, these were not created during head removal but were formed as slicing movements in planes distinct from the chopping marks present on the cervical vertebrae. They consist of two well-defined subparallel cut marks that course for approximately two centimeters along the mylohyoid line. Each line consists of small incisive marks created as a finely sharpened implement was worked back and forth. There are no such marks on the left ramus.

No cut marks near the mandibular condyles and no attempt to remove the mandible from the cranium indicate glossectomy; rather the tongue was removed by severing one side of the mylohyoid muscle insertion, pulling the tongue inferiorly (relative to the mandible) and then removing it by cutting only through soft tissues near the left mandibular body. This manner of tongue removal is similar to what Willey (1990) reported from the late prehistoric Crow Creek massacre site in South Dakota and is apparently consistent with tongue removal from butchered animals.

Additional traumata on this skeleton include a possible blunt-force wound on the occiput and punctures to some ribs. The occipital injury is obscured by taphonomic fracturing but is distinct because of its semicircular shape and internally beveled margin. The wound gives rise to two significant radiating fractures whose exposed inner cortex and trabeculae are as dark as the external table of the skull. The taphonomically caused fractures all have much lighter coloring of their exposed cortices and trabeculae. The few puncture marks suggest that some stabs were made to the chest, but the extremely fragmentary nature of the ribs makes it difficult to determine exactly where those stabs took place. Most likely they occurred on the ventro-lateral aspect of the thorax since they are not adjacent to the sternal or vertebral ends.

There is significant carnivore damage on this skeleton. The right humerus is absent and the right scapula and clavicle both have evidence of carnivore chewing. In fact, most of the right arm is missing; only the distal third of the arm remains, including the entire hand and the complete distal aspects of the radius and ulna. The damage happened well after the body was buried leaving the head and hand undisturbed in the ground when the humerus and parts of the radius and ulna as well as part of the pelvis were dragged away.

12FL73

Located in Floyd County not far from 12HR6, this site sits on the Ohio River bank. It is part of an enormous site, or collection of contemporaneous sites, that stretch for over half a mile. At least two distinct shell middens, each over 10 centimeters thick, are exposed in the river bank, about one and two meters below the current ground surface, respectively. Rick Burdin of the University of Kentucky is leading the research on this area and has concluded that 12FL73 is likely an early Late Archaic site (Stafford and Cantin, 2009). Crews from the University of Indianapolis excavated the burials in 2001.

Most of the site is not a mortuary, but the locality deemed 12FL73 has produced nine burials eroding from the riverbank. Most of these burials were located after exposure by erosion, so the skeletons were generally very incomplete. Still, some mortuary evidence could be gleaned. For example, at least three burials were flexed and laying on their right sides. Both males and females are present and there are very few subadult remains. No attempt was made to find burials that were not already exposed because the erosion was undercutting a narrow riparian forest that was along a state highway, making any effort to work deeply into the riverbank very hazardous.

No decapitated or dismembered burials were found at this site. Instead, one adult male burial was accompanied by 10 forearms. Study of the bone morphology indicates that left and right elements can be matched up for all of the forelimbs. Thus, both the left and right forelimbs were taken from five individuals. The 20 forelimb long bones (10 radii and 10 ulnae) were stacked in a cache placed beside the flexed burial. A few metacarpals also were present and at least one scaphoid was in articulation with one of the radii suggesting that at least one of the forelimbs had its hand with it at the time of burial. Some of the radii and ulnae were paired with their natural mates (i.e., they came from the same individual) but others were separated. This implies that some of the forelimbs had soft tissue present at the time of burial, while others did not.

All the bones were arranged so that their distal ends faced out of the riverbank and the proximal ends faced into it. In that orientation, the hands, or at least hand-end of the bones, were pointing in the opposite direction of the head of the individual with whom they were buried. Measurements of the bones indicate that they are all likely from males and they all have complete fusion of their epiphyses. None of the bones have cut marks on them. At least two decorated bone pins were found with the cache, but no other artifacts were found. Burials that include trophy limbs have been found at Green River sites including Indian Knoll (Snow 1948), but no other site has as many forelimbs buried with a single individual.

Implications of Cut Marks

Perhaps the first consideration to make when encountering such a frequency of cut marks is bone cleaning. Numerous Native American populations would clean soft tissues off the bones of the recently deceased as part of a large mortuary ritual where all who had died for the last several years were buried in a common grave, or ossuary, (e.g., Schmidt and Larsen 2002). Thus, all individuals entered the ground in a skeletal condition, even if they had only died recently (Ubelaker 1974). Olsen and Shipman (1994) established base criteria for elucidating evidence for bone cleaning in ancient skeletons. Skeletons that have been cleaned of their flesh tend to have cut marks throughout, especially in those places where large tendons cross joints, such as around the shoulders and hips.

Similar cut marks are common on the skull and mandible. Cuts may be accompanied by scrape marks, which are wide scratches where the cutting implement was used to undercut and peel away soft tissue, rather than to slice it. Although widespread, the cut and scrape marks are usually patterned, appearing bilaterally and adjacent to joints. Defleshing can also lead the burial of partial or tightly bundled skeletons.

While sharing some of these bone-cleaning characteristics (i.e., having cut marks), the skeletons discussed here do not have distinct evidence of bone cleaning. The cut marks are not widespread; none are at significant joint locations, such as around the pelvis; there are no scrape marks; and neither the Bluegrass nor the Firehouse individual is tightly bundled. Thus, bone cleaning seems to be an unlikely explanation.

Ancestor veneration involves the removal of body parts, such as skulls and limbs, from a decomposed or decomposing body that has been placed within a burial crypt or structure. Thus, removal of bony elements should not require significant cutting. Since the Bluegrass and Firehouse sites heads were taken perimortem, as evidenced by significant cervical cut marks, such an explanation is untenable.

Another consideration is that these individuals are victims of warfare. Osteological evidence indicates that warfare was an important component of life in pre-Columbian North America (e.g., Lambert 2002, 2007, 2008; LeBlanc 1999; Milner 1995; Milner et al. 1991; Walker 2001; Willey 1990). In addition to projectile-point injury, traumata used to argue for the presence of warfare include scalping, trophy taking of limbs and heads, other body mutilation (i.e., removal of soft-tissue trophies such as noses and ears), perimortem blunt-force trauma, puncture and stab traumata, and carnivore damage created once the dead are left on the battleground. Victims of warfare are commonly buried in shared graves and are not afforded the same ritualized placement as those who die during times of peace (e.g., Lambert 1997; Milner et al. 1991; Willey 1990). There are a number of documented instances of pre-Columbian warfare from the Plains and Eastern Woodlands of the United States that can be used for direct osteological comparison including Crow Creek, Norris Farms, Orendorf, Mobridge, Fay Tolton, Larson, Cahokia, and Koger Island (Bridges 1996; Emerson 2007; Holliman and Owsley 1994; Milner et al. 1991; Owsley et al. 1977; Steadman 2008; Willey 1990).

Perhaps the best described are those from Norris Farms in Illinois (Milner 1995) and Crow Creek in South Dakota (Willey 1990). Although distinct in the manners in which their cemeteries formed, both of these sites display overwhelming evidence for warfare. Crow Creek is a massacre site, a cemetery of nearly 500 men, women, and children who died in a single attack. Roughly 90 percent of the victims were scalped and others were mutilated in numerous fashions, leaving large wounds on the faces and cranial vaults. The dead were left for some time, as seen by scavenger damage to a significant number of bones. In contrast, the Norris Farms site is not a massacre; the cemetery is an accumulation of victims of periodic attacks. The makeup of the cemetery, however, is strikingly similar to that of Crow Creek. A majority of the victims were scalped, blunt-force trauma was rampant, and carnivore damage was present. Another commonality between these two sites is the manner in which the dead were disposed of. At both, the victims were placed in shared graves—in the case of Norris Farms, a series of shared graves. The warfare victims were not afforded standard burials as seen at other contemporaneous sites.

The Firehouse and Meyer sites burials have explicit evidence for violent deaths. The Firehouse male has projectile-point damage and punctures on his thorax. The young male from Meyer site likely has perimortem blunt-force trauma and stab marks. All the sites have some example of trophy taking. At Firehouse and at Bluegrass the mutilated individual lost his or her head and a single forearm. At 12HR6 at least one person lost a scalp and forearm and at Meyer the victim seems to have lost his tongue. The site 12FL73 gives some insight into what happened to people who lost their forearms. In at least five instances, men lost both of their forearms and those limbs were eventually buried with another man.

Except for those at 12HR6, the burials do not have evidence of scalping. Scalping, however, was present during the Late Archaic in the Tennessee River valley and at Green River sites including Indian Knoll (Smith 1993, 1997; Snow 1948; Webb 1946; Christopher W. Schmidt and Rachel Lockhart Sharkey, personal observations of skeletons at Indian Knoll not previously described). If we see scalping as a convenient means of taking a trophy from the head, then perhaps the removal of the Bluegrass and the Firehouse heads can be seen as essentially serving the same purpose as scalping. Taking heads as war trophies was rather common in the New World, particularly in Central and South America (e.g., Chacon and Dye 2007; Tung 2008).

It may be thought that the osteological expression of warfare and the mortuary treatment of the dead are functions of population size. Crow Creek and Norris Farms were late prehistoric sites that probably consisted of several hundred people, especially in the case of Crow Creek. This may make these groups seem too large for comparison to Middle and Late Archaic populations that likely lived in less densely populated villages. To investigate how scale affects the osteological evidence of warfare in late prehistoric sites, we studied a cemetery from northwestern Indiana, the Bicycle Bridge site (12C335). Although cemetery size does not always equate with population size, in conjunction with overall land use and settlement of the landscape, it can provide some input regarding relative population density (e.g., Jefferies et al. 2005). The Bicycle Bridge cemetery is small and the known surrounding sites are modest in size and usually were briefly occupied. Thus, while the site is roughly contemporary to Norris Farms, it is minute by the standards of late prehistory. The cemetery was found in Carroll County, Indiana, and was excavated by Purdue University in 1999. Although part of the site was disturbed by construction work, it was intact enough for osteological study at the University of Indianapolis.

Bicycle Bridge has a minimum of 19 individuals (11 adults) and is clearly much smaller than Norris Farms ($n = 264$) and Crow Creek ($n = 486$). However, the evidence for conflict is unmistakable. One individual has a projectile point embedded in an upper thoracic vertebra, the wound unhealed (Figure 8-7). There also is a small shared grave of two females and a male, all of whom were scalped. Nearly half of all the adults bear evidence of perimortem violence. Healed depressed fractures of the skull are found on at least three of the adults. The num-

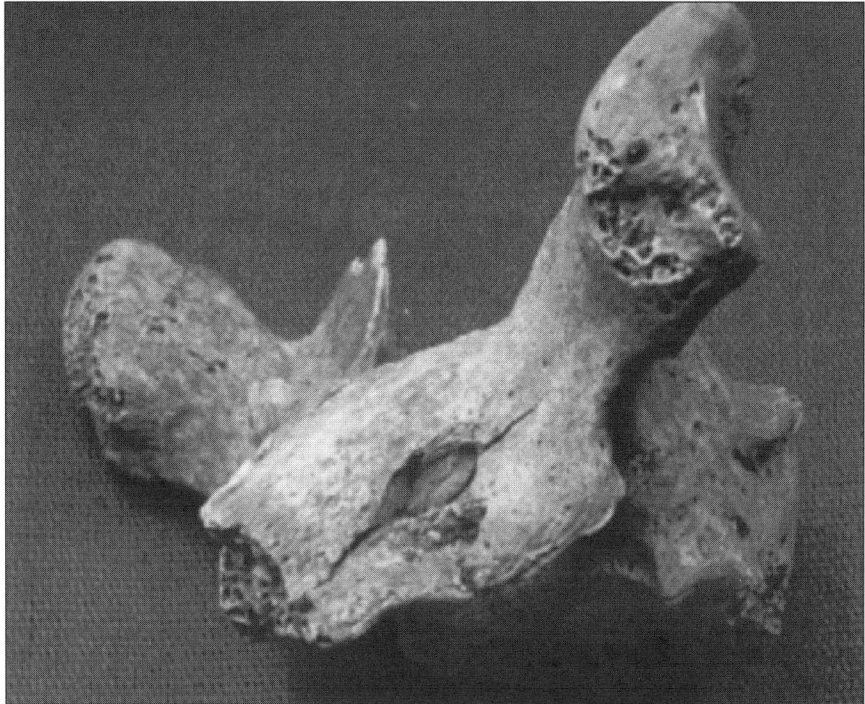

Figure 8-7. *Projectile point embedded in an upper thoracic vertebra from the late prehistoric Bicycle Bridge site (12C335). (Photograph by Rachel Lockhart Sharkey.)*

ber of injuries and the presence of a shared grave are strikingly similar to Norris Farms, despite the fact that the site is miniscule by comparison. The implication here is that the osteological profile indicating warfare is not strictly a function of site size—even small late prehistoric sites can harbor the indicators of warfare that are found at larger sites.

Thus, during the late prehistoric, large and small sites shared similar osteological patterns of violence. The remaining question is why the Middle to Late Archaic burials look so different. Why didn't warfare during this period produce osteological manifestations similar to those of late prehistory? Clearly, at least one person from each of four of the Indiana sites (Bluegrass, Meyer, 12HR6, and Firehouse) died a violent death. Each of these victims, in turn, lost a part of his or her body, presumably to attackers. But there are details in the Archaic skeletons that differentiate them from those of the late prehistoric. First, the means of removing the limbs is remarkably consistent over time and space. Second, there is little evidence for carnivore activity. Finally, the mortuary treatment of the victims is not at all like that seen at Crow Creek, Norris Farms, or Bicycle Bridge.

The strategy for removing the forelimbs during the late Middle Archaic through the Late Archaic changed little. The tools used and the strokes used to slice through the soft tissues are almost identical when comparing the woman from the Bluegrass site (ca. 6,000 years old) to the young man from the Firehouse site (no more than 4,000 years old). The cut marks on both consist of nearly parallel transverse incisions just above coronoid fossa as well as deeper cuts or small chops on the epicondylar ridges above each epicondyle. No other cuts are present on the rest of the humeri. The morphology of the 12HR6 cuts is identical as well; however, since only the distal end remains, it is impossible to determine if cuts were present elsewhere on the bone.

To better comprehend the forelimb removal process, two authors (Rachel Lockhart Sharkey and Christopher W. Schmidt) conducted experimental forearm removals on cadavers housed at the University of Indianapolis. They used slate and steel blades to cut through the soft tissues in an attempt to replicate the cuts made in antiquity. While cutting through the muscle and tendons was fairly easy, separating the ulna from the humerus was very difficult due to the tough fibrous joint capsule. One way to dislodge the ulna from the distal humerus was to twist it. Unfortunately, in each attempt at using this method the ulna fractured, a condition not seen in the ulnae from the 12FL73 cache. Our cut marks differed from the archaic by being far shallower and occurring in more places, especially on the articular surfaces of the distal humerus—the articular cartilage minimized most of this damage, however. Despite our relative ineptitude, removal of a forelimb occurred in only a few minutes. Although preliminary, our cadaver study indicates that the precise and consistent manner in which the forelimbs were apparently removed was not required for swift forelimb removal. It indicates that a tradition was in place whereby there was a "right way" to perform limb removal. The motivations for such attention to detail are unclear, but it is possible that, as in present-day populations who take human trophies, they may have believed that failure to remove and handle the trophy in a particular way might lead to the victim's spirit harming the trophy taker (e.g., Harner 1972).

Such precision in limb removal is not always seen in other North American populations that took trophies. Andrushko and colleagues (2005) describe a population in California dating to around 2,000 years ago that also removed forearms from victims. Unlike the material from Indiana, the humeral cut marks were not limited to a few locations on the distal end. In fact, cut marks were found all over the humeri as well as on other bones of the skeleton. The authors concluded that the cut marks followed the same distribution as seen on butchered seals. Again, motivation for this method of limb removal is unclear, but the authors feel that it was associated with intergroup conflict.

Patterning of cut marks on late prehistoric warfare victims indicates that precision in cut mark morphology was less vital. Although certain types of cut marks are consistent, as in those indicating scalping, tremendous variability in scalping cut mark expression can exist within a single cemetery. Willey (1990) describes several "styles" of scalping at Crow Creek and even the Bicycle Bridge site scalping victims have cut marks in different locations. Thus, one small late prehistory site has more cut mark variability than is expressed over a period of 2,000 years from the Middle to the Late Archaic.

Another osteological deviation from the late prehistory warfare model is seen in the treatment of the dead after the conflict. Shared graves found at Crow Creek (which, in terms of its size, probably had a true mass grave), Norris Farms, Koger Island, and Bicycle Bridge are generally less stylized means of disposal than the graves of those who did not die in battle. War dead are collected from the surface and placed into a large pit. At times these individuals are put into a nearly flexed or extended position, but usually they are not accompanied by significant grave offerings or buried individually. In contrast, the victims from the Middle/Late Archaic are no less stylized in their burials than their compatriots, and sometimes they appear to be even more so. The burial at Firehouse was less flexed than the rest of the skeletons at that site, but was located near others in the cemetery and oriented in a similar manner. This is true for the Bluegrass female whose burial is consistent with everyone else in the cemetery. The subadult at Meyer has his extended arm holding his head, a condition that needed at least as much time, if not more, to arrange compared to the other skeletons in his cemetery. In fact, no other burial at Meyer comes close to this level of symbolic treatment: They are all flexed or tightly flexed.

It appears, then, that during the Middle/Late Archaic the victim was not treated with any less care than the rest of the population. It also is curious that each site has a single individual (although Bluegrass may have two) missing part of his or her body. Is it possible that the mutilated body in each cemetery is part of a laying-in ceremony (e.g., Byers 2005)? Was such a ritual necessary to properly dedicate a burial ground? Or are the mutilations evidence of human sacrifice? Although the Meyer site subadult has a blow to his head that is reminiscent of cranial trauma seen on sacrifice victims in South America, it is impossible to support or refute this supposition. What is known is that the victims in each cemetery are morphologically similar to those with whom they are buried and have virtually identical stable carbon and nitrogen isotopic signatures (Chambers et al. 2010). They do not appear to be "outsiders" or to have consumed anything that distinguishes them from the others in their respective cemeteries. Further study in this regard, however, is under way, including studies of oxygen isotopes and DNA.

The final significant discrepancy between late prehistory and the Middle to Late Archaic is in carnivore damage. Nearly every late prehistoric cemetery shows some evidence of carnivore damage, a result of bodies being left on the surface. At Norris Farms such damage is extensive, even though the attacks apparently did not kill everyone in the group: Somebody came back and buried the victims in shared graves several times over. For the mutilation victims during the Middle and Late Archaic, carnivore damage is almost completely absent. The exception to this is the subadult burial at the Meyer site. This burial shows carnivore damage to the right clavicle, scapula, right ribs, and right pelvis. There are no carnivore marks on the neck or the skull. Although it is possible that the boy was killed and left on the surface for some time, the carnivore damage indicates that it took place after the body was buried. Based on direct observations of carnivore damage at modern forensic sites (e.g., Schmidt and Greene 1999), it is difficult to imagine how such a localized swath of body removal could take place on a fleshed victim. For example, how would the right os coxa be pulled away in its entirety without damaging the sacrum on a body that had all of its muscles, tendons, and ligaments intact? The absence of carnivore damage on the rest of the Archaic mutilation victims indicates that the bodies were not left on the surface for long after they were killed. This is similar to examples from the ethnographic record of extant "head-hunting" groups. For example, the Jivaro of South America routinely ambush victims at night in or near their homes (Harner 1972). Such attacks leave the victims' bodies in domiciles where they are readily found by other members of the community.

Is This Warfare?

In his 2007 article on Hopewell trophy heads, Seeman argues that, although the taking of heads during the Middle Woodland was likely a violent act, it "must be a different kind of war." This is precisely the sentiment we wish to engender here. The debate, as we see it, is not whether or not violent acts occurred during the Middle and Late Archaic periods in Indiana. They did. The real questions regard the context of these violent acts and why the Middle/Late Archaic violent acts yield osteological phenomena (i.e., cut mark morphology and mortuary treatment) different from those produced a few millennia later during the Late Woodland and Mississippian periods. The comparison of the late prehistoric warfare victims to those from the Middle and Late Archaic has allowed us to draw certain conclusions.

First, the process of trophy taking changed from the Archaic to the late prehistoric. At first, the process itself seems to be an important part of the trophy taking ritual. Care was taken to make sure that the limb (or limbs) and head were taken in a particular manner. This practice was passed down for many generations and became standardized along the entire length of the Indiana aspect of the Ohio River and even along the Green River. Second, the scale of killing changed through time. Archaic trophy victims seem to be fairly modest in number with just one or perhaps a few victims present in each cemetery. By late prehistory, large numbers of people might be killed in a single attack. Given that Middle/Late Archaic trophies appear to be taken via attacks that focused on just a few victims at a time, it may be that the trophy taking was an important reason for the killings, much like what has been documented for modern "headhunters." During the late prehistoric, however, trophy taking may have been secondary to the act of killing. This does not mean that warfare, even violence like that seen during the late prehistoric, was absent during the Middle and Late Archaic. Jacobi (2007) documents evidence of shared graves in the Southeast, and shared graves may be present at some of the Green River sites (e.g., Mensforth 2001, 2007). What is evident is that at least some of the violence seen during the Middle/Late Archaic is associated with a particular means of trophy removal from just one or a few victims at a time; it does not appear to stem from violent acts perpetrated to kill as many people as possible (like at Crow Creek).

Interestingly, some aspects of bone removal changed little from the Archaic to the late prehistoric. From the beginning, heads and forelimbs were important trophies, and that continued into the post-contact era. There does not seem to be a transition from taking whole heads to taking just scalps over time. Scalping is present during the Late Archaic at the same time as head taking, and during late prehistory complete heads were still taken. In addition, throughout time, taking trophies from men, women, and subadults was acceptable. There does not seem to be a historic moment when only a particular demographic group was targeted; almost everyone was considered fair game for both the early and the later populations.

Although there are idiosyncratic components to the evidence of violence seen among the Indiana sites discussed here (e.g., the glossectomy), similar patterns of violence have been documented at other Middle to Late Archaic sites, especially in Kentucky and Tennessee. Evidence of decapitation, scalping, and dismemberment can be seen at Indian Knoll (Mensforth 2001, 2007; Snow 1948; Webb 1946), Carlston Annis (Webb 1950), and Ward (Mensforth 2001, 2007) from along the Green River in Kentucky. Likewise, there are documented cases of decapitation at the Robinson and Collegedale sites in Tennessee (Ross-Stallings 2007; Smith 1993) as well as scalping and trophy taking at Eva (Smith 1993). These sites (see Figure 8-8) tend to have more than one skeleton showing evidence of violence, but the relative frequency and types of traumata in these populations are consistent with those in Indiana. The similarities indicate that Middle/Late Archaic violence in Indiana is part of a regional phenomenon that will likely increase in area as researchers become more attuned to cut marks within Middle/Late Archaic skeletal assemblages from the Eastern Woodlands and the Midcontinent.

Conclusion

The traumata seen in the Middle/Late Archaic cemeteries from Indiana resulted from violence that targeted specific trophy elements. These elements were removed in particular ways that changed little over millennia. The bodies of the victims were afforded normal burials in their cemeteries and show no evidence of being left on the ground surface for any length of time after being killed. This profile of violence and victim burial is different from what is seen during the late prehistoric, when trophies were taken but the dead were often placed in shared graves and the bodies were ravaged by carnivores. The idiosyncrasies of the Middle/Late Archaic violence are not simply an artifact of scale, as even small late prehistoric sites have patterns of mutilation and burial that look like larger late prehistoric sites where warfare is clearly documented. It may have been that the focus of certain Archaic violence was the trophy taking effort; mass killing was not a concern. Thus, the Middle/Late Archaic violence produced osteologically distinct patterns evident in the traumata and means of burial associated with the mutilation victims that should be seen in its own context and not presumed to be generated under the same circumstances as violent events later in prehistory.

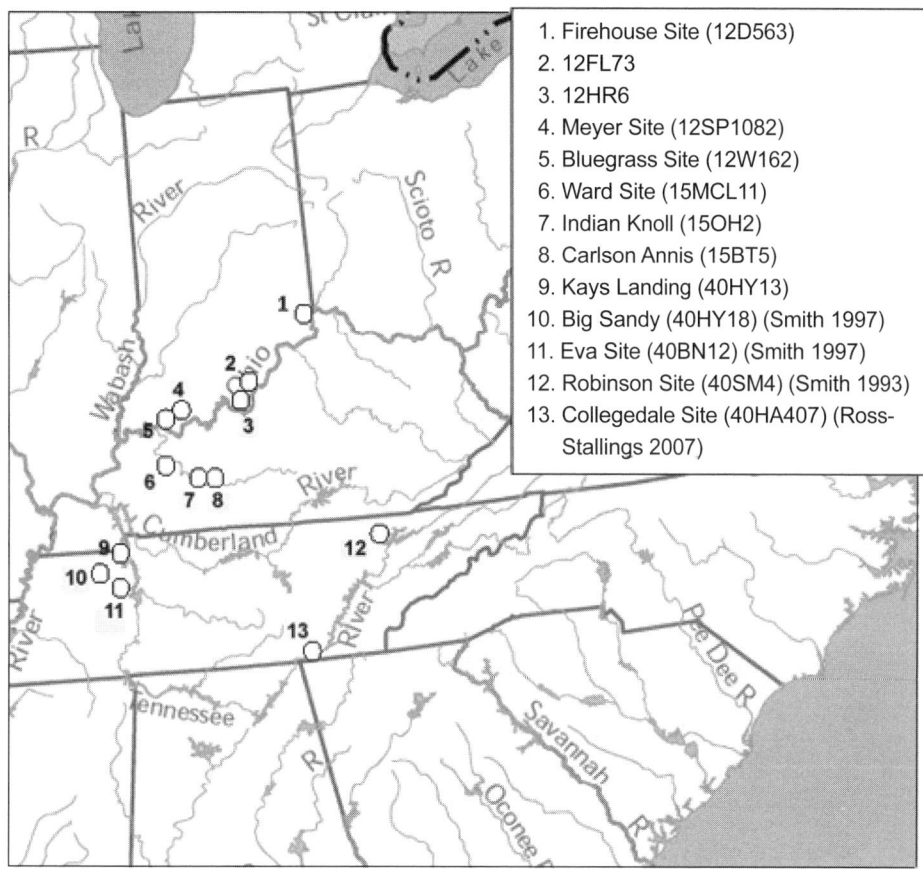

1. Firehouse Site (12D563)
2. 12FL73
3. 12HR6
4. Meyer Site (12SP1082)
5. Bluegrass Site (12W162)
6. Ward Site (15MCL11)
7. Indian Knoll (15OH2)
8. Carlson Annis (15BT5)
9. Kays Landing (40HY13)
10. Big Sandy (40HY18) (Smith 1997)
11. Eva Site (40BN12) (Smith 1997)
12. Robinson Site (40SM4) (Smith 1993)
13. Collegedale Site (40HA407) (Ross-Stallings 2007)

Figure 8-8. *Distribution of Middle/Late Archaic sites with evidence of decapitation, scalping, and/or dismemberment in the Eastern Woodlands.*

References

Andrushko, Valerie A., Kate Latham, Diane Grady, Allen Pastron, and Phillip Walker
    2005   Bioarcheological Evidence for Trophy-Taking in Prehistoric Central California. *American Journal of Physical Anthropology* 127:375–384.
Berryman, Hugh E.
    1980   The Averbuch Skeletal Series: A Study of Biological and Social Stress at a Late Mississippian Period Site from Middle Tennessee. Unpublished Ph.D. dissertation, Department of Anthropology, University of Tennessee, Knoxville.
Bridges, Patricia S.
    1996   Warfare and Mortality at Koger's Island, Alabama. *International Journal of Osteoarchaeology* 6:66–75.
Buikstra, Jane E., and Douglas H. Ubelaker
    1994   *Standards for Data Collection from Human Skeletal Remains*. Arkansas Archeological Survey, Fayetteville.

Byers, A. Martin
    2005   The Mortuary "Laying-In" Crypts of the Hopewell Site: Beyond the Funerary Paradigm. In *Interacting with the Dead: Perspectives on Mortuary Archeology for the New Millennium,* edited by Gordon F. M. Rakita, Jane E. Buikstra, Lane A. Beck, and Sloan R. Williams, pp.124–141. University of Florida Press, Gainesville.

Chacon, Richard J.
    2007   Seeking the Headhunter's Power: The Quest for *Arutam* among the Achuar of the Ecuadorian Amazon and the Development of Ranked Societies. In *The Taking and Displaying of Human Body Parts as Trophies by Amerindians,* edited by Richard J. Chacon and David H. Dye, pp. 523–546. Springer, New York.

Chacon, Richard J., and David H. Dye (editors)
    2007   *The Taking and Displaying of Human Body Parts as Trophies by Amerindians.* Springer, New York.

Chambers, Erica N., Christopher W. Schmidt, and Mark R. Schurr
    2010   Oxygen, Carbon, and Nitrogen Analysis of Middle/Late Archaic Populations from the Lower Ohio River Valley. *American Journal of Physical Anthropology* 50:S47–S48.

Ditch, Larry E., and Jerome C. Rose
    1972   A Multivariate Dental Sexing Technique. *American Journal of Physical Anthropology* 37:61–64.

Ember, Carol R.
    1978   Myths about Hunter-Gatherers. *Ethnology* 17:439–448.

Ember, Carol R., and Melvin Ember
    1997   Violence in the Ethnographic Record: Results of Cross-Cultural Research on War and Aggression. In *Troubled Times: Violence and Warfare in the Past,* edited by Debra L. Martin and David W. Frayer, pp. 1–20. Gordon and Breach, Amsterdam.

Emerson, Thomas E.
    2007   Cahokia and the Evidence for Late-Columbian War in the North American Midcontinent. In *North American Indigenous Warfare and Ritual Violence,* edited by Richard J. Chacon and Ruben G. Mendoza, pp. 129–148. University of Arizona Press, Tucson.

Harner, Michael J.
    1972   *The Jivaro: People of the Sacred Waterfalls.* University of California Press, Berkeley.

Holliman, Sandra E., and Douglas W. Owsley
    1994   Osteology of the Fay Tolton Site: Implications for Warfare During the Initial Middle Missouri Variant. In *Skeletal Biology of the Great Plains: Migration, Warfare, Health, and Subsistence,* edited by Douglas W. Owsley and Richard L. Jantz, pp. 344–345. Smithsonian Institution, Washington, D.C.

Jacobi, Keith P.
    2007   Disabling the Dead: Human Trophy Taking in the Prehistoric Southeast. In *The Taking and Displaying of Human Body Parts as Trophies by Amerindians,* edited by Richard J. Chacon and David H. Dye, pp. 299–338. Springer, New York.

Jefferies, Richard W., Victor D. Thompson, and George R. Milner
    2005   Archaic Hunter-Gatherer Landscape Use in West-Central Kentucky. *Journal of Field Archaeology* 30:3–23.

Lambert, Patricia M.
    1997   Patterns of Violence in Prehistoric Hunter-Gatherer Societies of Coastal Southern California. In *Troubled Times: Violence and Warfare in the Past,* edited by Debra L. Martin and David W. Frayer, pp. 77–110. Gordon and Breach, Amsterdam.

2002   The Archaeology of War: A North American Perspective. *Journal of Archaeological Research* 10:207–241.

2007   Ethnographic and Linguistic Evidence for the Origins of Human-Trophy Taking in California. In *The Taking and Displaying of Human Body Parts as Trophies by Amerindians,* edited by Richard J. Chacon and David H. Dye, pp. 65–89. Springer, New York.

2008   The Osteological Evidence for Indigenous Warfare in North America. In *North American Indigenous Warfare and Ritual Violence,* edited by Richard J. Chacon and Ruben G. Mendoza, pp. 202–221. University of Arizona Press, Tucson.

LeBlanc, Steven A.

1999   *Prehistoric Warfare in the American Southwest.* University of Utah Press, Salt Lake City.

Lockhart, Rachel A., and Christopher W. Schmidt

2007   Evidence of Decapitation and 'Trophy Taking' During the Late Archaic in Southern Indiana. Abstract in *American Journal of Physical Anthropology* 44:S158.

2008   Patterns in Head and Forearm Removal Traumata During the Late Archaic in Southern Indiana. Abstract in *American Journal of Physical Anthropology* 46:S141.

Lockhart, Rachel A., Christopher W. Schmidt, and Stephen A. Symes

2009   Understanding Middle and Late Archaic Forearm Removal. *American Journal of Physical Anthropology* 48:S177.

Lovisek, Joan A.

2007   Aboriginal Warfare on the Northwest Coast: Did the Potlatch Replace Warfare? In *North American Indigenous Warfare and Ritual Violence,* edited by Richard J. Chacon and Ruben G. Mendoza, pp. 58–73. University of Arizona Press, Tucson.

Maschner, Herbert D. G., and Katherine L. Reedy-Maschner

2007   Heads, Women, and the Baubles of Prestige: Trophies of War in the Arctic and Subarctic. In *The Taking and Displaying of Human Body Parts as Trophies by Amerindians,* edited by Richard J. Chacon and David H. Dye, pp. 32–44. Springer, New York.

Mays, Leigh Ann

1997   The Bluegrass Site (12W162): Bioarchaeological Analysis of a Middle-Late Archaic Mortuary Site in Southeastern Indiana. Unpublished Master's thesis, Department of Anthropology and Sociology, University of Southern Mississippi, Hattiesburg.

Mensforth, Robert P.

2001   Warfare and Trophy Taking in the Archaic Period. In *Archaic Transitions in Ohio and Kentucky Prehistory*, edited by Olaf H. Prufer, Sara E. Pedde, and Richard S. Meindl, pp. 110–140. Kent State University Press, Kent, Ohio.

2007   Human Trophy Taking in Eastern North America During the Archaic Period: The Relationship to Warfare and Social Complexity. In *The Taking and Displaying of Human Body Parts as Trophies by Amerindians,* edited by Richard J. Chacon and David H. Dye, pp. 222–277. Springer, New York.

Miller, Elizabeth

1994   Evidence for Prehistoric Scalping in Northeastern Nebraska. *Plains Anthropologist* 39:211–219.

Milner, George R.

1995   An Osteological Perspective on Prehistoric Warfare. In *Regional Approaches to Mortuary Analysis,* edited by Lane Anderson Beck, pp. 221–244. Plenum Press, New York.

2007   Warfare, Population, and Food Production in Prehistoric Eastern North America. In *North American Indigenous Warfare and Ritual Violence,* edited by Richard J. Chacon and Ruben G. Mendoza, pp. 182–201. University of Arizona Press, Tucson.

Milner, George R., Eve Anderson, and Virginia G. Smith
1991   Warfare in Late Prehistoric West-Central Illinois. *American Antiquity* 56:581–603.

Olsen, Sandra L., and Pat Shipman
1994   Cutmarks and Perimortem Treatment of Skeletal Remains on the Northern Plains. In *Skeletal Biology of the Great Plains: Migration, Warfare, Health, and Subsistence,* edited by Douglas W. Owsley and Richard L. Jantz, pp. 377–387. Smithsonian Institution, Washington, D.C.

Owsley, Douglas W.
1982   Dental Discriminate Sexing of Arikara Skeletons. *Plains Anthropologist* 27:165–169.

Owsley, Douglas W., and Hugh E. Berryman
1975   Ethnographic and Archaeological Evidence of Scalping in the Southeastern United States. *Tennessee Archaeologist* 31:41–58.

Owsley, Douglas W., Hugh E. Berryman, and William M. Bass
1977   Demographic and Osteological Evidence for Warfare at the Larson Site, South Dakota. *Plains Anthropologist Memoir* 13:119–131.

Owsley, Douglas W., Karin S. Bruwelheide, Laurie E. Burgess, and William T. Billeck
2007   Human Finger and Hand Bone Necklaces from the Plains and Great Basin. In *The Taking and Displaying of Human Body Parts as Trophies by Amerindians,* edited by Richard J. Chacon and David H. Dye, pp. 124–166. Springer, New York.

Ross-Stallings, Nancy
2007   Trophy Taking in the Central and Lower Mississippi Valley. In *The Taking and Displaying of Human Body Parts as Trophies by Amerindians,* edited by Richard J. Chacon and David H. Dye, pp. 339–370. Springer, New York.

Schaafsma, Polly
2007   Documenting Conflict in the Prehistoric Pueblo Southwest. In *North American Indigenous Warfare and Ritual Violence,* edited by Richard J. Chacon and Ruben G. Mendoza, pp. 114–128. University of Arizona Press, Tucson.

Schmidt, Christopher W., and Tammy R. Greene
1999   Excavation of Human Remains. Report submitted to the Morgan County Coroner's Office and the Morgan County Sheriff's Department, Mooresville, Indiana.

Schmidt, Christopher W., and Clark S. Larsen
2002   Demographic and Health Reconstruction of the Santa Catalina de Guale Ossuary, Amelia Island, Florida. *American Journal of Physical Anthropology* 34:S136.

Scott, Gary T., and Kenneth R. Parham
1979   Multivariate Sexing: Discrimination of the Sexes Within an East Tennessee Mississippian Skeletal Sample. *Tennessee Anthropologist* 4:189–198.

Seeman, Mark F.
2007   Predatory War and Hopewell Trophies. In *The Taking and Displaying of Human Body Parts as Trophies by Amerindians,* edited by Richard J. Chacon and David H. Dye, pp. 167–189. Springer, New York.

Smith, Maria O.
1993   A Probable Case of Decapitation at the Late Archaic Robinson Site (40SM4), Smith County, Tennessee. *Tennessee Anthropologist* 18:131–142.
1995   Scalping in the Archaic Period: Evidence from the Western Tennessee Valley. *Southeastern Archaeology* 14:60–68.

1997   Osteological Indications of Warfare in the Late Archaic Period of the Western Tennessee Valley. In *Troubled Times: Violence and Warfare in the Past*, edited by Debra L. Martin and David W. Frayer, pp. 241–265. Gordon and Breach, Amsterdam.

Snow, Charles E.
1948   *Indian Knoll Skeletons of Site Oh2, Ohio County, Kentucky.* Reports in Anthropology, Vol. 4, No. 3, Pt. 2. Department of Anthropology, University of Kentucky, Lexington.

Snow, Dean R.
2007   Iroquois-Huron Warfare. In *North American Indigenous Warfare and Ritual Violence,* edited by Richard J. Chacon and Ruben G. Mendoza, pp. 149–159. University of Arizona Press, Tucson.

Stafford, C. Russell, and Mark Cantin
2009   Archaic Period Chronology of the Hill Country of Southern Indiana. In *Archaic Societies: Diversity and Complexity Across the Midcontinent*, edited by Thomas E. Emerson, Dale L. McElrath, and Andrew C. Fortier, pp. 287–316. State University of New York Press, Albany.

Stafford, C. Russell, Ronald L. Richards, and C. Michael Anslinger
2000   The Bluegrass Fauna and Changes in Middle Holocene Hunter-Gatherer Foraging in the Southern Midwest. *American Antiquity* 65:317–336.

Steadman, Dawnie Wolfe
2008   Warfare Related Trauma at Orendorf, a Middle Mississippian Site in West-Central Illinois. *American Journal of Physical Anthropology* 136:51–64.

Tung, Tiffiny A.
2008   Dismembering Bodies for Display: A Bioarchaeological Study of Trophy Heads from the Wari Site of Conchopata, Peru. *American Journal of Physical Anthropology* 136:294–308.

Ubelaker, Douglas H.
1974   *Human Skeletal Remains: Excavation, Analysis, Interpretation*. Smithsonian Institution, Washington, D.C.

Walker, Phillip L.
2001   A Bioarchaeological Perspective on the History of Violence. *Annual Review of Anthropology* 30:573–596.

Webb, William S.
1946   *Indian Knoll Site Oh2 Ohio County, Kentucky*. Reports in Anthropology and Archaeology, Vol. 4, No. 3, Pt. 1, pp. 115–356. Department of Anthropology, University of Kentucky, Lexington.
1950   *The Carlston Annis Mound*. Reports in Anthropology and Archaeology, Vol. 7, No. 4, pp. 266–354. Department of Anthropology, University of Kentucky, Lexington.

Willey, Patrick S.
1990   *Prehistoric Warfare on the Great Plains: Skeletal Analysis of the Crow Creek Massacre Victims*. Garland, New York.

Williamson, Ron
2007   "Otinontsiskiaj ondaon" ("The House of Cut-Off Heads"): The History and Archaeology of Northern Iroquoian Trophy Taking. In *The Taking and Displaying of Human Body Parts as Trophies by Amerindians*, edited by Richard J. Chacon and David H. Dye, pp. 190–221. Springer, New York.

9. Population History of the Moquegua Valley, Far South Coast of Peru

Ken-ichi Shinoda, Sonia Guillén, and Izumi Shimada

Summary Statement: This study examines the changing population composition of the Moquegua Valley, which has long served as a major corridor for coast-highland interaction in the South-Central Andes. Through analysis of mitochondrial DNA from 14 individuals from the Formative period (3450–1850 B.P.) and 27 from the late pre-Hispanic Chiribaya culture (1050–600 B.P.), we tentatively define the timing, direction, and intensity of population movements that largely shaped the population composition of the Moquegua Valley. During the Formative era, haplogroup A predominated, but over time, particularly from the Middle Horizon (1450–1000 B.P.) to the Late Intermediate period (1000–550 B.P.), haplogroup B increased notably in its frequency. During this same period, genetic drift appears to have been a significant force and, combined with significant admixture with population(s) intruding from the adjacent highland area, also altered haplogroup frequencies. This influx increased during the Inka and colonial periods, seemingly replacing much of the indigenous population. In general, our genetic study supports archaeological inferences regarding the extent of colonization by highland populations, although the small sample size and the bidirectionality of biological and cultural interactions must be kept in mind.

Introduction

Prior to the availability of ancient DNA analysis, inferences regarding the genealogical lineage(s) of human skeletal remains excavated from archaeological sites, kinship ties with neighboring groups, and even kinship among skel-

Human Variation in the Americas: The Integration of Archaeology and Biological Anthropology, edited by Benjamin M. Auerbach. Center for Archaeological Investigations, Occasional Paper No. 38. © 2010 by the Board of Trustees, Southern Illinois University. All rights reserved. ISBN 978-0-88104-095-1.

etons from single sites were mostly based on morphological study of the skeletons themselves. Considerable expertise is required, however, to distinguish between genetically and environmentally influenced morphological features and a certain degree of error inevitably figures in to the conclusions reached through such observations. Advances in molecular biology over the last 20 years now enable analysis of DNA extracted from ancient bone samples, making it possible to obtain information on the origins, movements, and relationships of ancient human populations with significantly higher probabilities of accuracy compared with those based on morphological observations alone (Maca-Meyer et al. 2005; Melchior et al. 2008; Shinoda and Kanai 1999).

In our study, the aforementioned ancient DNA method was used on samples from excavated human burials to obtain data that would illuminate long-standing debate as to the nature, timing, and direction of coast-highland interaction and its long-term consequences in the Moquegua Valley on the far south coast of Peru and the adjacent highland plateau around Lake Titicaca. The valley has long served as a major corridor and elucidation of its population history is critical to understanding various issues in Andean archaeology, including those related to the inferred coastward expansion (1350–950 B.P.) of the Tiwanaku state and Aymara speakers as well as the late pre-Hispanic Chiribaya ethnogenesis (1050–600 B.P.) in the valley. Our study points to a notable shift in mitochondrial DNA haplogroup frequency from a predominance of haplogroup A during the Formative era (3450–1850 B.P.) to the predominance of haplogroup B, particularly from the Middle Horizon (1450–1000 B.P.) to the Late Intermediate period (1000–550 B.P.). During the last two periods, genetic drift appears to have been a significant force and, combined with significant admixture with population(s) intruding from the adjacent highland area, also altered haplogroup frequencies. The inferred influx from the highlands appears to have increased during the Inka and colonial periods, nearly replacing the indigenous population. While our sample is relatively small and results cannot be considered conclusive, a diachronic ancient DNA study in a nearby region reached quite similar conclusions (Moraga et al. 2005). In general, our genetic study effectively supports archaeological inferences regarding the extent of colonization by highland populations, although how this process relates to the Chiribaya ethnogenesis vis-à-vis biological and cultural hybridization remains to be elucidated via integration of multiple lines of evidence.

It is generally agreed that most mitochondrial DNA (mtDNA) from South American Amerindians can be traced to one of four matrilineal lineages present in the founders of New World populations and that these lineages can be defined by three restriction-site polymorphisms and a 9-bp deletion (Wallace et al. 1985). Torroni and colleagues (1992) labeled these lineages A through D. Haplogroup distribution defined through mtDNA analysis of modern Amerindian populations has accordingly opened a productive avenue for reconstructing the timing and geographical routes of prehistoric migrations, as well as the number and genetic character of founders who populated the New World thousands of years ago (Eshleman et al. 2003; Merriwether et al. 1995; Schurr 2004; Torroni et al. 1993).

Most of the research on local population events in South America is based on modern DNA analysis (Bert et al. 2001; Fuselli et al. 2003; Rodriguez-Delfin et al. 2001). Projecting backward in time from this modern genetic composition and distribution has a number of inherent limitations. Relatively short-term and local biological and cultural processes—such as epidemics, ritual sacrifice from select groups (e.g., young female virgins and male warriors), conquests and associated casualties and/or executions, and forced relocation—are all well documented, particularly for the late pre-Hispanic era (e.g., Benson and Cook 2001; Rowe 1982; Stanish 1992), and must be taken into account.

DNA analysis of ancient materials is currently based mainly on mtDNA, which has a high copy number and a fast mutation rate. This approach, however, only illuminates matrilineal genetic linkages and depends on successful extraction and analysis of minute and all too easily contaminated quantities of mitochondrial DNA. In spite of these difficulties, mtDNA analysis of teeth from pre-Hispanic individuals recovered from well-defined contexts by regionally based archaeological projects offers an effective means for understanding local and/or regional population history and dynamics (e.g., Reed 2005; Shimada et al. 2004, 2005; Shinoda et al. 2006). Such information, combined with the results of archaeological investigations, allows us to put forth and test new theories.

Individuals analyzed in our study were derived from 11 sites dating to the Formative (3450–1850 B.P.) and late pre-Hispanic Chiribaya (1050–600 B.P.) periods in the Moquegua Valley on the far south coast of Peru. Not only does the hyperaridity of this area favor preservation of human remains, but also intense and largely concurrent archaeological investigations since the mid-1980s both on the coast and in the adjacent highlands, particularly the altiplano region around Lake Titicaca, offer rich contextual details essential to our efforts to elucidate the pre-Hispanic population history and dynamics of southern coastal Peru. In fact, ancient DNA studies represent an effective means of testing various competing models of coast-highland cultural interactions and population movements that were stimulated by the pioneering verticality hypothesis of John V. Murra to be described below.

During the 1960s, the influential ethnohistorical studies of John V. Murra (e.g., 1968, 1972) documented the phenomenon of "verticality" or simultaneous, direct control by highland enclaves of multiple altitudinally differentiated, productive [ecological] zones by the same ethnic population during the immediately pre– and post–Spanish Conquest eras. Instead of communities occupying just one or a few ecological zones and focusing on local products to be traded or exchanged via markets with other similarly specialized communities occupying different altitudes, the verticality model (and its varied derivatives) argues that a given community established its own colonies at different productive zones to cultivate different crops for their own consumption. This economic arrangement was based on the concepts of reciprocity and resource sharing so that any important productive zone might be occupied by colonies of multiple ethnic groups in addition to the indigenous group.

The particular alpine ecology of the South-Central Andes engenders verticality. The extensive altiplano around Lake Titicaca, the highest freshwater lake at approximately 4,000 m asl, has long been recognized to have been the breadbasket and heartland of the Inka and earlier Tiwanaku empires. In spite of the tremendous potential for agricultural cultivation and large-scale herding of domesticated Andean camelids, the high elevation is constrained by as many as 300 days per year with frost, not to mention hail and snow (Mujica 1985). It is not surprising then that, both in the past and today, among altiplano inhabitants there has been strong interest in the adjacent river valleys on the Pacific side of the Andes where the warmer climate allows cultivation of complementary crops such as maize, coca, and ají. Abundant marine resources—including fish and seaweed (Masuda 1985), as well as guano on offshore islands (Julien 1985)—are additional attractions.

Murra's verticality model influenced many subsequent archaeological and bioarchaeological studies that focused on the population history, composition, and dynamics of the Moquegua Valley, which served as a major communication route connecting the snow-covered highland headwaters region and altiplano with the Pacific coast, a distance of around 140 km (Goldstein 1989; Owen 1993; Rice et al. 1989; Stanish 1992).

It has been postulated that the first systematic colonization of the warm lower elevation zones of the Moquegua Valley occurred during the Formative period by the Pukara culture, which flourished in the northern and northwestern portions of the altiplano (Focacci 1983; Kolata 1983; Rivera 1984). Their basic intent was to cultivate warm-climate crops (such as maize and ají peppers) that would effectively complement high-altitude crops such as tubers. Others (Mujica 1985; Stanish 2003) contend, however, that available evidence is inadequate to support the hypothesized establishment of Pukara colonies on the Pacific coast. Thus, one specific aim of our study is to shed light on this controversy by bringing to bear ancient DNA analysis of excavated burials dating to the Formative era.

The Tiwanaku polity centered on the southeastern shore of Lake Titicaca is believed to have initiated a second and much larger wave of colonization early in the seventh century (e.g., Goldstein 1989, 2005; Goldstein and Owen 2001). This lasted until the end of the tenth century, roughly coinciding with the emergence of the Chiribaya culture, whose population occupied both the warm and agriculturally productive middle (ca. 2,000–1,000 m asl) and lower portions of the Moquegua drainage, which encompassed freshwater springs, a river valley, *loma* vegetation that depends on condensation of dense winter fog, and the Pacific littoral. The Chiribaya culture dates from about 1050–600 B.P. and extended southward to the far north coast of Chile (Buikstra 1995; Goldstein 2005; Lozada and Buikstra 2002; Owen 1993). The principal settlements and cemeteries that have provided the data that define this culture as well as the burial samples used in our study derive from the lower portions of the coastal valleys.

Who were the Chiribaya people: descendants of a Tiwanaku colony, an indigenous population governed by ethnic chiefs (*señoríos*) and perhaps present as early as the Formative, or a hybrid population of both? Ghersi (1956) argued that the Chiribaya were a pre-Inka culture that was influenced by a highland tradition, while Lumbreras (1974) saw them clearly as a product of hybridization among various altiplano and coastal traditions. More recently, various archaeologists (Goldstein 2005; Owen 1993; Stanish 1989, 1992) have proposed that the Chiribaya culture resulted from the coastward diaspora of mid-valley colonies of the Tiwanaku empire that collapsed around 1050 B.P. with resulting ethnogenesis accompanied by both biological and cultural hybridization.

Various analytical methods and diagnostics such as dental caries, wear, and traits (Godoy 2005; Lozada 1998; Lozada and Buikstra 2002; Sutter 1997), cranial deformation (Blom et al. 1998), and architectural and funerary features (e.g., Bawden 1989; Buikstra 1995; Goldstein 1989) have been used to address these and related questions. Recent bioarchaeological studies based on nonmetric traits of excavated burials showed a statistically significant difference between lower valley Chiribaya and mid-valley Tiwanaku samples, suggesting a coastal origin of Chiribaya (Lozada 1998:176; Lozada and Buikstra 2002).

Some of these approaches suffer from the uncertainties inherent in statistical analysis. Stylistic or other diagnostics, because they are often transitory and/or situationally contingent, have their own limitations. While ancient mtDNA analysis reveals biological linkages only along matrilineal lines and is susceptible to contamination and degeneration of genetic materials, it nonetheless can offer precise and direct indications of biological relationships among individuals being compared. In this sense, our study also serves as an independent test of earlier inferences regarding regional population histories and relationships.

Materials and Methods

Archaeological Site and Specimens

The DNA samples used in this study were extracted from the teeth and bones of 76 individuals representing both sexes and varied adult ages excavated at 11 archaeological sites in the lower reaches of the Moquegua Valley on the far south coast of Peru (Figure 9-1). Twenty-five individuals date to the Formative and 51 to the Chiribaya culture of the Late Intermediate period.

Tooth enamel forms a natural barrier to exogenous DNA contamination; additionally, the DNA recovered from teeth appears to lack most of the inhibitors to the enzymatic amplification of ancient DNA (Woodward et al. 1994). Therefore, in many cases, tooth samples were used in the present analysis. When tooth samples were not available, cortical bones of the femur were used. A list of all the samples used in this study is presented in Tables 9-1 and 9-2.

Authentication Methods

When ancient DNA is analyzed, contamination with contemporary DNA can yield false-positives and is a serious concern. In order to ensure the accuracy and reliability of results, standard contamination precautions were employed in the present study: separation of pre- and post-PCR experimental areas, use of disposable laboratory wares and filter-plugged pipette tips, treatment with DNA contamination removal solution (DNA-OFF^{TM}; TaKaRa, Otsu, Japan), UV irradiation of equipment and benches, negative extraction controls, and negative PCR controls. Other rigorous authentication methods were employed through-

Map of archaeological sites

Figure 9-1. *Locations of the archaeological sites of Moquegua Valley mentioned in the text.*

out the DNA-based analyses as described elsewhere (Shinoda et al. 2006). Tooth preparation, DNA extraction, and PCR amplification were carried out in a physically separate lab room dedicated to the study of ancient DNA.

Extraction and Amplification of DNA

In order to prevent contamination from post-excavation handling, tooth samples were rinsed with DNA-decontamination agents and then washed thoroughly with distilled water before drying. Next, the samples were crushed into powder using a Multi-beads Shocker (Yasui Kikai Corporation). DNA was extracted from .5 g of the powder for each sample using a commercial DNA extraction kit (MO BIO Lab.).

We analyzed a segment of hypervariable region I (HV1) in the D-loop and the 9-bp repeat variation that defines haplogoup B. More specifically, mtDNA segments that cover nucleotide positions 16121–16238, and 16209–16402, relative to the revised Cambridge reference sequence (Andrews et al. 1999), were sequenced for all samples.

Five-microliter aliquots of the extracts were used as the templates for PCR. Amplifications were carried out in a total reaction volume of 25 μl containing one unit of Taq DNA polymerase (HotStarTaq^{TM} DNA polymerase; QIAGEN), .2 μM of each primer, and 200 μM of dNTPs in the 1× PCR buffer provided by the

Table 9-1. *Nucleotide Changes Observed in the Samples from the Formative Sites Analyzed in the Present Study*

Site	No.	Material	D-loop Sequence (16000+)	Haplogroup
			Formative Period	
Descanso	2177	Maxilla Left M3	223,290,319,362	A
	2179	Maxilla Left M3	223,290,319	A
	2180	Maxilla Left M3	223,319,362	A
	2184	Maxilla Left PM2	217,311,362*	B
	2238	Maxilla Left M2	223,290,319	A
Roca Verde	2373	Mandible Right M2	223,298,311,325,327	C
	2459	Mandible Right M3	223,290,319,362	A
	2538	Isolated tooth	223,298	C
La Cruz	1397	Mandible Left M3	217,311,355,368*	B
	2714	Mandible Right I2	223,290,319,362	A
	2854	Mandible Right M3	223,290,319,362	A
	4195	Mandible Left M3	217*	B
	4249	Mandible Right M3	223,290,319	A
	34-359	Carpal bone	223,290,319,362	A

Note: All polymorphic sites are numbered according to the revised Cambridge reference sequence (Andrews et al. 1999).
* Segments 16121–16141 cannot be amplified.

manufacturer. The conditions for PCR were as follows: incubation at 95°C for 15 min; 40 cycles at 94°C for 20 s, 46°C–56°C for 20 s, 72°C for 15 s; and final extension at 72°C for 1 min.

The primers used to amplify the regions described above are:

L16120 5′-TTACTGCCAGCCACCATGAA-3′
H16239 5′-TGGCTTTGGAGTTGCAGTTG-3′
L16208 5′-CCCCATGCTTACAAGCAAG-3′
H16403 5′-TTGATTTCACGGAGGATGGTG-3′

The PCR products were filtered using Centricon-100 spin columns (Amicon), and the filtrates were prepared for sequencing using forward and reverse primers and a BigDye Cycle Sequencing Kit (Applied Biosystems, Foster City, CA, USA). All sequencing reactions were analyzed using a model 3130 DNA Sequencer with SeqEd software, and the sequence of each region that did not contain primer regions was determined and compared with the revised Cambridge reference sequence (CRS).

Table 9-2. *Nucleotide Changes Observed in the Samples from the Chiribayan Sites Analyzed in the Present Study*

Site	No.	Material	D-loop Sequence (16000+)	Haplogroup
		Chiribaya Culture		
Loreto Viejo	1472	C3 (Child Mummy)	217*	B
	1474	Maxilla Left M2	189,209	B
	1475	Maxilla Right M3	223,298,325,327,362	C
	1476	Maxilla Left M3	189,209	B
	1479	Isolated tooth	223,290,319,362	A
Chiribaya Baja	2029	Isolated tooth	223,291,298,325,327	C
	2030	Mandible Right M3	223,278,311,362	D
	2038	Mandible Right M3	223,234,316,362	D
	2043	Isolated tooth	189,209	B
	2044	Mandible Right M2	223,311,343,362	D
	2147	Maxilla Left M3	223,298,325,327,362	C
Chiribaya Baja PVT	1498	Maxilla Right Canine	217,256,291*	B
	1510	Maxilla Left M3	223,319,362	A
	1527	Maxilla Right PM1	223,325,362	D
San Geronimo	3003	Maxilla Right M3	189,209	B
	3017	Isolated tooth	223,298,343,362	C
	3020	Mandible Right PM1	189,209	B
	3022	Maxilla Left Canine	223,290,319,362	A
	3023	Maxilla Left PM2	223,311,362	D
	3030	Isolated tooth	223,298,362	C
Algodonal	1010	Maxilla Left M2	223,325,362,370	D
	1239	Maxilla Left M2	217,311*	B
	1246	Maxilla Left M1	217,311*	B
	1271	Maxilla Left M3	223,362	D
	1275	Maxilla Left M3	223,290,319,362	A
Casa Vieja	3223	Mandible Left M2	223,362,390	D
	3254	Maxilla Left M2	223,325,362,370	D

Note: All polymorphic sites are numbered according to the revised Cambridge reference sequence (Andrews et al. 1999).
*Segments 16121–16141 cannot be amplified.

A DNA fragment encompassing the 9-bp repeat variation in the noncoding cytochrome oxidase II/tRNA^{Lys} intergenic region was amplified with primers 9-bpF (5′-ACAGTTTCATGCCCATCGTC-3′) and 9-bpR (5′-CTAAGTTAGCTT-TACAGTGGG-3′), which specify a product of 143 or 152 bp for deleted or non-deleted samples, respectively. The constitution of the PCR reaction mixture was the same as described above. The thermal conditions were as follows: incubation at 95°C for 15 min; 40 cycles at 94°C for 10 s, 52°C–54°C for 10 s, 72°C for 5 s; and final extension at 72°C for 1 min. PCR products were electrophoresed on 4 percent agarose gels and the bands visualized under ultraviolet light after staining with ethidium bromide.

Data Analysis

With recent advances in our understanding of global mtDNA phylogeny, control and coding region motifs have been identified for a majority of the major haplogroups and their subhaplogroups of Native Americans (Achilli et al. 2008; Fagundes et al. 2008; Kitchen et al. 2008; Tamm et al. 2007). Accordingly, we assigned each mtDNA to a haplogroup according to the HV1 and coding region data (9-bp deletion).

Several genetic parameters were computed on the sequence and the haplogroup level. Nucleotide diversity, Tajima's D value, and mean number of pairwise differences of mitochondrial D-loop sequences were computed in the Arlequin version 3.1 program (Excoffier et al. 2005), using Tamura and Nei distances and a gamma-parameter value of .26 (Meyer et al. 1999). Point mutation at HV1 np 16183 was excluded from analysis because it is dependent on the presence of a T-C transition at np 16189. A population differentiation test (Raymond and Rousset 1995) between the Moquegua Valley populations and comparative populations was also computed by means of the Arlequin program. Neighbor-joining (NJ) trees based on pairwise Fst were constructed using Mega 3.0 (Kumar et al. 2004) to study the relationships between the Moquegua Valley and other populations. Suspected false-positive results stemming from contamination with contemporary DNA and other questionable data were omitted from this study.

Results and Discussion

Genetic Characteristics of the Moquegua Population

We were able to successfully retrieve 280-bp HV1 sequences with different variations and 9-bp repeat variation in the noncoding cytochrome oxidase II/tRNA^{Lys} intergenic region from 41 out of 76 samples collected from the Moquegua Valley. It was possible to determine the HV1 sequence of 46 individuals. However, suspected false-positives stemming from contamination with contemporary DNA and other questionable data were detected among them, and thus five of the sequences were omitted from further consideration. Haplogroup distribution among the total sample was as follows: 31.7 percent haplogroup A, 29.3 percent haplogroup B, 17.1 percent haplogroup C, and 22.0 percent haplogroup D. Haplogroup frequency distributions, however, were different among populations.

In the Formative population, DNA was successfully amplified for 14 of 25 samples. The remaining 11 samples either failed to yield a product on amplification or produced several ambiguous positions during the sequencing reaction, possibly due to misincorporations by the DNA polymerase during PCR. Therefore, the success rate was 56 percent. Because of the often poor quality of the mtDNA extracted from ancient materials, it was not possible to amplify all the samples. It is known from past studies that the success rate for DNA analysis of ancient human remains is 50–70 percent at best, even using well-preserved samples and mtDNA, which is relatively easy to amplify. Our results suggest that the preservation conditions of DNA in the Formative samples are relatively good.

The base sequence and haplogroup of each individual sample of Formative remains is shown in Table 9-1. These individuals can be classified into eight haplotypes according to their sequences; mutations were observed in 11 portions. The largest haplotype group is composed of five individuals; the next largest, three; and the remaining seven haplotypes correspond to one individual each (Table 9-3).

As the source cemeteries were used by different cultures over a long period, there was the possibility of matrilineal relationships across generations at the same site. The high frequency of one haplotype (3 out of 6 samples) at the La Cruz site may denote just such shared matrilineal kinship.

Among the 51 Chiribaya individuals, ancient DNA was successfully amplified from 27 samples, an approximately 53 percent success rate, which is a slightly lower rate compared with that of the Formative samples. A sequence comparison helped identify 18 mitochondrial haplotypes defined on the basis of 18 segregating sites. Table 9-2 shows the positions at which the sequences of the Chiribaya population differed from the reference sequence described by Andrews and colleagues (1999). Sequence haplotypes of both Formative and Chiribaya populations are summarized in Table 9-3.

One of the main purposes of analyzing DNA from ancient burial sites is to determine whether the human remains represent related or unrelated individuals. An exact test of differentiation based on pairwise Fst values between Formative and Chiribayan groups revealed that these differences are not statistically significant. It should be considered, however, that the nonsignificance is due to the small sample size of Formative population. As shown in Tables 9-1, 9-2, and 9-3, at least 8 and 18 matrilineal lines were distinguished at the Formative and Chiribaya sites, respectively. They shared only three matrilineal lines. Those 3 lines, however, account for 50 percent of the Formative and 18.5 percent of the Chiribaya individuals.

In recent years, regional studies based on mtDNA HV1 sequences of Andean populations have begun to investigate genetic diversity in depth and their data (Fuselli et al. 2003; Lewis et al. 2005, 2007a) can be compared with our results. Basic statistics for our data and reference populations are summarized in Table 9-4. The haplotype diversities were .857 and .954 for the Formative and the Chiribaya, respectively. The nulceotide diversities were .018 and .021, respec-

Table 9-3. *mtDNA Sequence Variation in the Formative and Chiribayan Populations*

Haplotype	n (Total)	Formative	Chiribaya	D-loop Sequence (16000+)	9-bp deletion[a]	Haplogroup
Type 1	8	5	3	223,290,319,362	-	A
Type 2	3	3	0	223,290,319	-	A
Type 3	2	1	1	223,319,362	-	A
Type 4	1	1	0	217,311,362*	+	B
Type 5	1	1	0	217,311,355,368*	+	B
Type 6	2	1	1	217*	+	B
Type 7	5	0	5	189,209	+	B
Type 8	1	0	1	217,256,291*	+	B
Type 9	2	0	2	217,311*	+	B
Type 10	1	1	0	223,298,311,325,327	-	C
Type 11	1	1	0	223,298	-	C
Type 12	2	0	2	223,298,325,327,362	-	C
Type 13	1	0	1	223,291,298,325,327	-	C
Type 14	1	0	1	223,298,343,362	-	C
Type 15	1	0	1	223,298,362	-	C
Type 16	2	0	2	223,325,362,370	-	D
Type 17	1	0	1	223,278,311,362	-	D
Type 18	1	0	1	223,234,316,362	-	D
Type 19	1	0	1	223,311,343,362	-	D
Type 20	1	0	1	223,325,362	-	D
Type 21	1	0	1	223,311,362	-	D
Type 22	1	0	1	223,362	-	D
Type 23	1	0	1	223,362,390	-	D

Note: Sites are numbered according to the revised Cambridge reference sequence (CRS) of Andrews and colleagues (1999).

[a] + and - denote the presence of the 9-bp (CCCCCTCTA) deletion and nondeletion (i.e., two repeats of the 9-bp fragment) in the COII/tRNA^{Lys} intergenic region, respectively.

*Segments 16121–16141 cannot be amplified.

Table 9-4. *mtDNA HV1 Haplotype Diversity Parameters of the Analyzed Pre-Hispanic Populations and the Contemporary Reference Populations*

Population	n	Number of haplotypes	Haplotype diversity	Nucleotide diversity	Reference
Formative	14	8	.857	.018	This study
Chiribaya	27	18	.954	.021	This study
Ancash	34	27	.979	.016	Lewis et al. 2005
Arequipa	22	18	.978	.014	Fuselli et al. 2003
Tayacaja	60	42	.967	.017	Fuselli et al. 2003
Puno (Quechua)	30	23	.974	.018	Lewis et al. 2007a
Mapuche	34	8	.834	.019	Moraga et al. 2000

tively. These diversity estimates are congruent with those from other Andean populations. Tajima's D values for Formative and Chiribaya populations were not significantly different from 0 at a confidence level of .05 (D = .028, -.521 respectively), indicating no sign of demographic expansion.

Diachronic Changes in Mitochondrial DNA Haplogroup Frequencies in the Andean Coastal Region

To assess the population history of this area, a haplogroup frequency comparison was made among the analyzed Formative and Chiribaya individuals and a sample of modern Aymara (Merriwether et al. 1995). As shown in Figure 9-2, the frequency of haplogroup B tends to increase over time, whereas A tends to decrease during the same time interval considered. In fact, among contemporary Aymara now living in the same geographic area, B is by far the most frequent haplogroup, followed by D, while C and A occur in lower frequencies. Fst P-values among these populations are statistically significant (Significant level = .05); that is, there is a significant difference in the haplogroup frequencies among them. Since genetic variation is inherited from a group's ancestors, frequencies of genetic markers in the contemporary population would be expected to be similar to those in their ancestors. The sampled Formative and Chiribaya populations, however, have very different frequencies of genetic markers and the modern Aymara sampled may not be closely related to our ancient samples, although effects of genetic drift or selective forces cannot be ruled out (see below).

To clarify further the genetic characteristics of the regional population, our mtDNA data were compared with those of previous work conducted in a nearby region. It is noteworthy that our analytical results closely parallel those reported by Moraga and colleagues (2005) and Lewis and associates (2007b) for the far

frequency comparisons

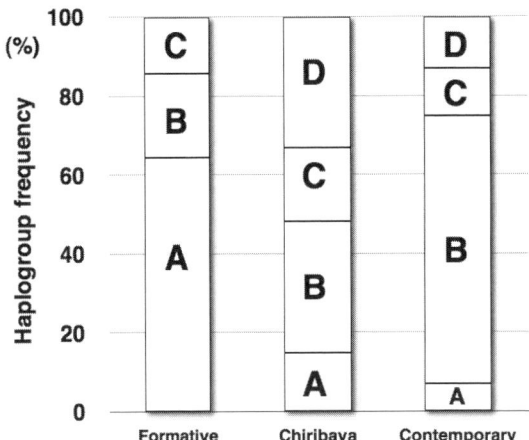

Figure 9-2. *Distribution of mtDNA haplogroup frequencies in Moquegua Valley and contemporary highland Aymara populations.*

north coast of Chile south of the Moquegua Valley (Table 9-5). The percentage of haplogroup A in the Late Archaic and Formative periods of both areas is higher than that of later periods, whereas haplogroup B is more frequent after the Middle Horizon (1450–1000 B.P.). To investigate the relationships among major cultural divisions in the Moquegua Valley and the far north coast of Chile, phylogenetic analysis was carried out. Genetic affinities among populations were visualized in a NJ tree (Figure 9-3). Analysis of genetic distances based on haplogroup frequencies revealed two distinct, temporally separated clusters.

MtDNA haplogroup frequencies in the ancient human remains from the far north coast of Chile and the Moquegua Valley change simultaneously over time. These similarities may be attributed to gene flow between the two populations. Combining their and our results allows us to consider the population history and dynamics of the broader region over a longer time span. The distribution of mtDNA haplogroups among these populations suggests that the formation of the Chiribaya population was not the result of a simple increase in number of individuals. Even though there is a diachronic change in the haplogroup frequencies, no increase of overall mitochondrial genetic variability can be observed in the HV1 sequence data.

Recent research using a simulation model to estimate population variation over time in South-Central Andean populations, however, suggests local genetic drift could explain the mtDNA haplogroup frequency changes between contemporary and prehistoric populations (Lewis 2009). Thus, genetic drift may have also been a significant force that altered haplogroup frequencies in this region over time.

The population history of the Moquegua Valley in general agrees with the well-established geographical distribution of mtDNA haplogroups on the

Table 9-5. *Distribution of mtDNA Haplogroup Frequencies in the Populations Analyzed in This Study and Other Pre-Hispanic and Contemporary Andean Populations Reported in Other Studies*

Population	n	Haplogroup (%)				Haplogroup diversity	Chronological age	Reference
		A	*B*	*C*	*D*			
Late Archaic	14	50.0	35.7	7.1	7.1	.66		Moraga et al. 2005
Formative	14	64.3	21.4	14.3	0	.56	3900 B.P.–2500 B.P.	This study
Chen Chen	23	39.1	39.1	17.4	4.3	.69	1215 B.P.–1000 B.P.	Lewis et al. 2007b
Middle Horizon	19	31.6	42.1	26.3	0	.69	1650 B.P.–1000 B.P.	Moraga et al. 2005
Chiribaya	27	14.3	33.3	18.5	33.3	.75	1100 B.P.–650 B.P.	This study
Late Intermediate	15	20.0	53.3	20.0	6.7	.68	1000 B.P.–500 B.P.	Moraga et al. 2005
Contemporary Aymara	172	7.0	68.0	11.9	13.1	.51	present	Merriwether et al. 1995

South American continent. Haplogroup A is dominant among contemporary North Andean Amerindians, while hapologroup B is relatively high in modern Central Andean populations, and high frequencies of haplogroups C and D are characteristic of South Andeans (Bert et al. 2001; Keyeux et al. 2002; Melton et al. 2007; Merriwether et al. 1995; Moraga et al. 2000; Rodriguez-Delfin et al. 2001). It should be kept in mind that these modern haplogroup frequencies are based predominantly on highland (as opposed to coastal) populations of each region.

Sampling Issues

It is obvious that the behavioral and social significance of genetic information gained from the study of ancient DNA rests largely on the size, context, and representativeness of the analyzed sample. Elsewhere, we noted that "[m]itochondrial DNA analysis is most valuable if it is effectively integrated in a sustained, interdisciplinary, regional archaeological investigation with a specific research agenda" and that "it is incumbent on the archaeologist to define worthwhile research issues and aims and accordingly collect relevant samples of *well-preserved and well-dated burials with good, regional contextual understanding*" (emphasis added; Shimada et al. 2005:81–82).

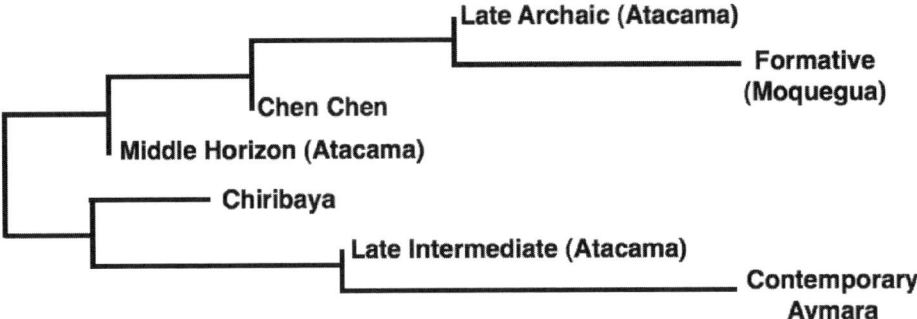

Figure 9-3. *Phylogenetic relationship between seven ancient and modern populations determined by neighbor-joining using the Fst values based on haplogroup frequencies.*

The archaeological reality, however, is that burial excavations are often the result of unexpected encounters during fieldwork conducted for different purposes or the result of rescue operations following looting that revealed the presence of burials. Large coherent corpora of ancient burials with well-preserved, well-dated, and well-understood regional contexts ideally suited for ancient DNA study are quite rare. Unlooted cemeteries are just as rare. Further, too often time, manpower, and other logistical constraints do not allow thorough contextual documentation or recovery of many burials. Indeed, such was the case for the burials recorded and recovered by Guillén in our study.

In Peru and elsewhere, the archaeologist finds it nearly impossible to determine the representativeness of a given sample of excavated burials and to define the composition and organization of associated burials. As Drennan (1996:86) and others have noted, the notion of representativeness in archaeological samples is "slippery" and to a large degree unachievable since the target population is not accessible due to a variety of archaeological and cultural formation processes, among other reasons.

Until large corpora of appropriately contextualized burials become available through sustained interdisciplinary regional studies, the results of our study should be seen as preliminary. In this regard, we are cautiously encouraged that there is a good degree of agreement between the results of our study and those of studies of nearby areas.

Archaeological Ramifications and Conclusions

The high frequency of mitochondrial DNA haplogroup A observed among the Formative samples was consistent with that of the equatorial coastal region during the Formative era. There is, however, a notable long-term increase in the relative proportion of haplogroup B, particularly after the onset of the Middle Horizon around 1400–1350 B.P. when the Central and South-Central Andes suffered severe droughts and a mega–El Niño (Binford et al. 1997; Kolata 2000;

Shimada et al. 1991; Thompson and Mosley-Thompson 1987). It is starting in this time period that we have clear indications of the establishment of intrusive enclaves pertaining to the Tiwanaku Empire. The observed trend suggests that over the centuries increasing numbers of one or more highland populations migrated to the lower reaches of the Moquegua and adjacent coastal valleys to the south, interbreeding to varying degrees with local population(s). This influx accelerated during the Inka and colonial periods, seemingly replacing much of the indigenous population of this area.

The above interpretation, however, rests on untested assumptions of a random mating pattern among indigenous and intrusive highland populations and the representativeness of our burial sample of local resident individuals. Post-marital residence rules of the indigenous and intrusive populations and our understanding of the representativeness of the analyzed sample are not known. Furthermore, it is important to keep in mind that the coast-highland interaction was most likely both biologically and culturally bidirectional and that our mtDNA study has revealed only maternal linkages of ancient regional populations.

The inference that there were various waves of population movement from the highlands to the adjacent coast, however, is supported by the limited number of shared haplotypes: just 3 among the 23 haplotypes observed in the samples from the Formative and the Chiribaya sites.

The ethnic or linguistic identities of the highland populations that migrated to the coast are not readily apparent. While many archaeologists working in the South-Central Andes hold the view that an Aymara ethnic population or Aymara speakers were the builders of the pre-Inkaic Tiwanaku Empire (e.g., Browman 1984; Kolata 1993, 2004), this is by no means certain. Certainly, from the time of Spanish Conquest of the Inka Empire to the present, the ethnic Aymara have been the dominant resident population of the altiplano of the South-Central Andes. Cerrón-Palomino (1989, 2000) and other leading Andean linguists (e.g., Hardman 2000; Torero 2002) believe that Aymara speakers expanded southward out of their original homeland in the central highlands of Peru (Yauyos Valley, Department of Lima), reaching the altiplano or Tiwanaku territory relatively late. It is likely that the ethnic Aymara population formed a major component of the inferred highland population(s) that settled in the Moquegua Valley during the late pre-Hispanic and colonial eras. One thing that is certain at present is that archaeologists cannot readily assume concordance among linguistic, genetic, and archaeological data.

Acknowledgments

This study was supported by a Grant-in-Aid for Scientific Research 19405016 from the Ministry of Education, Science, Sports, and Culture, Japan. The sampled teeth were exported from Peru to the National Museum of Natural History and Science in Tokyo with the authorization of the National Institute of Culture of Peru. We are grateful to Melody Shimada for her helpful comments and editorial assistance. Gabriela Cervantes kindly provided us with some Peruvian publications and theses that were difficult to access.

References

Achilli, Alessandro, Ugo A. Perego, Claudio M. Bravi, Michael D. Coble,
Qing-Peng Kong, Scott R. Woodward, Antonio Salas, Antonio Torroni, and
Hans-Jürgen Bandelt
    2008   The Phylogeny of the Four Pan-American mtDNA Haplogroups: Implications for Evolutionary and Disease Studies. *PLoS ONE* 3(3):e1764. doi: 10.1371/journal .pone.0001764.

Andrews, Richard M., Iwona Kubacka, Patrick Chinnery, Robert N. Lightowlers,
Douglass M. Turnbull, and Neil Howell
    1999   Reanalysis and Revision of the Cambridge Reference Sequence for Human Mitochondrial DNA. *Nature Genetics* 23:147.

Bawden, Garth
    1989   The Tumilaca Site. In *Ecology, Settlement and History in the Osmore Drainage, Peru,* edited by Don S. Rice, Charles Stanish, and Phillip R. Scarr, pp. 287–302. BAR International Series 545. British Archaeological Reports, Oxford.

Benson, Elizabeth P., and Anita Cook (editors)
    2001   *Ritual Sacrifice in Ancient Peru*. University of Texas Press, Austin.

Bert, Francesc, Alfons Corella, Manel Gené, Alejandro Pérez-Pérez, and
Daniel Turbón Borrega
    2001   Major Mitochondrial DNA Haplotype Heterogeneity in Highland and Lowland Amerindian Populations from Bolivia. *Human Biology* 73:1–16.

Binford, Michael W., Alan L. Kolata, Mark Brenner, John W. Janusek, Matthew T. Seddon,
Mark Abbott, and Jason H. Curtis
    1997   Climate Variation and the Rise and Fall of an Andean Civilization. *Quaternary Research* 47:235–248.

Blom, Deborah E., Benedikt Hallgrimsson, Linda Keng, Maria C. Lozada C., and
Jane E. Buikstra
    1998   Tiwanaku "Colonization": Bioarchaeological Implications for Migration in the Moquegua Valley, Peru. *World Archaeology* 30:238–261.

Browman, David L.
    1984   Prehistoric Aymara Expansion, the Southern Altiplano and San Pedro de Atacama. *Estudios Atacameños* 7:236–252.

Buikstra, Jane
    1995   Tombs for the Living or for the Dead: The Osmore Ancestors. In *Tombs for the Living: Andean Mortuary Practices*, edited by Tom Dillehay, pp. 229–280. Dumbarton Oaks, Washington, D.C.

Cerrón-Palomino, Rodolfo
    1989   *Lengua y sociedad en el Valle del Mantaro.* Instituto de Estudios Peruanos, Lima
    2000   *Lingüística aimara.* C.E.R.A. Bartolomé de Las Casas, Cuzco.

Drennan, Robert D.
    1996   *Statistics for Archaeologists: A Commonsense Approach.* Plenum Press, New York.

Eshleman, Jason A., Ripan S. Malho, and David Glenn Smith
    2003   Mitochondrial DNA Studies of Native Americans: Conceptions and Misconceptions of the Population Prehistory of the Americas. *Evolutionary Anthropology* 12:7–18.

Excoffier, Laurent, Guillame Laval, and Stefan Schneider
    2005   Arlequin ver. 3.0: An Integrated Software Package for Population Genetics Data Analysis. *Evolutionary Bioinformatics Online* 1:47–50.

Fagundes, Nelson J. R., Ricardo Kanitz, Roberta Eckert, Ana C. S. Valls,
Mauricio R. Bogo, Francisco M. Salzano, David Glenn Smith, Wilson A. Silva Jr.,
Marco A. Zago, Andrea K. Ribeiro-dos-Santos, Sidney E. B. Santos,
Maria Luiza Petzl-Erler, and Sandro L. Bonatto
    2008   Mitochondrial Population Genomics Supports a Single Pre-Clovis Origin with a Coastal Route for the Peopling of the Americas. *American Journal of Human Genetics* 82:583–592.

Focacci, Guillermo
    1983   El Tiwanaku clásico en el valle de Azapa. In *Asentamientos aldeanos en los valles costeros de Arica*, edited by Iván Muñoz and Julia Cordova, pp. 94–113. Documento de Trabajo no. 3, Universidad de Tarapacá, Tarapacá, Chile.

Fuselli, Silvia, Eduardo Tarazona-Santos, Isabelle Dupanloup, Alonso Soto,
Donata Luiselli, and Davide Pettener
    2003   Mitochondrial DNA Diversity in South America and the Genetic History of Andean Highlanders. *Molecular Biology and Evolution* 20:1682–1691.

Ghersi, Humberto
    1956   Informe sobre las Excavaciones en Chiribaya. *Revista del Museo Nacional, Lima-Perú* 25:89–119.

Godoy, María C.
    2005   Tiwanaku and Chiribaya: Diet and Dental Diseases During the Middle Horizon and Late Intermediate Period in the Lower Osmore Valley, Southern Peru. Unpublished Master's thesis, Institute of Archaeology, University College London, London.

Goldstein, Paul S.
    1989   Omo, a Tiwanaku Provincial Center in Moquegua, Peru. Unpublished Ph.D. dissertation, Department of Anthropology, University of Chicago, Chicago.
    2005   *Andean Diaspora: The Tiwanaku Colonies and the Origins of South American Empire*. University of Florida Press, Gainesville.

Goldstein, Paul S., and Bruce Owen
    2001   Tiwanaku en Moquegua: Las colonias altipánicas. In *Huari y Tiwanaku: Modelos vs. evidencias*, edited by Peter Kaulicke and William H. Isbell, pp. 139–168. Boletín de Arqueología PUCP 5. Fondo Editorial de la Pontificia Universidad Católica del Perú, Lima.

Hardman, Martha J.
    2000   *Jaqaru*. Lincom Europa, München.

Julien, Catherine J.
    1985   Guano and Resource Control in Sixteenth-Century Arequipa. In *Andean Ecology and Civilization*, edited by Shozo Masuda, Izumi Shimada, and Craig Morris, pp. 185–231. University of Tokyo Press, Tokyo.

Keyeux, Genoveva, Clemencia Rodas, Nancy Gelvez, and Dee Carter
    2002   Possible Migration Route into South America Deduced from Mitochondrial DNA Studies in Colombian Amerindian Populations. *Human Biology* 74:211–233.

Kitchen, Andrew, Michael M. Miyamoto, and Connie J. Mulligan
    2008   A Three-Stage Colonization Model for the Peopling of the Americas. *PLoS ONE* 3(2):e1596. doi:10.1371/journal.pone.0001596.

Kolata, Alan L.
    1983   The South Andes. In *Ancient South Americans*, edited by Jesse D. Jennings, pp. 240–285. W. H. Freeman, San Francisco.
    1993   *The Tiwanaku: Portrait of an Andean Civilization*. Basil Blackwell, Oxford, United Kingdom.

2000   Environmental Thresholds and the 'Natural History' of an Andean Civiliza-
tion. In *Environmental Disaster and the Archaeology of Human Response*, edited by
Garth Bawden and Richard Martin Reycraft, pp. 163–178. Anthropological Papers
No. 7, Maxwell Museum of Anthropology, University of New Mexico Press.

2004   The Flow of Cosmic Power: Religion, Ritual, and the People of Tiwanaku. In
Tiwanaku: Ancestors of the Inca, edited by Margaret Young-Sanchez, pp. 97–125. Den-
ver Art Museum and the University of Nebraska Press, Lincoln.

Kumar, Sudhir, Tamura Kouichiro, and Nei Masatoshi
2004   MEGA3: Integrated Software for Molecular Evolutionary Genetics Analysis
and Sequence Alignment. *Briefings in Bioinformatics* 5:150–163.

Lewis, Cecil M., Jr.
2009   Difficulties in Rejecting a Local Ancestry with mtDNA Haplogroup Data in the
South-Central Andes. *Latin American Antiquity* 20:76–90.

Lewis, Cecil M., Jr., Jane E. Buikstra, and Anne C. Stone
2007b   Ancient DNA and Genetic Continuity in the South Central Andes. *Latin American Antiquity* 18:145–160.

Lewis, Cecil M., Jr., Beatriz Lizárraga, Raúl Y. Tito, Paul W. López, Gian Carlo Iannacone,
Angel Medina, Rolando Martínez, Susan I. Polo, Augusto F. De La Cruz,
Angela M. Cáceres, and Anne C. Stone.
2007a   Mitochondrial DNA and the Peopling of South America. *Human Biology*
79:159–178.

Lewis, Cecil M., Jr., Raúl Y. Tito, Beatriz Lizárraga, and Anne C. Stone
2005   Land, Language, and Loci: mtDNA in Native Americans and the Genetic History of Peru. *American Journal of Physical Anthropology* 127:351–360.

Lozada, María C.
1998   The Señorío of Chiribaya: A Bio-Archaeological Study in the Osmore Drainage
of Southern Peru. Unpublished Ph.D. dissertation, Department of Anthropology,
University of Chicago.

Lozada, María C., and Jane E. Buikstra
2002   El *Señorío de Chiribaya en la Costa Sur del Peru*. Instituto de Estudios Peruanos,
Lima.

Lumbreras, Luis G.
1974   Los Reinos post-Tiwanaku en el Área altiplanica. *Revistadel Museo National,
Lima-Perú* 15:56–85.

Maca-Meyer, Nicole, Vicente M. Cabrera, Matilde Arnay, Carlos Flores, Rosa Fregel,
Ana M. González, and José M. Larruga
2005   Mitochondrial DNA Diversity in 17th–18th Century Remains from Tenerife
(Canary Islands). *American Journal of Physical Anthropology* 127:418–426.

Masuda, Shozo
1985   Algae Collectors and Lomas. In *Andean Ecology and Civilization*, edited by Shozo Masuda, Izumi Shimada, and Craig Morris, pp. 233–250. University of Tokyo
Press, Tokyo.

Melchior, Linea, Marcus Thomas Pius Gilbert, Toomas Kivisild, Niels Lynnerup, and
Jorgen Dissing
2008   Rare mtDNA Haplogroups and Genetic Differences in Rich and Poor Danish
Iron-Age Villages. *American Journal of Physical Anthropology* 135:206–215.

Melton, Phillip E., Ignacio Briceño, A. Gómez, Eric J. Devor, J. E. Bernal, and
Michael H. Crawford
2007   Biological Relationship Between Central and South American Chibchan Speak-

ing Populations: Evidence from mtDNA. *American Journal of Physical Anthropology* 133:753–770.

Merriwether, D. Andrew, Francisco Rothhammer, and Robert E. Ferrell
  1995   Distribution of the Four Founding Lineage Haplotypes in Native Americans Suggests a Single Wave of Migration for the New World. *American Journal of Physical Anthropology* 98:411–430.

Meyer, Sonja, Gunter Weiss, and Arndt von Haeseler
  1999   Pattern of Nucleotide Substitution and Rate of Heterogeneity in Hypervariable Region I and II of Human mtDNA. *Genetics* 152:1103–1110.

Moraga, Mauricio, Paola Rocco, Juan F. Miquel, Fravio Nervi, Elena Llop, Ranajit Chakraborty, Francisco Rothhammer, and Ranajit Carvallo
  2000   Mitochondrial DNA Polymorphisms in Chilean Aboriginal Populations: Implications for the Peopling of the Southern Cone of the Continent. *American Journal of Physical Anthropology* 113:19–29.

Moraga, Mauricio, Calogero M. Santoro, Vivien G. Standen, Pilar Carvallo, and Francisco Rothhammer
  2005   Microevolution in Prehistoric Andean Populations: Chronologic mtDNA Variation in the Desert Valleys of Northern Chile. *American Journal of Physical Anthropology* 127:170–181.

Mujica, Elias
  1985   Altiplano-Coast Relationships in the South-Central Andes: From Indirect to Direct Complementarity. In *Andean Ecology and Civilization*, edited by Shozo Masuda, Izumi Shimada, and Craig Morris, pp. 103–140. University of Tokyo, Tokyo.

Murra, John V.
  1968   An Aymara Kingdom in 1567. *Ethnohistory* 15:115–151.
  1972   El "control vertical" de un máximo de pisos ecológicos en la economía de las sociedades andinas. In *Visita de la provincia de León de Huánuco 1562,* Tomo II, edited by John V. Murra, pp. 27–468. Universidad Nacional Hermilio Valdizán, Huánuco, Peru.

Owen, Bruce
  1993   A Model of Multiethnicity: State Collapse, Competition, and Social Complexity from Tiwanaku to Chiribaya in the Osmore Valley, Peru. Unpublished Ph.D. dissertation. Department of Anthropology, University of California, Los Angeles.

Raymond, Michel, and François Rousset
  1995   An Exact Test of Population Differentiation. *Evolution* 49:1280–1283.

Reed, David (editor)
  2005   *Biomolecular Archaeology: Genetic Approaches to the Past.* Occasional Paper No. 32. Center for Archaeological Investigations, Southern Illinois University, Carbondale.

Rice, Don S., Charles Stanish, and Phillip R. Scarr (editors)
  1989   *Ecology, Settlement and History in the Osmore Drainage.* British Archaeological Reports, Oxford.

Rivera, Mario
  1984   Altiplano and Tropical Lowland Contacts in Northern Chilean Prehistory: Chinchorro and Alto Ramírez Revisited. In *Social and Economic Organization in the Prehispanic Andes*, edited by David L. Browman, Richard L. Burger, and Mario A. Rivera, pp. 143–160. BAR International Series 194. British Archaeological Reports, Oxford.

Rodriguez-Delfin, Luis A., Verónica Rubin-de-Celis, and Marco A. Zago
  2001   Genetic Diversity in an Andean Population from Peru and Regional Migration

Patterns of Amerindians in South America: Data from Y Chromosome and Mitochondrial DNA. *Human Heredity* 51:97–106.

Rowe, John H.
    1982   Policies and Institutions Relating to the Unification of the Empire. In *The Inca and Aztec States, 1400–1800*, edited by George Allen Collier, Renato I. Rosaldo, and John D. Wirth, pp. 93–118. Academic Press, New York.

Schurr, Theodore G.
    2004   The Peopling of the New World: Perspectives from Molecular Anthropology. *Annual Review of Anthropology* 33:551–583.

Shimada, Izumi, Crystal Schaaf, Lonnie G. Thompson, and Ellen Mosley-Thompson
    1991   Cultural Impacts of Severe Droughts in the Prehistoric Andes: Application of a 1,500-Year Ice Core Precipitation Data. *World Archaeology* 22:247–270.

Shimada, Izumi, Ken-ichi Shinoda, Steve Bourget, Walter Alva, and Santiago Uceda
    2005   mtDNA Analysis of Mochica and Sicán Populations of Pre-Hispanic Peru. In *Biomolecular Archaeology: Genetic Approaches to the Past*, edited by David Reed, pp. 61–92. Occasional Paper No. 32. Center for Archaeological Investigations, Southern Illinois University, Carbondale.

Shimada, Izumi, Ken-ichi Shinoda, Julie Faunum, Robert Corruccini, and Hirokatsu Watanabe
    2004   An Integrated Analysis of Prehispanic Mortuary Practices: A Middle Sicán Case Study. *Current Anthropology* 45:369–402.

Shinoda, Ken-ichi, Noboru Adachi, Sonia Guillén, and Izumi Shimada
    2006   Mitochondrial DNA Analysis of Ancient Peruvian Highlanders. *American Journal of Physical Anthropology* 131:98–107.

Shinoda, Ken-ichi, and Satoshi Kanai
    1999   Intracemetery Genetic Analysis at the Nakazuma Jomon Site in Japan by Mitochondrial DNA Sequencing. *Anthropological Science* 107:129–140.

Stanish, Charles
    1989   An Archaeological Evaluation of an Ethnohistorical Model in Moquegua. In *Ecology, Settlement and History in the Osmore Drainage*, edited by Don S. Rice, Charles Stanish, and Phillip R. Scarr, pp. 303–322. BAR International Series 545. British Archaeological Reports, Oxford.
    1992   *Ancient Andean Political Economy*. University of Texas Press, Austin.
    2003   *Ancient Titicaca: The Evolution of Complex Society in Southern Peru and Northern Bolivia*. University of California Press, Berkeley and Los Angeles.

Sutter, Richard C.
    1997   Dental Variation and Biocultural Affinities among Prehistoric Populations from the Coastal Valleys of Moquegua. Unpublished Ph.D. dissertation, Department of Anthropology, University of Missouri, Columbia.

Tamm, Erika, Toomas Kivisild, Mauro Roidla, Mait Metspalu, David Glenn Smith, Connie J. Mulligan, Claudio M. Bravi, Olga Rickards, Cristina Martinez-Labarga, Elsa K. Khusnutdinova, Sardana A. Fedorova, Maria V. Golubenko, Vadim A. Stepanov, Marina A. Gubina, Sergey I. Zhadanov, Ludmila P. Ossipova, Larisa Damba, Mikhail I. Voevoda, Jose E. Dipierri, Richard Villems, and Ripan S. Malhi
    2007   Beringian Standstill and Spread of Native American Founders. *PLoS ONE* 2(9):e829. doi:10.1371/journal.pone.0000829.

Thompson, Lonnie G., and Ellen Mosley-Thompson
    1987   Evidence of Abrupt Climatic Change During the Last 1,500 Years Recorded in Ice Cores from the Tropical Quelccaya Ice Cap, Peru. In *Abrupt Climatic Change:*

Evidence and Implications, edited by Wolfgang H. Berger and Laurent D. Labeyrie, pp. 99–110. D. Reidel, Dordrecht, Holland.

Thompson, Lonnie G., Ellen Mosley-Thompson, John F. Bolzan, and Bruce R. Koci
    1985   A 1500-Year Record of Tropical Precipitation in Ice Cores from the Quelccaya Ice Cap, Peru. *Science* 229:971–973.

Torero, Alfredo
    2002   *Idiomas de los Andes: Lingüística e Historia*. Instituto Francés de Estudios Andinos, Lima.

Torroni, Antonio, Theodore G. Schurr, Margaret F. Cabell, Michael D. Brown,
James V. Neel, Merethe Larson, David G. Smith, Carlos M. Vullo, and Douglas C. Wallace
    1993   Asian Affinities and Continental Relation of the Four Founding Native American mtDNAs. *American Journal of Human Genetics* 53:563–590.

Torroni, Antonio, Theodore G. Schurr, Chi-Chuann Youg, Emoke J. E. Szathmary,
Robert C. Williams, Moses S. Shanfield, Gary A. Troup, William C. Knowler,
Dale N. Lawrence, Kenneth M. Weiss, and Douglas C. Wallace
    1992   Native American Mitochondrial DNA Analysis Indicates That the Amerind and the Nadene Population Were Founded by Two Independent Migrations. *Genetics* 130:153–162.

Wallace, C. Douglas, Katherine Garrison, and William C. Knowler
    1985   Dramatic Founder Effects in Amerindian Mitochondrial DNAs. *American Journal of Physical Anthropology* 68:149–155.

Woodward, Scott R., Marie J. King, Nancy M. Chiu, Marvin J. Kuchar, and
C. Wilfred Griggs
    1994   Amplification of Ancient Nuclear DNA from Teeth and Soft Tissues. *PCR Methods and Applications* 3:244–247.

10. Climate Variation, Biological Adaptation, and Postcranial Metric Variation in Precontact North America

Kathryn A. King

Summary Statement: This chapter explores questions about human biological variation in response to climate variation. Twenty-five prehistoric and protohistoric populations from North America are included in this study. The skeletal remains of 854 individuals from these populations were measured to assess the influence of climate variation on body form.

Among the climate factors examined, temperature had the strongest impact on body form. Long-bone lengths in particular were influenced by temperature, with colder temperatures being related to shorter long-bone lengths. This effect was particularly pronounced in the tibia and radius. The dimensions of the joint surfaces were also related to temperature, with populations in colder climates exhibiting wider epiphyseal breadths than those from warmer climates. The breadth of the long-bone shafts were most commonly correlated with annual precipitation, suggesting that factors other than temperature influence variation in body form, including the possible relationships among nutrition, rainfall, and the biomechanical loads associated with procuring food in various environments.

These results imply that studies of human biological diversity in North America cannot ignore the corresponding effects of variation in climate and lifeways of the study populations. Similarly, archaeological studies should appreciate the inherent variation in body form that occurs simultaneously with cultural variation.

Human Variation in the Americas: The Integration of Archaeology and Biological Anthropology, edited by Benjamin M. Auerbach. Center for Archaeological Investigations, Occasional Paper No. 38. © 2010 by the Board of Trustees, Southern Illinois University. All rights reserved. ISBN 978-0-88104-095-1.

Introduction

Modern American Indians and First Peoples are the descendants of immigrants who likely derived from a single group or several geographically proximate groups in Asia. They were subject to similar climate-related selective pressures during the migration process, regardless of whether their ancestors' route was coastal or terrestrial. From this founding population(s), a multitude of groups emerged and settled every habitat in the Americas, from deserts to rain forests to polar regions. This population expansion resulted in cultural and biological diversification. It is the biological component of this diversification that this work seeks to explore, particularly those features associated with climatic adaptation in the postcranial skeleton.

The hypothesis of this work is that indigenous groups have inhabited the New World for sufficient time to have adapted to specific climates. This idea will be explored through the analysis of postcranial osteometric data in relation to a variety of climate variables. Significant positive relationships between increased skeletal robusticity and shortened limb length in colder climates support this hypothesis.

The theoretical basis for the study of ecogeographical patterning in humans and other warm-blooded animals derives from Carl Bergmann's study on the relationship between increasing body size and decreasing mean annual temperature within polytypic warm-blooded species (Bergmann 1847, as translated in Katzmarzyk and Leonard 1998). A complementary study by Joel Allen (1877) argues that warm-blooded organisms with longer extremities are better adapted to hot climates than those with shorter or smaller extremities are, and vice versa.

The mechanism that governs both of these rules is the relationship between heat loss and heat production (Schreider 1951, 1975). In situations where a warm-blooded organism is exposed to external temperatures significantly lower than its body temperature, heat loss through radiation needs to be minimized while heat production through metabolic processes should be maximized. In climates where external temperatures often exceed body temperature, heat loss should be maximized while heat production should be kept to a minimum (Allee and Schmidt 1951). This is achieved through the maximizing or minimizing of an organism's thermolytic surfaces relative to its body mass through changes in the overall body size and/or length and breadth of the appendages (Allee and Schmidt 1951; Schreider 1975).

In humans, the forearm and lower leg lose heat more rapidly than the upper arm and thigh do due to the formers' larger surface area to volume ratio, so relatively shorter distal limb segments would be expected in cold adapted populations (Holliday 1999). This pattern has been documented in a variety of modern and prehistoric populations (Hall et al. 2004; Porter 1999; Yamaguchi 1989).

A multitude of climatic variables have been examined as possible correlates of ecogeographical patterns. While many studies examine ecogeographical patterning in terms of large-scale climate differences, many studies have noted correlations between specific variables and body proportions. Significant associations between body size and/or proportions have been noted for precipitation (Crognier

1981; Hall and Hall 1995; Yom-Tov and Geffen 2006), humidity (Hiernaux and Froment 1976), temperature during the hottest months (Crognier 1981; Hiernaux and Froment 1976; Schreider 1951), temperature during cold months (Johnston and Selander 1971; Newman 1960; Newman and Munro 1955), mean annual temperature (Katzmarzyk and Leonard 1998; Murphy 1985; Newman 1953, 1960, 1962; Newman and Munro 1955; Roberts 1953, 1978; Stinson and Frisancho 1978), and seasonal differences in temperature and precipitation (Murphy 1985). Similar patterns have also been noted for latitude (Holliday 1999; Jacobs 1985; Johnston and Selander 1973; Murphy 1985; Ruff 1991), a variable that implies differences in a number of specific measures. The variety of factors that have been found to be associated with variation in body proportions suggests that such relationships exist in humans but that these patterns are not easily correlated with one specific measure. Human morphology responds to multiple stresses simultaneously, making it impossible to name a single causative factor for ecogeographical patterning in humans.

It can be difficult to separate the influence of genetic and nongenetic factors that affect human morphological plasticity (Meadows Jantz and Jantz 1999). This study seeks to minimize the impact of nonclimatic environmental factors by focusing on skeletal remains from prehistoric and protohistoric times. This approach eliminates the need to consider secular change, gene flow from European and African populations, the impacts of Old World technology, and any possible effects of the continent-wide epidemics that accompanied the first Old World explorers and colonists as causative agents of biological change.

The possible influences of an environment on growth and development are more difficult to remove. The samples used in this study derive from groups that differed in both ecological settings and cultural practices, which likely resulted in differences in subsistence strategy, disease load, and habitual activity patterns. Using only osteometric variables, it is difficult if not impossible to account for potential differences in nutritional, health, and activity status among these samples. In an attempt to minimize the influence of disease and poor nutrition, specimens with notable pathologies were excluded from this analysis. Activity-related remodeling of the skeleton is difficult to address in this context, but intergroup differences should be negligible as all samples used in this study were engaged in some form of subsistence production or acquisition that relied on lithic technology and human labor. Even in socially stratified prehistoric societies, there likely was little difference between the elites and the rest of the populace in habitual activity patterns as can be inferred from the development of muscle insertion sites (Harle and King 2004), though it should be noted that other authors have argued for status-based differences in long-bone morphology based on cortical bone thickness (Hatch et al. 1983).

Homo sapiens in the Old World represent the maximum amount of ecogeographic variation seen in modern humans. Humans have inhabited Africa, Europe, and Asia millennia longer than humans have lived in the Americas. Old World populations demonstrate the extremes of human adaptation to climate, from hot and dry sub-Saharan Africa to extremely cold regions of northern Europe and Asia. Significant anthropometric and/or osteometric variation in the Old World related to ecogeographical patterning has been documented by Crognier

(1981), Hiernaux and Froment (1976), Holliday (1997a), Jacobs (1985, 1993), Katzmarzyk and Leonard (1998), Roberts (1953, 1978), Ruff (1991, 1994), and Yamaguchi (1989). Similar, though often less pronounced, results using indigenous New World populations have also been documented (Hall and Hall 1995; Katzmarzyk and Malina 1999; Lazenby and Smashnuk 1999; Newman 1953, 1960, 1962).

The relatively recent colonization of the Americas by *Homos sapiens* presents a unique natural experiment in which to examine human adaptation to climate. Significant populations of humans were likely dispersed throughout the Americas by 11,500 RCYBP as evidenced by the establishment of the widespread, archaeologically visible Clovis sites (Haynes 2002). The indigenous populations of the Americas demonstrate the amount of biological adaptation that a group can undergo in less than 13,500 years.

Have the Americas been populated long enough for ecogeographical patterns to have evolved? The studies discussed above document that clear clinal patterns are visible in the Old World. Such patterns also are argued for in much of the literature on New World samples, but the relationships between body proportions and climate variables may be more ambiguous than those in the rest of the world. North American house sparrows have experienced significant morphological differentiation rapidly, in a span of 50 to 115 generations (Johnston and Selander 1971). Are humans capable of adapting as rapidly, or does our species evolve at a slower rate?

Gradual genetic-based change in morphology in response to climate does not appear to occur among humans in time spans less than a millennium, or 25 to 50 generations—assuming generations to be between 20 and 40 years in length. This excludes situations in which major selection events resulted in a drastically altered and reduced gene pool, as is suspected to have occurred during the peopling of the Pacific islands. Body proportions seem to be a conservative trait, as demonstrated by differences in the average sitting height to stature ratio between African American children and European American children (Martorell et al. 1988). Similar ancestry-related patterns can be seen in Chile (Palomino et al. 1979) and on the Croatian island of Hvar (Rudan et al. 1986).

While the pace of human adaptation may not occur at detectable levels within a few centuries, it still may proceed at a somewhat rapid pace, taking only a few thousand years to evolve. Change in body proportions from the seemingly warm-adapted anatomically modern humans of the European Early Upper Paleolithic to the more cold-adapted forms of modern Europeans occurred in 15,000 to 20,000 years, with modern patterns of body proportions emerging during the Middle Upper Paleolithic (Holliday 1997b; Trinkaus 1997).

Newman (1962) argues that ecogeographical patterning is present in the New World, citing the range in mean male statures as being too wide to be indicative of anything but genetic adaptation. Pearson and Millones (2005) argue that the settlement of the Americas occurred long enough ago for adaptation to climate to have taken place in the locations most distant from Beringia: Tierra del Fuego. The authors demonstrate that two Fuegan groups have very robust limb bones and relatively broad pelves, with values similar to those found in Inuit and Sami samples.

Time depth does affect the strength of the statistical relationships among climate variables and morphological traits (Roberts 1978). Roberts noted that anthropometric measures are more strongly correlated with climate factors among Old World populations, which have resided in their local areas for a significant length of time, as compared with Native American populations, which by comparison immigrated more recently to their respective regions. A similar pattern also explains the lack of robust statistical results in Newman and Munro's (1955) study of climate and body size in U.S. males compared to their European counterparts.

Morphological variation among prehistoric American Indians and First Nations peoples should demonstrate some relationship with climate variables. These relationships may not be as strong or as obvious as those seen in Old World populations, depending on the length of time necessary for adaptation to climate to reach the levels seen in Africa, Asia, and Europe. Patterns may be related to heat stress, cold stress, precipitation, annual temperature range, latitude, or any combination of these variables. Abbreviation of the distal segments of the limbs may or may not be present depending on the time depth necessary for these patterns to develop and on whether or not this feature is a true ecogeographical correlate.

The Origins of the Indigenous Populations of the Americas

In order to explore questions about the biological adaptations of modern American Indians, it is necessary to investigate the processes of the peopling of the Americas. There is much debate about the number of migrations and the timing of those migrations. In the context of this work, it is necessary to establish a feasible time line for the latest possible entry into the Americas, as this would establish the minimum amount of time indigenous groups have been subjected to the climatic patterns of the New World.

The earliest evidence of human habitation in the Americas comes from a handful of pre-Clovis sites, all of which have produced radiocarbon dates earlier than 11,500 RCYBP (Dillehay 1997; Goodyear 2004). The antiquity of one of these sites, Monte Verde in Chile, which dates to 12,500–12,000 cal B.P., has been evaluated and accepted by a team of prehistorians. A late Pleistocene occupation of Monte Verde, 16,000 kilometers from Beringia, implies an earlier colonization of the Americas than was previously accepted (Dillehay 1989, 2000; Meltzer et al. 1997). Further support for a pre-Clovis presence in the Americas comes from 14,000-year-old coprolites that contain human DNA from Paisley Caves in Oregon (Gilbert et al. 2008).

In the context of biological anthropology in the Americas, the proposed antiquity of the pre-Clovis sites is irrelevant if the early colonizers were not the ancestors of the later American populations. The low archaeological visibility of pre-Clovis sites (Beaton 1991) suggests that these sites might reflect early "failed migrations" into the North and South American continents by groups too small or too geographically distant to establish large population levels (Meltzer 1989). A more conservative estimate for the immigration of the founding populations of the Americas would be 11,500 RCYBP, when the first Clovis sites were occupied and shortly before the widespread dispersal of Clovis technology over the North American continent (Haynes 2002), which would have been possible only if a dramatic increase in population occurred around this time or if there was a large population previously residing in North America.

For the purposes of investigating the biological adaptation to climate seen in American Indians, we must be concerned not with the earliest occupation of the Americas but with the earliest occupation that led to population expansion and divergence. Considering the archaeological, biological, and linguistic data available, I have decided to use the date 11,500 RCYBP, or approximately 13,500 cal B.P. (Stuiver et al. 2005), the approximate date of the earliest Clovis culture sites (Fiedel 2004; Humphrey and Ferring 1994), as the starting point for the large-scale settlement of the Americas. While it is noted that there are very few human biological remains associated with the Clovis period (Jones 1996), the extensive collection of unambiguous artifacts and sites associated with this period strongly suggests the presence of a large, widespread population. No assumption will be made about the climatic adaptations seen in the first migrants to the Americas, as the founding population(s) may have been indigenous to either the cold climate of Siberia and Beringia of the late Pleistocene or the temperate or tropical climates, following Hall and colleagues (2004). Evidence of the development of morphology associated with climate stresses specific to the region from which a sample derives will be considered support of biological adaptation to climate.

Bioarchaeological Samples and Climate Data

The samples used in this study are from two sources: those measured by the investigator at four institutions and those collected by the Office of Repatriation at the Smithsonian Institution National Museum of Natural History. Sites were chosen in an effort to sample from a wide geographic range. Sites were also temporally restricted to strengthen the validity of intersite comparisons. Table 10-1 provides the approximate dates or time periods for the sites used in this analysis. All sites used in this analysis date within a 3,500-year range, with the majority (21 of 25) having no component older than 1,500 years. The majority are prehistoric, with the latest component of any site dating to 120 B.P. Figure 10-1 illustrates the locations of these sites.

Only skeletally mature individuals were used in this study. An individual was excluded from the study if the proximal epiphysis of the humerus was not, at the minimum, in the process of fusing to the diaphysis. Several individuals whose humeri showed complete fusion had unfused sternal epiphyses of the clavicles. In such cases the bones of the limbs were measured, but the clavicles were not.

The sex of the individuals measured by the investigator was estimated from gross morphological features of the os coxae (Phenice 1969). If the os coxa was unavailable or poorly preserved or the features were ambiguous, sex was assigned using the vertical head diameter of the humerus or the epicondylar breadth of the humerus. In the rare event that neither the os coxa nor the humerus was available or well preserved, gross morphological features of the cranium and mandible were used to estimate sex (Bass 1995). If none of these elements were present, the individual was not used in this study.

Table 10-1. *Bioarchaeological Samples*

State or province	Site[a]	Approximate dates or time periods
Alaska	Kauwerak[1]	Late Prehistoric
	Mummy Caves[2]	Late Prehistoric/Early Historic
	St. Michael[3]	Late Prehistoric/Early Historic
	Umnak Island[4]	3000 B.P.–Early Historic
British Columbia	Prince Rupert Harbour[5]	3550 B.P.–1500 B.P.
Illinois	Steuben[6]	Middle Woodland (Hopewell)
Manitoba	Fort Prince of Wales[7]	Historic
	Souris Valley[8]	Late Prehistoric–Protohistoric
Michigan	Bussinger[9]	Early-Late Woodland
	Huron Village[10]	Early-Late Woodland
	Juntunen[11]	850–300 B.P.
Mississippi	Edwards Mound[12]	Late Prehistoric/Protohistoric
	Lake George[13]	1250–750 B.P.
New Jersey	Bell-Philhower[14]	Late Prehistoric–Early Historic
New Mexico	Hawikku[15]	750–350 B.P.
	Kwastiyukwa[16]	Late Prehistoric–Early Historic
	Pueblo Bonito[17]	1060–810 B.P.
Nunavut	Native Point[18]	1900 B.P.–Historic
Ohio	Madisonville[19]	950–250 B.P.
	Turpin[20]	950–700 B.P.
South Dakota	Sully[21]	350–120 B.P.
Tennessee	Dr. Jarman[22]	Mississippian
	Thompson Village[23]	Late Mississippian
	Toqua[24]	650–350 B.P.
Washington	Berrians Island[25]	200–140 B.P.

[a] Numbers correspond to locations in Figure 10-1.

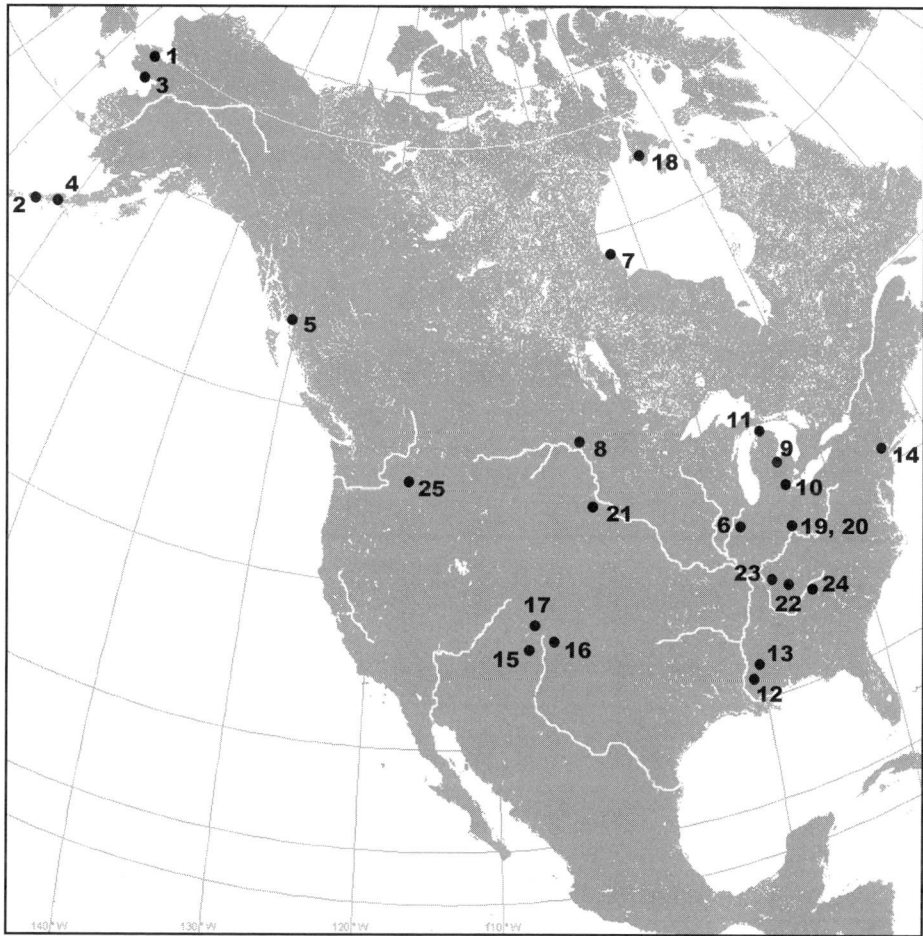

Figure 10-1. *Locations of archaeological sites. Numbers correspond to the sites listed in Table 10-1.*

Most of the skeletons used in this study were from individual burials. Multiple burials that contained more than one adult were not used in this study. If a multiple burial included one adult and any number of subadults, the adult was considered for inclusion. Elements that showed significant pathologies that would influence measurements—including osteomyeletis, poorly set antemortem fractures, and bowed long bones—were not used in this analysis.

The skeletal materials available for study were often subject to preservation issues. Measurements were only taken if no exfoliation of the cortical bone had occurred. Fragmentary bones were used only if they could be reassembled and no parts essential to taking the measurement were absent. Due to these restrictions, it was not possible to take all 29 bilateral measurements on any of the skeletons used.

Table 10-2 gives sample sizes for the 25 sites used in this analysis, categorized by state or province. Two of these samples, Souris Valley and Prince Rupert Harbour, are composites of several smaller sites from the same geographic area that date to the same time periods (Capes 1963; Cybulski 1992).

Table 10-3 lists the postcranial measurements used in this analysis. Most of the measurements were taken as described in the *Standards for Data Collection from Human Skeletal Remains* (Buikstra and Ubelaker 1994). Three other measurements were taken but dropped from the analysis: bi-iliac breadth and radius distal epiphyseal breadth, which were not available in the data obtained from the Smithsonian Institution's Office of Repatriation, and wrist breadth, which was not measurable on most skeletons as it requires a complete paired ulna and radius and was not readily replicable. The original data collected for this study are on file at the individual institutions where the data was collected. No attempt was made to address interobserver error between data collected by the author and data obtained from the Office of Repatriation because of the pending or completed repatriation of most of the skeletal material measured through that office.

The 29 measurements used in this analysis were selected because of their potential to be informative about ecogeographical adaptation. Measures of the epiphyseal size, such as head diameters and epicondylar breadths, are associated with body mass (Auerbach and Ruff 2004). More robust limbs have a reduced surface area to volume ratio compared with slender limbs. Diaphyseal diameters may be associated with limb thickness, though diaphyseal shape is also associated with habitual activity patterns (Ruff et al. 1984) and may not correlate as clearly with climate variables as epiphyseal dimensions do. Measures of the length of the limb bones are related to the length of the segments of the extremities. Longer extremities increase the surface area to volume ratio of the limbs, providing more thermolytic surfaces through which to facilitate heat dissipation. This is particularly true for the distal segments of the limbs, which are more slender than the proximal segments. An increase in the length of the distal segments increases the surface area to volume ratio more rapidly than an equivalent increase in the proximal segments does.

The above measurements were analyzed in conjunction with climate variables from the 25 archaeological sites used in this analysis. All of the sites in this study date to the last third of the Holocene. While there were minor temperature fluctuations in this epoch, these dates are recent enough that modern climate data is valid (Jacobs 1985). Information for sites in the United States was obtained from the National Climatic Data Center (National Oceanic and Atmospheric Administration et al. 2002). Climate information for the Canadian sites was obtained from the Meteorological Service of Canada's National Climate Archive (Meteorological Service of Canada, Environment Canada 2002). These publications provided data for mean January temperature, mean July temperature, mean annual temperature, and mean annual precipitation data compiled between 1971 and 2000. Annual temperature range was calculated as the difference between mean July temperature and mean January temperature. Values given in Fahrenheit and inches were converted to Celsius and millimeters. Climate data for each site are provided in Table 10-4.

Table 10-2. *Sample Sizes and Sex Distribution*

Site	Total individuals	Males	Females
Kauwerak	21	13	8
Mummy Caves	37	20	17
St. Michael	27	14	13
Umnak Island	20	11	9
Prince Rupert Harbour	35	28	7
Steuben	17	11	6
Fort Prince of Wales	2	1	1
Souris Valley	19	15	4
Bussinger	8	5	3
Huron Village	5	4	1
Juntunen	23	15	8
Edwards Mound	4	3	1
Lake George	3	2	1
Bell-Philhower	34	18	16
Hawikku	137	56	81
Kwastiyukwa	72	23	49
Pueblo Bonito	33	11	22
Native Point	52	28	24
Madisonville	64	32	32
Turpin	10	7	3
Sully	148	83	65
Dr. Jarman	19	10	9
Thompson Village	8	3	5
Toqua	33	16	17
Berrians Island	23	14	9
Totals	854	443	411

Results

Correlation and Regression Analysis

The relationship between surface area and body size in a living individual will be reflected in the morphology of the skeleton, as it is the frame upon which thermolytic and thermogenic soft tissues are situated. Theoretically, the long, thin extremities and torso that are typical features of warm-adapted populations would be reflected in long limb bones, small epiphyseal breadths in the long bones, and short clavicles. In cold-adapted populations with short, robust

Table 10-3. *Postcranial Measurements*

Element	Measurement
Humerus	Maximum length
	Maximum diameter at midshaft
	Minimum diameter at midshaft
	Maximum vertical head diameter
	Epicondylar breadth
Radius	Maximum length
	Medial-lateral diameter at midshaft
	Anterior-posterior diameter at midshaft
	Maximum head breadth
Clavicle	Maximum length
	Superior-inferior diameter at midshaft
	Anterior-posterior diameter at midshaft
Femur	Maximum length
	Bicondylar length
	Anterior-posterior diameter at midshaft
	Medial-lateral diameter at midshaft
	Circumference at midshaft
	Maximum head breadth
	Anterior-posterior subtrochanteric diameter
	Medial-lateral subtrochanteric diameter
	Epicondylar breadth
Tibia	Maximum length
	Maximum proximal epiphyseal breadth
	Maximum distal epiphyseal breadth
	Maximum diameter at nutrient foramen
	Medial-lateral diameter at nutrient foramen
	Circumference at nutrient foramen
Os Coxa	Maximum length
	Iliac breadth

Table 10-4. *Climate Data*

Site	Mean annual temp (°C)	Mean January temp (°C)	Mean July temp (°C)	Mean annual precip. (mm)	Annual temp range (°C)
Kauwerak	-5.9	-17.6	8.6	289.8	26.2
Mummy Caves	3.6	-2.1	10.3	1023.1	12.4
St. Michael	-2.7	-15.9	13.1	316.0	15.8
Umnak Island	4.5	-.3	10.5	1455.7	10.8
Prince Rupert Harbour	7.1	1.3	11.1	2593.6	9.8
Steuben	11.1	-4.7	24.3	962.7	29.0
Fort Prince of Wales	-6.9	-26.7	12.0	431.6	38.7
Souris Valley	2.5	-16.0	16.3	516.2	32.3
Bussinger	9.5	-4.8	22.8	801.6	27.6
Huron Village	9.4	-4.8	22.6	897.9	27.4
Juntunen	6.3	-7.5	19.7	673.9	13.4
Edwards Mound	16.9	4.7	27.8	1377.7	23.1
Lake George	18.1	7.2	27.8	1517.4	20.6
Bell-Philhower	9.1	-4.2	21.8	1220.0	26.0
Hawikku	10.6	0	22.1	324.6	22.1
Kwastiyukwa	7.8	-3.4	19.6	330.2	23.0
Pueblo Bonito	9.9	-2.2	22.6	241.8	24.8
Native Point	-11.6	-30.0	9.3	285.7	39.3
Madisonville	12.6	-.7	24.7	1070.6	25.4
Turpin	12.6	-.7	24.7	1070.6	25.4
Sully	8.2	-7.9	23.7	379.5	31.6
Dr. Jarman	13.9	1.7	25.2	1380.0	23.5
Thompson Village	13.9	1.0	25.4	1363.5	24.4
Toqua	14.1	2.5	24.9	1483.1	22.4
Berrians Island	11.6	.6	22.9	191.8	22.3

limbs and wide or deep torsos, the expectation would be for short long bones with broad epiphyses and longer clavicles.

Multiple statistical tests were run on the data to detect if ecogeographic patterns are present within these samples. All statistical tests were performed using NCSS statistical software (NCSS 2004). All measurements are from left elements unless the left was poorly preserved or missing, in which case measurements from the corresponding right element were used.

Pearson correlations were calculated for each measurement and each climate variable. Raw measurements for males and females were analyzed separately.

The results for females are given in Table 10-5 and for males in Table 10-6. Each skeletal measurement was regressed on each climatic variable using ordinary least-squares regression. Raw measurements for each sex were analyzed. T-tests for each regression were performed to investigate if the slopes of the regression lines were significantly different from a slope of zero.

The series of correlations and linear regressions show several patterns between climate variables and measurements. In regard to bone length, the four limb bones behaved in a similar manner in both males and females. In females, the maximum lengths of the humerus, radius, femur, and tibia were all correlated more strongly with mean July temperature than with any of the other climate variables. A similar pattern is seen in the male subsample. In the combined sample, the values for the correlations between mean July temperature and maximum limb-bone lengths were intermediate between the values calculated for the sexes separately. The correlations between radius and tibia lengths and higher temperature are somewhat stronger than those of humerus and femur lengths. This suggests that the bones of the distal limb segments elongate disproportionately in relation to increased temperature than the bones of the proximal limb segments.

The four limb-bone length measurements were more weakly correlated with mean annual or mean January temperature than they were with mean July temperature. However, most of the limb-bone lengths were significantly correlated with mean January temperature; only femur length does not produce a slope significantly different from zero when it is regressed on mean January temperature. The pattern is present in both sexes, though is more pronounced in females ($p = .944$) compared with males ($p = .629$).

Maximum length of the clavicle was moderately correlated with mean annual temperature, mean January temperature, and mean July temperature in both sexes. Mean annual temperature showed the highest correlation in both sexes.

Many of the epiphyseal measurements show strong correlations with mean January temperature and mean annual precipitation. Males and females continue to exhibit similar patterns of correlation. In the female subsample, mean January temperature was the strongest climate correlate for five of the seven epiphyseal measurements. Among the males, mean annual precipitation was the strongest correlate for three of the seven measurements; mean January temperature was the climate variable most strongly correlated with two measurements. The combined sample not unexpectedly shows a combination of the two patterns, with mean annual precipitation and mean January temperature being the strongest correlates for six of the seven measurements. The seventh, proximal epiphyseal breadth of the tibia, has a slightly higher correlation with annual temperature range ($p = .227$) than with mean January temperature ($p = .223$).

In both sexes, most measures of epiphyseal size produce slopes significantly different from zero when regressed on mean annual temperature. In the female subsample, linear regression produced significant results for all epiphyseal measurements and mean January temperature, with the exception of epicondylar breadth ($p = .089$). Among males, regression with mean January temperature only produced significant results for humerus vertical head diameter, femur epicondylar breadth, and tibia distal epiphyseal breadth. All other climate variables produced inconsistent results for epiphyseal measurements.

Unlike limb-bone length and epiphyseal measurements, measures of diaphyseal size do not show a clear association with any one climate variable. The male and female subsamples show somewhat different patterns, though there is some overlap between the sexes. Among females, 10 of the 14 diaphyseal measurements are significantly and positively correlated with mean annual precipitation. Humerus maximum diameter at midshaft and radius anterior-posterior diameter at midshaft are most strongly correlated with mean January temperature. Both the superior-inferior and anterior-posterior midshaft measurements of the clavicle and medial-lateral subtrochanteric diameter of the femur are most strongly correlated with mean annual precipitation.

In the male subsample, diaphyseal measurements of the humerus, radius, and femur are strongly correlated with either mean annual temperature or mean July temperature. Radius anterior-posterior diameter at midshaft, superior-inferior and anterior-posterior midshaft diameters of the clavicle, and medial-lateral subtrochanteric diameter of the femur are most strongly correlated with mean annual precipitation. Anterior-posterior midshaft diameter of the femur is most highly correlated with mean January temperature, although it is not strongly correlated with any climate variable.

Regression analysis showed no clear patterns for diaphyseal measurements. Many of them produce significant results with multiple climate variables. In females, both diaphyseal clavicle measurements only produce significant results when regressed on mean annual precipitation and maximum diameter at the nutrient foramen is only significant for annual temperature range. Similar patterns are seen in the male subsample, with the exception of no significant results for clavicle superior-inferior midshaft diameter (though approaching significance at the $\alpha = .05$ level with mean annual precipitation, $p = .055$) and maximum diameter at the nutrient foramen producing significant results with mean annual temperature and mean July temperature.

In both sexes, maximum length of the os coxa and iliac breadth were most strongly correlated with mean annual temperature and mean January temperature. In females, regression analysis yielded statistically significant results for all variables except mean July temperature and mean annual precipitation. The male subsample was statistically significant for all variables except mean annual precipitation. The total sample followed the same trend as that of males, with all climate variables except mean annual precipitation producing statistically significant results.

Multiple regression analysis was used in order to examine the complex relationships among the climate variables and each measurement. Mean annual temperature, mean January temperature, mean July temperature, mean annual precipitation, and annual temperature range were regressed onto each measurement. The sexes were analyzed separately. The models for each measurement and the p-values and adjusted r^2 for the models can be found in Tables 10-7 and 10-8. Climate variables that were not found to significantly influence the slope of the regression when all other climate variables were excluded are italicized.

Table 10-5. *Pearson Correlations* (r), *Climate Variables and Individual Raw Data, Females*

Bone	Measurement	Annual temp	January temp	July temp	Annual precip.	Annual temp range
Humerus	Max length	.246	.129	.317	.208	.108
	Max midshaft diam.	-.100	-.079	-.136	.101	-.005
	Min midshaft diam.	-.141	-.195	-.080	.217	.154
	Max vertical head diam.	-.171	-.245	-.100	.214	.219
	Epicondylar breadth	-.078	-.106	-.084	.221	.065
Radius	Max length	.486	.327	.561	.203	.101
	ML midshaft diam.	-.209	-.222	-.195	.229	.113
	AP midshaft diam.	.157	.057	.235	.087	.127
	Max head breadth	-.026	-.122	.057	.137	.230
Clavicle	Max length	.304	.242	.293	.204	-.036
	SI midshaft diam.	.026	-.034	.067	.206	.105
	AP midshaft diam.	.092	.055	.089	.223	.000
Femur	Max length	.158	.003	.307	.141	.255
	Bicondylar length	.150	.004	.289	.157	.234
	AP midshaft diam	-.212	-.227	-.110	.090	.242
	ML midshaft diam.	-.230	-.274	-.179	.182	.199
	Midshaft circumference	-.269	-.341	-.168	.152	.290
	Max head breadth	-.182	-.302	-.043	.200	.307
	AP subtroch diam.	-.310	-.374	-.194	.088	.236
	ML subtroch diam.	.122	.006	.195	.239	.204
	Epicondylar breadth	-.244	-.321	-.140	.112	.267
Tibia	Max length	.314	.137	.471	.067	.235
	Max proximal epiphyseal Breadth	-.233	-.376	.062	.003	.362
	Max distal epiphyscal br.	-.328	-.409	-.205	.159	.325
	Max diam. at NF	.004	-.080	.088	.081	.207
	ML diam. at NF	-.080	-.198	.041	.237	.289
	Circumference at NF	-.022	-.146	.106	.154	.291
Os Coxa	Max length	-.183	-.243	-.107	.106	.191
	Iliac breadth	-.235	-.312	-.144	.123	.254

Note: Shaded cells indicate that OLS regression produced a slope significantly different from zero at α = .05 level.

Table 10-6. *Pearson Correlations (r), Climate Variables and Individual Raw Data, Males*

Bone	Measurement	Annual temp	January temp	July temp	Annual precip.	Annual temp range
Humerus	Max length	.239	.124	.289	.069	.065
	Max midshaft diam.	-.281	-.127	-.461	.334	-.244
	Min midshaft diam.	-.336	-.234	-.406	.243	-.08
	Max vertical head diam.	-.141	-.143	-.180	.258	.005
	Epicondylar breadth	-.145	-.090	-.246	.259	-.113
Radius	Max length	.500	.354	.493	.129	.021
	ML midshaft diam.	-.264	-.177	-.378	.293	-.110
	AP midshaft diam.	-.033	-.011	-.142	.390	-.091
	Max head breadth	-.106	-.108	-.176	.289	-.029
Clavicle	Max length	.215	.168	.209	.060	-.030
	SI midshaft diam.	.083	.042	.077	.235	.019
	AP midshaft diam.	.009	.035	-.054	.239	-.051
Femur	Max length	.158	-.026	.338	-.093	.304
	Bicondylar length	.170	-.008	.342	-.087	.285
	AP midshaft diam.	-.116	-.129	-.065	.010	.079
	ML midshaft diam.	-.272	-.233	-.282	.161	.039
	Midshaft circumference	-.243	-.225	-.209	.059	.073
	Max head breadth	-.300	-.311	-.241	.100	.140
	AP subtroch diam.	-.274	-.252	-.222	-.039	.031
	ML subtroch diam.	.054	-.053	.086	.243	.210
	Epicondylar breadth	-.214	-.215	-.199	.146	.061
Tibia	Max length	.351	.137	.531	-.139	.272
	Max proximal epiphyseal Breadth	-.039	-.123	.049	.039	.135
	Max distal epiphyseal br.	-.287	-.342	-.208	.150	.233
	Max diam. at NF	.115	.066	.112	.083	.024
	ML diam. at NF	-.197	-.227	-.139	.095	.139
	Circumference at NF	.055	-.029	.107	.096	.128
Os Coxa	Max length	-.298	-.316	-.262	.003	.172
	Iliac breadth	-.257	-.304	-.172	-.060	.230

Note: Shaded cells indicate that OLS regression produced a slope significantly different from zero at $\alpha = .05$ level.

Table 10-7. *Multivariate Regression Models for Females*

Bone	Measurement	Adjusted r^2	Model[a]
Humerus	Max length	.216	108.092 - .619*ATR[1] + 6.104*MAT[2] - 4.235*MJANT[3] - 1.302*MJULT[4] + 5.527 E-03*MAP[5]
	Max midshaft diam.	.027	42.267 - 5.187 E-03*ATR + .405*MAT - .215*MJANT - .265*MJULT + 1.515 E-04*MAP
	Min midshaft diam.	.164	19.244 - 3.118 E-02*ATR + .139*MAT - .195*MJANT + 2.951 E-02*MJULT + 1.237 E-03*MAP
	Max vertical head diam.	.252	82.348 - .082*ATR + 1.713*MAT - 1.101*MJANT - .774*MJULT + 1.151 E-03*MAP
	Epicondylar breadth	.161	113.374 - .089*ATR + 2.362*MAT - 1.333*MJANT - 1.236*MJULT + 2.948 E-04*MAP
Radius	Max length	.431	-94.523 - .148*ATR + 8.550*MAT - 4.737*MJANT - 2.763*MJULT + 2.521 E-03*MAP
	ML midshaft diam.	.156	29.253 - 4.238 E-02*ATR + .176*MAT - .198*MJANT - 3.948 E-02*MJULT + 1.033 E-03*MAP
	AP midshaft diam.	.125	7.849 - 3.538 E-02*ATR + .594*MAT - .355*MJANT - .233*MJULT - 6.277 E-05*MAP
	Max head breadth	.181	35.792 + 7.175 E-03*ATR + 1.096*MAT - .601*MJANT - .554*MJULT + 1.156 E-04*MAP
Clavicle	Max length	.144	90.843 - .175*ATR + 4.553*MAT - 2.455*MJANT - 1.935*MJULT + 1.407 E-03*MAP
	SI midshaft diam.	.097	8.990 - 3.119 E-02*ATR + .348*MAT - .242*MJANT - .112*MJULT + 6.632 E-04*MAP
	AP midshaft diam.	.088	8.908 - 5.646 E-02*ATR + .357*MAT - .252*MJANT - .105*MJULT + 5.245 E-04*MAP
Femur	Max length	.227	61.145 - 1.163*ATR + 6.240*MAT - 5.757*MJANT + .557*MJULT + 1.015 E-02*MAP
	Bicondylar length	.254	67.463 - 1.136*ATR + 4.778*MAT - 4.911*MJANT + 1.141*MJULT + 1.126 E-02*MAP
	AP midshaft diam.	.149	36.483 - 6.984 E-02*ATR + .175*MAT - .309*MJANT + 8.075 E-02*MJULT + 1.325 E-03*MAP
	ML midshaft diam.	.209	61.017 - .058*ATR + .883*MAT - .620*MJANT - .401*MJULT + 1.080 E-03*MAP

Midshaft		
Circumference	.252	163.082 - .161*ATR + 1.932*MAT - 1.614*MJANT - .657*MJULT + 3.958 E-03*MAP
Max head breadth	.334	61.718 - 9.638 E-02*ATR + 1.148*MAT - .915*MJANT - .328*MJULT + 1.965 E-03*MAP
AP subtroch diam.	.246	45.776 - .154*ATR + .337*MAT - .506*MJANT + 7.307 E-02*MJULT + 1.402 E-03*MAP
ML subtroch diam.	.215	25.261 - 5.708 E-02*ATR + 1.659*MAT - 1.022*MJANT + .637*MJULT + 1.139 E-03*MAP
Epicondylar breadth	.179	131.888 - 4.375 E-02*ATR + 1.765*MAT - 1.167*MJANT - .825*MJULT + 1.247 E-03*MAP
Tibia		
Max length	.363	-35.542 - .844*ATR + 10.987*MAT - 7.347*MJANT - 2.452*MJULT + 4.431 E-04*MAP
Max proximal Epiphyseal breadth	.282	221.791 + 9.303 E-02*ATR + 5.400*MAT - 2.856*MJANT - 3.082*MJULT - 2.631 E-03*MAP
Max distal epiphyseal Breadth	.297	140.316 - 2.514 E-02*ATR + 2.519*MAT - 1.560*MJANT - 1.307*MJULT + 1.198 E-03*MAP
Max diam. at NF	.075	28.648 + 5.518 E-03*ATR + .764*MAT - .496*MJANT - .272*MJULT + 7.354 E-04*MAP
ML Diam at NF	.272	20.363 - .048*ATR + .616*MAT - .532*MJANT - .102*MJULT + 1.796 E-03*MAP
Circumference at NF	.208	64.778 - .1077*ATR + 2.386*MAT - 1.775*MJANT - .600*MJULT + 3.864 E-03*MAP
Os Coxa		
Max length	.148	330.932 - .241*ATR + 4.682*MAT - 3.091*MJANT - 2.100*MJULT + 3.531 E-03*MAP
Iliac breadth	.284	305.931 - .548*ATR + 5.802*MAT - 3.962*MJANT - 2.426*MJULT + 2.469 E-03*MAP

[a]Variables in italics are not significant at α = .05

Table 10-8. *Multivariate Regression Models for Males*

Bone	Measurement	Adjusted r^2	Model[a]
Humerus	Max length	.223	118.130 - 1.467*ATR − 5.183*MAT - 4.374*MJANT - .179*MJULT + 3.531E-03*MAP
	Max midshaft diam.	.298	78.972 - .101*ATR + .472 *MAT - .345*MJANT - .323*MJULT + 8.344 E-04*MAP
	Min midshaft diam.	.235	42.554 - 5.499 E-02* ATR - .269*MAT + 4.812 E-03*MJANT + .166*MJULT + 1.378 E-03*MAP
	Max vertical head diam.	.220	85.017 - .115* ATR + .240*MAT - .861*MJANT - .532*MJULT + .001*MAP
	Epicondylar breadth	.183	106.827 - .227* ATR + .936*MAT - .755*MJANT - .350*MJULT + 1.331 E-03*MAP
Radius	Max length	.439	-19.437 - .911* ATR + 9.167*MAT - 5.362*MJANT - 2.897*MJULT - 1.120 E-03*MAP
	ML midshaft diam.	.236	45.109 - 6.384 E-02*ATR + .205*MAT - .209*MJANT - .106*MJULT + 9.638 E-04*MAP
	AP midshaft diam.	.252	17.369 - 5.292 E-02*ATR + .261*MAT - .225*MJANT - 6.159 E-02*MJULT + 9.040 E-04*MAP
	Max head breadth	.217	40.774 - 5.086 E-02*ATR + .699*MAT - .436*MJANT - .331 *MJULT + 5.194 E-04*MAP
Clavicle	Max length	.045	119.891 - .083*ATR + 3.385*MAT - 1.701*MJANT - 1.565*MJULT - 1.562 E-03*MAP
	SI midshaft diam.	.103	3.031 - 4.186 E-02*ATR + .180*MAT - .179*MJANT + 1.634 E-02*MJULT + 9.221 E-04*MAP
	AP midshaft diam.	.052	8.135 - 1.315 E-02*ATR - .110*MAT + 1.792 E-02*MJANT + 9.933 E-02*MJULT + 8.726 E-04*MAP
Femur	Max length	.207	115.696 - .750*ATR + 3.158*MAT - 3.434*MJANT + 1.272*MJULT + 6.298 E-03*MAP
	Bicondylar length	.198	106.205 - .743*ATR + 2.835* MAT - 3.172*MJANT + 1.364*MJULT + 6.008 E-03*MAP
	AP midshaft diam.	.032	30.231 - 9.109 E-02*ATR - .229*MAT - 5.145 E-02*MJANT + .268*MJULT + 8.147 E-04*MAP
	ML midshaft diam.	.145	53.140 - 2.592 E-02*ATR + 9.724 E-02*MAT - .171*MJANT - .029*MJULT + 1.145 E-03*MAP

	R^2	Equation
Midshaft circumference	.084	$137.603 - .166*ATR - .206*MAT - .305*MJANT + .313*MJULT + 2.492 E{-}03*MAP$
Max head breadth	.195	$80.258 - 6.500 E{-}02*ATR + .355*MAT - .405*MJANT - 8.496 E{-}02*MJULT + 1.444 E{-}03*MAP$
AP subtroch diam.	.116	$44.972 - .165*ATR - .215*MAT - .128*MJANT + .270*MJULT + 5.320 E{-}04*MAP$
ML subtroch diam.	.273	$37.153 + .014*ATR + 1.418*MAT - .885*MJANT - .572*MJULT + 1.914 E{-}03*MAP$
Epicondylar breadth	.119	$130.823 - 9.604 E{-}02*ATR + .7209*MAT - .636*MJANT - .279*MJULT + 2.093 E{-}03*MAP$
Tibia Max length	.366	$-120.028 - .968*ATR + 6.170*MAT - 4.606*MJANT + 4.315 E{-}02*MJULT - 7.183 E{-}04*MAP$
Max proximal Epiphyseal breadth	.045	$68.985 - .117*ATR + .486*MAT - .514*MJANT + 3.128 E{-}02*MJULT + 1.561 E{-}03*MAP$
Max distal epiphyseal Breadth	.253	$103.859 - 2.753 E{-}02*ATR + 1.016*MAT - .777*MJANT - .445*MJULT + 2.204 E{-}03*MAP$
Max diam. at NF	.040	$27.053 - 6.332 E{-}02*ATR + .665*MAT - .418*MJANT - .220*MJULT + 3.780 E{-}04*MAP$
ML diam. at NF	.107	$41.281 - 1.935 E{-}02*ATR + .110*MAT - .210*MJANT + 2.343 E{-}02*MJULT + 1.317 E{-}03*MAP$
Circumference at NF	.084	$68.317 - .132*ATR + 1.270*MAT - 1.026*MJANT - .185*MJULT + 2.869 E{-}03*MAP$
Os Coxa Max length	.169	$417.048 - .130*ATR + 4.137*MAT - 2.506*MJANT - 2.364*MJULT + 1.886 E{-}04*MAP$
Iliac breadth	.129	$291.847 - .169*ATR + 3.176*MAT - 2.108*MJANT - 1.567*MJULT + 7.993 E{-}04*MAP$

aVariables in italics are not significant at $\alpha = .05$

Several interesting relationships between the measurements and the climate variables emerge from the multivariate regressions. All of the measurements produced regressions that were significant at the $\alpha = .05$ level. Only two, maximum diameter of the humerus at midshaft in females and maximum proximal epiphyseal breadth of the tibia in males, produced regressions that were not significant at the $\alpha = .01$ level.

The four long-bone lengths overall had high r^2 values in both sexes. The models for these measurements produced r^2 greater than .200, with the tibia producing r^2 of .363 in females and .366 in males. Radius length showed an even stronger relationship, with r^2 of .431 in females and .439 in males. Among the long-bone lengths, mean January temperature had significant partial regression coefficients in all eight of the regressions, while annual temperature range and mean annual temperature were significant in seven of the eight models.

Results for the epiphyseal dimensions were somewhat less robust than those for the long-bone lengths. Some of the r^2 values for epiphyseal measurements were inconsistent between the sexes, the largest two discrepancies being femoral head breadth ($r^2 = .334$ in females, .195 in males) and maximum proximal epiphyseal breadth of the tibia ($r^2 = .282$ in females, .045 in males). All other r^2 values differed by less than .05 between the sexes. With the exception of maximum proximal epiphyseal breadth of the tibia in males, all r^2 values for epiphyseal measurements fell between .119 and .334. Mean January temperature had significant partial regression coefficients for all measures of epiphyseal breadth. Mean annual temperature and mean annual precipitation were significant for six of eight epiphyseal measurements.

Models and r^2 for the diaphyseal measurements were not similar between the sexes. There was very little consistency in which climate variables were associated with these measurements, with the exception of mean annual precipitation, which was significant in 12 of 14 measurements among males and 10 of 14 among females.

In both sexes the measurements of the os coxa were significantly associated with mean annual temperature, mean January temperature, and mean July temperature. The r^2 values for these measurements varied in size, from .129 for iliac breadth in males to .284 for iliac breadth in females.

Testing for Allometry: Analysis of Size and Shape Variables

The humerus, radius, femur, and tibia exhibited moderately strong correlations with mean July temperature. In both sexes, the tibia and the radius were more highly correlated with this variable than either the femur or the humerus was. The correlations and linear regressions suggest that with higher temperatures the bones of the distal limb segments are disproportionately longer than the proximal limb segments. To explore this question further, the relationships of all the limb bones and the climate variables will be analyzed using "size" and "shape" variables (Darroch and Mosimann 1985).

Size and shape variables were created using the natural logarithms of humerus maximum length, radius maximum length, femur maximum length, and tibia maximum length (Darroch and Mosimann 1985; Meadows Jantz and Jantz 1999). As the creation of the size variables requires the presence of maximum length measurements for all four limb bones used in this analysis, only individuals with all four of these elements were used. This reduced the sample sizes in both sexes by a considerable degree (see Table 10-9 for sample sizes by site). Despite the similarity of the relationships between limb-bone lengths and climate variables demonstrated previously, males and females were analyzed separately and no attempt was made to pool the sexes for this portion of the analysis.

This type of analysis allows for examination of changes in shape relative to overall size. Mosimann-type variables, as are used here, are superior to various residual analyses for identifying specimens of the same shape but of different sizes (Jungers et al. 1995). The first variable created is the "size" variable, the arithmetic mean of the logarithms of the measurements to be examined. This variable functions as a measure of isometric size from the four limb bones (Jungers et al. 1995). Each size variable is unique to the individual. The "shape" variables were created for each measurement by subtracting the log transformed size variable from the log of each measurement (Darroch and Mosimann 1985; Meadows Jantz and Jantz 1999).

To determine if the lengths of the humerus, radius, femur, and tibia were isometric with the "size" variable, each "shape" variable was correlated with the "size" variable. A significant positive correlation indicates that the long-bone length is increasing faster than the geometric mean of the lengths; a significant negative correlation indicates that the measurement is increasing at a slower rate than the geometric mean; a correlation of zero indicates isometry (Holliday 1997a).

As shown in Table 10-10, all "shape" variables with the exception of tibia "shape" in females were significantly correlated with the "size" variables at the $\alpha = .05$ level. Two patterns emerge in this data: The bones of the distal limb segments, the radius and tibia, are positively allometric with "size," meaning they increase in length at a faster rate than overall average limb-bone length does. Additionally, the bones of the proximal limb segments, the humerus and the femur, are negatively allometric with "size." This supports the results found by Holliday (1999) that also showed that the distal limb segments respond more rapidly to changes in temperature variables than the proximal limb segments do.

Each shape variable was regressed onto the five climate variables to determine if the bone lengths increase or decrease in relation to these climate variables. Significant results indicate that variation in bone length unrelated to size corresponds to changes in the climate factors. The slopes for the size-on-climate variables and shape-on-climate variables regressions are tabulated in Table 10-11 for females and Table 10-12 for males.

Size was significant for all climate variables except annual temperature range in the female subsample. Among males, size was significant only for mean annual temperature, mean January temperature, and mean July temperature. In both sexes the radius and femur produced significant results for all five climate variables. Radius shape increased in all cases. Femur shape decreased with increases in mean annual temperature, mean January temperature, and mean July

Table 10-9. *Sample Sizes for Mosimann Variables*

Site	Males	Females	Total
Kauwerak	6	3	9
Mummy Caves	11	9	20
St. Michael	7	5	12
Umnak Island	8	2	10
Prince Rupert Harbour	14	5	19
Steuben	6	2	8
Fort Prince of Wales	1	0	1
Souris Valley	8	3	11
Huron Village	2	1	3
Juntunen	8	4	12
Edwards Mound	2	1	3
Lake George	2	0	2
Bell-Philhower	3	6	9
Hawikku	30	44	74
Kwastiyukwa	3	5	8
Pueblo Bonito	3	9	12
Native Point	24	23	47
Madisonville	20	23	43
Turpin	4	3	7
Sully	24	12	36
Dr. Jarman	6	8	14
Thompson Village	1	3	4
Toqua	10	5	15
Berrians Island	6	3	9
Totals	209	179	388

Table 10-10. *Slopes and Standard Errors, Regressions of Log Transformed Bone Length on Size*

Sex	Hum. shape	Rad. shape	Fem. shape	Tib. shape
Females				
Slope	.895	1.188	.873	1.045
SE	.032	.038	.031	.030
Males				
Slope	.857	1.119	.889	1.135
SE	.028	.037	.029	.027

Note: Shaded cells contain results that are significantly different from a slope of 1 at the $\alpha = .05$ level.

Table 10-11. *Pearson Correlations for Size and Shape Variables Regressed on Climate Variables for Females*

Climate variable	Size	Hum. shape	Rad. shape	Fem. shape	Tib. shape
Mean annual temp	.379	-.282	.679	-.573	.009
Mean January temp	.262	-.175	.655	-.580	-.070
Mean July temp	.457	-.400	.583	-.457	.142
Mean annual precip.	.215	.044	.277	-.179	-.228
Annual temp range	-.022	-.080	-.369	.391	.165

Note: Shaded cells contain results that are significantly different from a slope of 1 at the α = .05 level.

Table 10-12. *Pearson Correlations for Size and Shape Variables Regressed on Climate Variables for Males*

Climate variable	Size	Hum. shape	Rad. shape	Fem. shape	Tib. shape
Mean annual temp	.371	-.265	.598	-.576	.090
Mean January temp	.202	-.109	.605	-.577	-.082
Mean July temp	.485	-.419	.383	-.403	.354
Mean annual precip.	-.014	.184	.314	-.231	-.366
Annual temp range	.066	-.244	-.422	.407	.373

Note: Shaded cells contain results that are significantly different from a slope of 1 at the α = .05 level.

temperature and increased with increases in mean annual precipitation and annual temperature range. In both sexes humerus shape decreased significantly with increases in mean annual temperature and mean July temperature. Humerus shape also decreased with increased annual temperature range in males. Tibia shape was not strongly related to many climate variables. Among females, tibia shape was only significant with mean annual precipitation and annual temperature range at the α = .05 level. The pattern was similar among males, with mean July temperature also being significant.

Discussion

The noteworthy correlation of limb-bone length and mean July temperature shows that as temperature increases, limb-bone length increases, as would be expected if Allen's rule were valid for these samples. Heat stress appears to be a driving factor, as the correlations with mean July temperature are stronger than those with mean annual temperature or mean January temperature. This is especially true for the femur, whose length is barely correlated with mean January temperature in either sex.

Maximum length of the clavicle does not seem to follow ecogeographic ex-pectations. It increases in size as temperature increases, suggesting that the upper torso and shoulder girdle become wider as temperature increases. While this is not the pattern expected if the body were to decrease in width in response to increased temperature, this pattern is also seen in the Boas anthropometric data (Jantz et al., Chapter 11). These results may be influenced by the fact that clavicle length is not simply a linear measure of torso width but is also associated with torso depth.

The inconsistent correlations in radius head breadth and humerus epicon-dylar breadth between the sexes suggest that climate variables are not greatly or consistently influencing these measurements. As the elbow joint shows signifi-cant sexual size dimorphism, it perhaps provides little useful information about climate adaptation, as potential adaptations may be masked by a stronger dimor-phic component. Epiphyseal size, especially femoral head diameter, may also be related to gross differences in body size (Auerbach and Ruff 2004). It should be noted that femoral head diameter is negatively correlated with increased tem-perature and positively with body mass, which supports the results found in previous studies (Ruff 1994).

The lack of consistency observed between the sexes for the diaphyseal mea-surements suggests that they are being influenced by variables other than cli-mate. Antemortem biomechanical stresses have caused cortical remodeling in some of the dimensions measured. Among these measurements, diaphyseal size may be more influenced by the function of the long bone as a structural support for the body than it is influenced by ecogeographical factors (Ruff et al. 1984).

The analysis of the os coxa produced very similar results in both sexes, sug-gesting that the dimorphic features of this element do not greatly influence its relationship with climate. The general trend of longer ossa coxae and broader ilia with increasing latitude may be correlated to a worldwide trend toward increase in bi-iliac breadth with increase in latitude (Ruff 1991), a trend that would be expected if ecogeography was a significant influence on the morphology of this element.

Multivariate analysis revealed that variation in temperature and precipitation does explain a considerable degree of postcranial metric variation in this sample; however, this set of climate variables can only partially explain the variation ob-served in these measurements. The highest r^2 values in both sexes are associated with radius length, suggesting that approximately 43 percent of the variation in this measurement can be attributed to variation in the climate variables. Tibia length shows a similar though less robust pattern, with approximately 36 percent of the variation in this measurement being attributable to variation in the climate variables. Humerus and femur length have more modest r^2 values, suggesting that these measurements are less strongly associated with these climate variables.

Many of the epiphyseal measurements have r^2 values greater than .200, which supports the hypothesis that there is some variation in these measure-ments attributable to climate variation, though the explanatory power of these variables is rather modest. There is great variation in the r^2 values of the diaphy-seal measurements, suggesting that there is not a clear, straightforward relation-ship among the diaphyseal dimensions in general and any combination of these climate variables. The measurements of the os coxa show a clear relationship with the temperature variables, although the r^2 values for these measurements are not particularly large.

The allometric relationships among the size variable and the four bone lengths support the hypothesis that the distal limb segments are disproportionately longer in relation to higher temperatures than the proximal segments. This pattern is in concordance with the ecogeographical expectations. The distal limb segments are thinner than the proximal ones and therefore have a higher surface area to volume ratio than the proximal segments (Holliday 1999). An increase in the length of the distal segments with greater temperature causes an increase in the surface area to volume ratio, which would be beneficial in climates in which efficient heat loss would be advantageous.

The positive slopes of all the significant size versus climate variable regressions support the hypothesis of climate related patterns in morphology. As the size variable is reflective of an increase in overall limb-bone length, an increase in size that corresponds to an increase in temperature indicates that the limbs become longer with increased temperature. While all measures of temperature show this relationship, annual temperature range and mean annual precipitation have not been significantly associated with overall limb-bone length.

It should be noted that the correlations reported here are not atypically low for studies of modern human morphology and climate variation. For example, Katzmarzyk and Leonard (1998) report similar r values for anthropometric data from a worldwide sample when compared to mean annual temperature. Among males in their study, the correlation between surface area to volume ratio and mean annual temperature was .288; among females, it was .341. They further reported correlations between mean annual temperature and body mass, body mass index, and relative sitting height as -.267, -.221, and -.369 for males and -.279, -.295, and -.456 for females respectively. While anthropometric and osteometric measurements are rarely directly comparable, these values suggest that similar relationships between temperature and body shape are present in modern humans regardless of the method of inquiry.

Conclusions

The goal of this study was to investigate if the patterns of postcranial variation in indigenous North American populations were related to adaptation to climate. Significant relationships between several osteometric measurements and climate variables were found, especially for long-bone length, epiphyseal size, and dimensions of the os coxa. This analysis supports the hypothesis that adaptation to climate has occurred in North America since it was first populated, though it is more apparent in some skeletal features than in others. Ultimately, it is difficult to separate out the individual effects of each climate variable because of their collinearity. However, overall relationships between the classes of measurements (epiphyseal, diaphyseal, and lengths) and the classes of climate variables (temperatures, precipitation, and seasonality) can be discussed.

Among the long bones, maximum femur length seems to deviate from the pattern of correlations seen in the other limb bones. Most of the limb-bone lengths are positively correlated with mean January temperature, but the femur is not. Perhaps the lack of relationship between femur length and mean January temperature suggests a biomechanical restraint in femur length, as it is highly correlated with stride length and locomotor efficiency (Wang and Crompton 2003; Witte et al. 1991). This also might explain why tibia length is less strongly correlated with mean January temperature than radius length is, despite their being analogous structures and the similar effects of their relative size on the surface area to volume ratio of the body.

Multivariate analyses demonstrate that limb-bone length and some measures of epiphyseal size are well explained by variation in temperature. Generally speaking, mean annual precipitation and annual temperature range are not as influential as are mean annual temperature, mean January temperature, and mean July temperature. All limb-bone length and most epiphyseal dimensions respond in a similar manner to climate variables, while diaphyseal measures vary greatly in their relationships with temperature, precipitation, and seasonality.

The analysis of the shape variables demonstrates that the radius and the tibia increase more rapidly with increased temperature than the humerus and the femur do. This is in concordance with Holliday's (1999) argument that elongation of the distal limb segments is more adaptive in warm climates than an overall increase in limb length is. Multiple regression analysis further confirmed the strong association between radius and tibia lengths and temperature.

Distal lengthening of the limbs supports the idea that the indigenous populations of North America have resided on the continent long enough to have undergone considerable adaptation to climate. If distal segment change occurs late in the process of adaptation to cold stress (Holliday 1999), then evidence of it suggests that earlier processes—such as the overall lengthening of the extremities, assuming more cold-adapted morphologies in "cold-filtered" ancestors—had previously occurred.

The "size" variable created from the four limb-bone lengths was significantly and positively correlated with mean annual temperature, mean January temperature, and mean July temperature, indicating that an increase in temperature is related to an increase in overall limb-bone length. The regressions of the individual shape variables confirm the results of the allometry test, indicating that relative humerus length is shorter with higher temperatures and the distal segments are disproportionately longer with increased temperature than proximal segment lengths.

The significant and mostly positive association between mean annual precipitation and all measurements is not consistent with any ecogeographic expectation. Increased precipitation is related to increased humidity, and if Ruff's (1993) argument—that humid conditions are best dealt with through a reduction in overall body size—holds true in this sample, then the osteometric variables should be negatively correlated with precipitation. In fact, the opposite pattern is seen.

It is possible that the relationship between increased precipitation and increased size is a nutritional effect. Rainfall is correlated with resource availability, especially in dry environments (Yom-Tov and Geffen 2006). The impact of increased nutrition on overall growth is well documented (i.e., Froehlich 1970; Greulich 1976; Ruff 2002) and can ultimately influence skeletal morphology. Other possible complex factors involving several of these components could be affecting these results. The potential interrelationships among rainfall, biomass, food availability, subsistence strategy, and the biomechanical demands of obtaining food in specific environments are areas that should be explored in future studies of this nature.

Are These Patterns Indicative of Biological Adaptation to Climate?

The analyses performed in this study demonstrate that there is an appreciable relationship between postcranial morphology and climate variables among the late prehistoric inhabitants of North America. However, correlation among sets of variables does not prove a causal relationship exists. Nor does it prove that the correlations are the product of biological adaptation, as Bergmann (1847) and Allen (1877) argued; or of physiological effects, as Serrat and colleagues (2007) claim is inducible in mice; or of stochastic processes working on a small founding population.

This study suggests that these patterns in postcranial morphology are strongly related to variation in temperature and precipitation, regardless of the mechanism. Due to the fact that similar patterns can be found in Old World human populations and in many nonhuman species, the possibility that these changes reflect biological adaptation cannot be excluded.

References

Allee, Warder C., and Karl P. Schmidt
    1951   *Ecological Animal Geography*, pp. 457–472. John Wiley and Sons, New York.
Allen, Joel
    1877   The Influence of Physical Conditions in the Genesis of Species. *Radical Review* 1:108–140.
Auerbach, Benjamin M., and Christopher B. Ruff
    2004   Human Body Mass Estimation: A Comparison of "Morphometric" and "Mechanical" Methods. *American Journal of Physical Anthropology* 125:331–342.
Bass, William M.
    1995   *Human Osteology: A Laboratory and Field Manual*. Missouri Archaeological Society, Columbia.
Beaton, John M.
    1991   Colonizing Continents: Some Problems from Australia and the Americas. In *The First Americans: Search and Research*, edited by Thomas D. Dillehay and David J. Meltzer, pp. 209–230. CRC Press, Boca Raton, Florida.
Bergmann, Carl
    1847   Increase in the Effectiveness of Heat-Conservation in Large Subjects. *Gottinger Studien* 3:595–708.
Buikstra, Jane E., and Douglas H. Ubelaker (editors)
    1994   *Standards for Data Collection from Human Skeletal Remains*. Arkansas Archeological Survey, Fayetteville.

Capes, Katherine H.

   1963   *The W. B. Nickerson Survey and Excavations, 1912–1915, of the Southern Manitoba Mounds Region.* National Museum of Canada, Ottawa, Ontario.

Crognier, Emile

   1981   Climate and Anthropometric Variations in Europe and the Mediterranean Area. *Annals of Human Biology* 8:99–107.

Cybulski, Jerome S.

   1992   *A Greenville Burial Ground: Human Remains and Mortuary Elements in British Columbia Prehistory.* Canadian Museum of Civilization, Hull, Quebec.

Darroch, John N., and James E. Mosimann

   1985   Canonical and Principal Components of Shape. *Biometrika* 72:241–252.

Dillehay, Thomas D.

   1989   *Monte Verde, a Late Pleistocene Settlement in Chile: Paleoenvironment and Site Context,* Vol. 1. Smithsonian Institution, Washington, D.C.

   1997   *Monte Verde, a Late Pleistocene Settlement in Chile: The Archaeological Context,* Vol. 2. Smithsonian Institution, Washington, D.C.

   2000   *The Settlement of the Americas: A New Prehistory.* Basic Books, New York.

Fiedel, Stuart J.

   2004   Clovis Age in Calendar Years: 13,500–13,000 CALYBP. In *New Perspectives on the First Americans,* edited by Bradley T. Lepper and Robsen Bonnichsen, pp. 73–80. Center for the Study of the First Americans, Texas A&M University Press, College Station.

Froehlich, Jeffery W.

   1970   Migration and Plasticity of Physique in Japanese-Americans of Hawaii. *American Journal of Physical Anthropology* 32:429–442.

Gilbert, M. Thomas P., Dennis L. Jenkins, Anders Götherstrom, Nuria Naveran, Juan J. Sanchez, Michael Hofreiter, Philip Francis Thomsen, Jonas Binladen, Thomas F. G. Higham, Robert M. Yohe II, Robert Parr, Linda Scott Cummings, and Eske Willerslev

   2008   DNA from Pre-Clovis Human Coprolites in Oregon, North America. *Science* 320:786–789.

Goodyear, Albert C.

   2004   Evidence of Pre–Clovis Sites in the Eastern United States. In *Paleoamerican Origins: Beyond Clovis,* edited by Robsen Bonnichsen, pp. 89–98. Center for the Study of the First Americans, Texas A&M University Press, College Station.

Greulich, William Walter

   1976   Some Secular Changes in the Growth of American-Born and Native Japanese Children. *American Journal of Physical Anthropology* 45:553–568.

Hall, Roberta L., and Don A. Hall

   1995   Geographic Variation of Native People Along the Pacific Coast. *Human Biology* 67:407–426.

Hall, Roberta, Diana Roy, and David Boling

   2004   Pleistocene Migration Routes into the Americas: Human Biological Adaptations and Environmental Constraints. *Evolutionary Anthropology* 13:132–144.

Harle, Michaelyn S., and Kathryn A. King

   2004   Skeletal Markers of Occupational Stress: Gender and Rank Based Division of Labor in a Late Mississippian Population. Poster presented at the Annual Meeting of the Southeastern Archaeological Conference.

Hatch, James W., Patrick S. Willey, and Edward E. Hunt Jr.

   1983   Indicators of Status-Related Stress in Dallas Society: Transverse Lines and Cortical Thickness in Long Bones. *Midcontinental Journal of Archaeology* 8:49–71.

Haynes, Gary
    2002   *The Early Settlement of North America: The Clovis Era*. Cambridge University Press, Cambridge, England.
Hiernaux, Jean, and Alain Froment
    1976   Correlations Between Anthropobiological and Climatic Variables in Sub-Saharan Africa—Revised Estimates. *Human Biology* 48:757–767.
Holliday, Trenton W.
    1997a   Postcranial Evidence of Cold Adaptation in European Neanderthals. *American Journal of Physical Anthropology* 104:245–258.
    1997b   Body Proportions in Late Pleistocene Europe and Modern Human Origins. *Journal of Human Evolution* 32:423–448.
    1999   Brachial and Crural Indices of European Late Upper Paleolithic and Mesolithic Humans. *Journal of Human Evolution* 36:549–566.
Humphrey, John D., and C. Reid Ferring
    1994   Stable Isotopic Evidence for Latest Pleistocene and Holocene Climatic Change in North-Central Texas. *Quaternary Research* 41:200–213.
Jacobs, Kenneth H.
    1985   Climate and Hominid Postcranial Skeleton in Würm and Early Holocene Europe. *Current Anthropology* 26:512–514.
    1993   Human Postcranial Variation in the Ukrainian Mesolithic-Neolithic. *Current Anthropology* 34:311–324.
Johnston, Richard F., and Robert K. Selander
    1971   Evolution in House Sparrow. 2. Adaptive Differentiation in North American Populations. *Evolution* 25:1–28.
    1973   Evolution in House Sparrow. 3. Variation in Size and Sexual Dimorphism in Europe and North and South America. *American Naturalist* 107:373–390.
Jones, J. Scott
    1996   The Anzick Site: Analysis of a Clovis Burial Assemblage. Master's thesis, Department of Applied Anthropology, Oregon State University, Corvallis.
Jungers, William L.
    1985   Body Size and Scaling of Limb Proportions in Primates. In *Size and Scaling in Primate Biology*, edited by William L. Jungers, pp. 345–381. Plenum Press, New York.
Jungers, William L., Anthony B. Falsetti, Christine E. Wall
    1995   Shape, Relative Size, and Size-Adjustments in Morphometrics. *American Journal of Physical Anthropology* 38:137–161.
Katzmarzyk, Peter T., and William R. Leonard
    1998   Climatic Influences on Human Body Size and Proportions: Ecological Adaptations and Secular Trends. *American Journal of Physical Anthropology* 106:483–503.
Katzmarzyk, Peter T., and Robert M. Malina
    1999   Body Size and Physique among Canadians of First Nation and European Ancestry. *American Journal of Physical Anthropology* 108:161–172.
Lazenby, Richard, and Amanda Smashnuk
    1999   Osteometric Variation in the Inuit Second Metacarpal: A Test of Allen's Rule. *International Journal of Osteoarchaeology* 9:182–188.
Martorell, Reynaldo, Robert M. Malina, Ricardo O. Castillo, Fernando S. Mendoza, and I. G. Pawson
    1988   Body Proportions in Three Ethnic Groups: Children and Youths 2–17 Years in NHANES II and HHANES. *Human Biology* 60:205–222.

Meadows Jantz, Lee, and Richard L. Jantz
    1999   Secular Change in Long Bone Lengths and Proportion in the United States, 1800–1970. *American Journal of Physical Anthropology* 110:57–67.
Meltzer, David J.
    1989   Why Don't We Know When the First People Came to North America? *American Antiquity* 54:471–490.
Meltzer, David J., Donald K. Grayson, Gerardo Ardila, Alex W. Barker, Dena F. Dincauze, C. Vance Haynes, Francisco Mena, Lautaro Núñez, and Dennis J. Stanford
    1997   On the Pleistocene Antiquity of Monte Verde, Southern Chile. *American Antiquity* 62:659–663.
Meteorological Service of Canada, Environment Canada
    2002   National Climate Archive. http://climate.weatheroffice.ec.gc.ca/climate_normals/results_e.html, accessed January 2007.
Murphy, Edward C.
    1985   Bergmann's Rule, Seasonality, and Geographic Variation in Body Size of House Sparrows. *Evolution* 39:1327–1334.
National Oceanic and Atmospheric Administration, National Environmental Satellite, Data, and Information Service, National Climatic Data Center
    2002   *Monthly Station Normals of Temperature, Precipitation, and Heating and Cooling Degree Days, 1971–2000.* Climatography of the United States, No. 81, National Climatic Data Center/NESDIS/NOAA, Asheville, North Carolina.
NCSS
    2004   Number Cruncher Statistical System 2004. NCSS, Kaysville, Utah.
Newman, Marshall T.
    1953   The Application of Ecological Rules to the Racial Anthropology of the Aboriginal New World. *American Anthropologist* 55:311–327.
    1960   Adaptations in the Physique of American Aborigines to Nutritional Factors. *Human Biology* 32:288–313.
    1962   Evolutionary Changes in Body Size and Head Form in American Indians. *American Anthropologist* 64:237–257.
Newman, Russell W., and Ella H. Munro
    1955   The Relation of Climate and Body Size in U.S. Males. *American Journal of Physical Anthropology* 13:1–17.
Palomino, Hernan, William H. Mueller, and William J. Schull
    1979   Altitude, Heredity, and Body Proportions in Northern Chile. *American Journal of Physical Anthropology* 50:39–50.
Pearson, Osbjorn M., and Mario Millones
    2005   Skeletal Features Showing Adaptation to Climate and Activity of Aboriginal Inhabitants of Tierra del Fuego. *Magallania* 33(1):37–50.
Phenice, Terrell W.
    1969   A Newly Developed Visual Method of Sexing the Os Pubis. *American Journal of Physical Anthropology* 30:297–301.
Porter, Alan M. W.
    1999   Modern Human, Early Modern Human and Neanderthal Limb Proportions. *International Journal of Osteoarchaeology* 9:54–67.
Roberts, Derek F.
    1953   Body Weight, Races, and Climate. *American Journal of Physical Anthropology* 11:533–558.
    1978   Migration and Genetic Change. *Human Biology* 60:521–539.

Rudan, Pavao, Derek F. Roberts, Branka Janicijevic, Nina Smolej, Lajos Szirovicza, and Andrija Kastelan

    1986   Anthropometry and the Biological Structure of the Hvar Population. *American Journal of Physical Anthropology* 70:231–240.

Ruff, Christopher B.

    1991   Climate and Body Shape in Hominid Evolution. *Journal of Human Evolution* 21:81–105.

    1993   Climatic Adaptation and Hominid Evolution: The Thermoregulatory Imperative. *Evolutionary Anthropology* 2:53–60.

    1994   Morphological Adaptation to Climate in Modern and Fossil Hominids. *American Journal of Physical Anthropology* 37:65–107.

    2002   Variation in Human Body Size and Shape. *Annual Review of Anthropology* 31:211–232.

Ruff, Christopher B., Clark Spencer Larsen, and Wilson C. Hayes

    1984   Structural Changes in the Femur with the Transition to Agriculture on the Georgia Coast. *American Journal of Physical Anthropology* 64:125–136.

Schreider, Eugéne

    1951   Anatomical Factors of Body-Heat Regulation. *Nature* 167:823–824.

    1975   Morphological Variations and Climatic Differences. *Journal of Human Evolution* 4:529–539.

Serrat, Maria A., Donna King, and C. Owen Lovejoy

    2009   Temperature Regulates Limb Length in Homeotherms by Directly Modulating Cartilage Growth. *Proceedings of the National Academy of Sciences of the United States of America* 105:19348–19353.

Stinson, Sara, and A. Roberto Frisancho

    1978   Body Proportions of Highland and Lowland Peruvian Quechua Children. *Human Biology* 50:57–68.

Stuiver, Minze, Paula J. Reimer, and Ron W. Reimer

    2005   CALIB 5.0.

Trinkaus, Erik

    1997   Appendicular Robusticity and the Paleobiology of Modern Human Emergence. *Proceedings of the National Academy of Sciences of the United States of America* 94:13367–13373.

Wang, Weijie J., and Robin H. Crompton

    2003   Size and Power Required for Motion with Implication for the Evolution of Early Hominids. *Journal of Biomechanics* 36:1237–1246.

Witte, Hartmut, Holger Preuschoft, and Sebastian Recknagel

    1991   Human Body Proportions Explained on the Basis of Biomechanical Principles. *Zeitschrift fur Morphologie und Anthropologie* 78:407–423.

Yamaguchi, Bin

    1989   Limb Segment Proportions in Human Skeletal Remains of the Jomon Period. *Bulletin, National Science Museum*, Series D, Anthropology 15:41–48.

Yom-Tov, Yoram, and Eli Geffen

    2006   Geographic Variation in Body Size: The Effects of Ambient Temperature and Precipitation. *Oecologia* 148:213–218.

11. Body Proportions in Recent Native Americans: Colonization History Versus Ecogeographical Patterns

Richard L. Jantz, Paul Marr, and Claire A. Jantz

Summary Statement: In this contribution we examine anthropometric variation among native North Americans and assess it in relation to Bergmann-
Allen ecogeographical rules. We use a data set collected by Franz Boas
during the late nineteenth and early twentieth centuries, containing body
measurements from 7,732 individuals distributed over 97 tribes. Measurements and indices used are height, sitting height, leg length, arm length,
shoulder breadth, cormic index, relative arm length, and relative shoulder
breadth. All absolute dimensions except sitting height have a positive correlation with mean annual temperature. Cormic index has a negative correlation with mean annual temperature, but relative arm length and relative
shoulder breadth are not correlated with temperature. The absolute dimensions show that larger, wider bodies are found in warmer temperatures, the
opposite of ecogeographical expectations. The finding of relatively longer
legs in warmer climates agrees with expectations. However, trend surface
analysis shows that cormic index increases from the southeast to the northwest. Body measurements provide little evidence of climate patterning.
Rather, the patterning is more likely to reflect colonization history of the
continent. The spatial patterning, along with high variability, is inconsistent
with single-migration origin models. Native American limb proportions are
more consistent with those of ancestors originating in the lower to mid latitudes of East and Southeast Asia.

Human Variation in the Americas: The Integration of Archaeology and Biological Anthropology, edited by
Benjamin M. Auerbach. Center for Archaeological Investigations, Occasional Paper No. 38. © 2010
by the Board of Trustees, Southern Illinois University. All rights reserved. ISBN 978-0-88104-095-1.

Introduction

The ecogeographical rules set forth over a hundred years ago by Bergmann and Allen have been widely applied to human populations. Correlations of body size and shape variables with climate have resulted in widespread agreement that humans generally conform to Bergmann-Allen predictions. Roberts's (1953, 1973) extensive survey of various relative and absolute body dimensions convincingly documented that they are correlated strongly with mean annual temperature (MAT). Recent secular changes in human populations since Roberts's 1953 paper have reduced but not eliminated correlations between climate and weight and surface area (Katzmarzyk and Leonard 1998). Climate–body size and climate-shape correlations have also been confirmed in hunter-gatherer populations (Taylor-Weale and Vinicius 2008). Newman (1953) explicitly examined how body size and body shape correlated with climate in native North and South Americans, concluding that the Bergmann-Allen ecogeographical rules account for the observed clines and that these clines probably represent change since arrival in the Americas.

Evidence for the adaptiveness of body size and shape is based primarily on correlation between body dimensions and climate. We are frequently warned that correlation is not causation, but correlations of body dimensions with climate are so strong and consistent that few doubt that selection has played a role in establishing them. Holliday (1997, 1999) has shown that early Upper Paleolithic Europeans have more-tropical limb proportions, which he considers consistent with an African origin. Late Upper Paleolithic and Mesolithic samples are more similar to modern Europeans, consistent with gradual adaptation to the colder conditions of Europe. The European results seem to indicate that the changes in limb proportions approach their adaptive values slowly, requiring 20,000 years to approximate modern Europeans. Hence body proportions may be used to infer population history and migrations. In the case of native North Americans, who are relatively recent arrivals, it may be instructive to ask what body proportions have to say about the climate that ancestors of Native Americans experienced prior to coming to America.

It is often argued that proportions are less subject to plastic change than general size is. What this argument overlooks is that size and shape are not necessarily independent. Leg length, and specifically distal leg length, is positively allometric with stature (Holliday and Ruff 2001; Meadows and Jantz 1995). In America over the past 150 years, tibia length has increased relatively faster than height, resulting in relatively longer tibiae and hence an increase in crural index. It is therefore necessary to control for variation in size when comparing variation in relative limb length.

In this paper an analysis of Franz Boas's anthropometric data will be carried out with three related aims: (1) evaluate the extent to which Native Americans' size and proportions are related to climate, which would suggest an evolutionary response in America, (2) to examine spatial patterning of data to ascertain whether it is suggestive of population history, and (3) to relate Native American body size and proportions to the larger world context as a way of suggesting climates of origin.

Samples and Methods

Boas's Data

All of the above-mentioned studies contain few Native American groups, relative to the rest of the world. To some extent the paucity of data has been remedied by the rediscovery of Boas's anthropometric data (Jantz 1995, 2003, 2006; Jantz et al. 1992). It may do well to briefly describe the various projects Boas put together to acquire the largest anthropometric data set of native North Americans in existence. Boas began collecting anthropometric data in 1890 and was able to string together projects in such a way that data collection continued almost without interruption until 1901. Table 11-1 shows the various projects and approximate number of subjects contained in each. The bulk of the data collection was supported by the World's Columbian Exposition (WCE)—the 1893 World's Fair in Chicago—which alone resulted in about 13,000 subjects. Other projects, such as British Association for the Advancement of Science and the Jesup North Pacific (JNP) expedition played into Boas's interest in the Northwest, while some others, such as the Huntington California Expedition, filled in gaps. Theoretical issues usually powered these projects. The WCE was designed to synthesize patterns of variation in North America and the JNP expedition was designed to explore cultural and biological similarities between western North America and eastern Siberia. Boas was able to hire some 50 anthropometrists who traveled to locations in the United States and Canada where Native Americans could be found, mostly at Indian agencies or Indian schools.

Measurements

Boas hired around 50 anthropometrists (Jantz and Spencer 1997) who were provided with instruments and instructions in an effort to standardize procedures as far as possible. The basic data set consisted of six body measurements: stature, sitting height, shoulder breadth (biacromion), span, shoulder height, and finger height. Some of these measurements can be converted into values that more directly reflect structures of interest. These along with relative measurements (see Jantz 1995 for more details) are:

> Height (direct measurement)
> Sitting height (direct measurement)
> Shoulder breadth (direct measurement)
> Arm length (= shoulder height − finger height)
> Leg length (= height − sitting height)
> Cormic index (= [sitting height / height] × 100)
> Relative arm length (= [arm length / height] × 100)
> Relative shoulder breadth (= [shoulder breadth / height] × 100)

It is obvious these measurements are somewhat limited. Weight is lacking, and shoulder breadth is the only transverse dimension. These data will primarily characterize absolute and relative longitudinal dimensions.

Table 11-1. *Boas's Native American Anthropometric Projects, 1890–1901*

Supporting Organization	No. Subjects
Bureau of American Ethnology	143
British Association for the Advancement of Science	703
World's Columbian Exposition	12,915
American Association for the Advancement of Science	195
Jesup North Pacific Expedition	854
Huntington California Expedition	203
Total	15,013

Samples

Table 11-2 presents the distribution of tribes and individuals by culture area, as defined by the Smithsonian Institution Handbook of North American Indians series. There is considerable variation among culture areas in numbers of tribes and individuals measured. This reflects population density and the attention Boas directed to different areas. There is considerable variation in sampling contexts. Most subjects were measured at Indian agencies with the agent in charge acting as contact. Subjects were also obtained at Indian schools or in the remaining intact communities, particularly in the Northwest. Boas personally measured about 3,000 subjects, indicating the extent of his personal investment in the projects.

It would have been close to impossible in this, as well as most anthropological samples, to obtain a random sample. It has been argued that the sampling from Indian agencies resulted in measurements of individuals who worked for or lived close to the agency. These individuals would have been the most acculturated segment of the population (Moore and Campbell 1995). Students in Indian schools would have had a diet comparable to the white majority of the time. The general level of acculturation varied greatly. Eastern tribes had been in contact with Europeans for over 200 years, while many tribes in the Plains and West had much less exposure. Admixture levels also vary greatly, some individuals in eastern tribes report their blood quantum in terms of thirty-seconds while in some western tribes one-half is the finest division, indicating the longer contact history of the former compared to the latter. The sample was limited to those with three-fourths or higher Native blood quantum. The position taken in this study is that the sampling bias will have little effect on the dimensions of interest.

Climate and Spatial Data

The climate variable used in this analysis is mean annual temperature (MAT). It is undoubtedly an oversimplification but has the advantage in that it is comparable to much of what has been done in the past. Mean annual tempera-

Table 11-2. *Distribution of Tribes/Individuals by Culture Area*

Culture Area	Males		Females	
	Tribes	Individuals	Tribes	Individuals
Arctic	3	72	3	50
Subarctic	5	172	5	123
Northwest Coast	16	434	16	237
Plateau	14	648	14	436
California	11	275	10	160
Great Basin	4	211	4	84
Southwest	5	290	4	42
Plains	16	1,391	15	522
Northeast	17	1,014	15	497
Southeast	6	798	6	276
Total	97	5,305	92	2,427

ture was obtained from the 5 arc-minute resolution WorldClim climate data set (Hijmans et al. 2005), which is based on a 1950–2000 time series of meteorological data. Because latitude and longitude were available for the aboriginal location of each tribe (Dillingham 2005), we were able to calculate MAT for each tribe's location by averaging the MAT within a 3 x 3 cell neighborhood centered on the geographic location of each tribe. Each cell is 100 km^2 in area.

Statistical and Geographical Analysis

Correlation of body dimensions with mean annual temperature was used to assess relationships with climate, including an analysis using a covariance adjustment of leg length based on allometric relationships we discovered in our data. We performed two additional analyses to investigate spatial patterns of body dimensions and mean annual temperature: (1) a least squares linear regression of cormic index on MAT was performed and the residuals of this analysis were mapped, and (2) linear trend surfaces for cormic index and shoulder breadth were mapped to evaluate any existing directional patterns. These surfaces were generated using x (east-west) and y (north-south) coordinates from the Lambert conformal conic projection as independent and body measurements as the dependent variables in a first-order linear polynomial regression, which would reveal direction trends in the data. Instead of using an unprojected coordinate system (i.e., latitude and longitude), the Lambert projection minimizes systematic decreases in the unit measurement of degrees longitude with increasing latitude.

Results

Climate Correlations

The results of the correlation of anthropometric variables with MAT are shown in Table 11-3. The two sexes show a similar pattern, which can be described as follows: (1) the direct measurements of height, shoulder breadth, arm length, and leg length have significant positive correlations with mean annual temperature, the strongest of which is leg length; (2) cormic index shows a negative correlation, but neither relative arm length nor relative shoulder breadth is correlated with mean annual temperature. The results indicate that relative leg length is the only aspect of shape that is correlated with climate. All other significant correlations are driven by size. All absolute dimensions except sitting height have significant correlations with MAT, and all but leg length disappear when expressed as a ratio to stature. The picture that emerges is that larger body size is associated with higher temperature, the reverse of what we would expect following ecogeographical rules, which predict smaller body sizes with higher temperature.

Allometry

Since so much of climate patterning is size related, allometric relationships with size may drive the correlation of cormic index with climate. To test that idea, we undertook a brief analysis of allometric relationships. Allometric coefficients of body dimensions with size were obtained as the least-squares regression of the natural log variable on log size—in this case height—as

$$\log(\text{variable}) = b \log(\text{height}) + \log(a).$$

Allometric coefficients (b) describe shape change with change in size. Positive allometry ($b > 1$) means a structure gets relatively larger as size increases, while negative allometry ($b < 1$) means a structure is relatively smaller with increasing size. Isometry ($b = 1$) means proportions are unchanged with changes in size. Table 11-4 shows the allometric coefficients for sitting height, shoulder breadth, arm length, and leg length. It shows that all variables except leg length are negatively allometric. The allometry coefficients for leg length and sitting height are necessarily complementary, since as the proportion of one increases the other must decrease. In this case, the allometry coefficients indicate taller people have relatively longer legs and shorter trunks. All allometry coefficients differ significantly from isometry.

It is therefore necessary to ask whether the correlation of cormic index with climate might be driven by size. We carried out a covariance adjustment of leg length, adjusting leg lengths to the same stature. Correlations with MAT for adjusted leg length were reduced slightly compared to raw leg length: .61 ver-

Table 11-3. *Regression and Correlation Statistics of Anthropometric Variables on Mean Annual Temperature*

Variable	Males				Females			
	b	SE	r	p	b	SE	r	p
Height	4.0217	.5951	.5698	<.001	2.8878	.4893	.5283	<.001
Sit Ht	-.3002	.3430	-.0894	ns	-.0148	.3006	-.0052	ns
Sh Br	.9556	.2419	.3757	<.001	.7906	.2687	.2962	<.01
Arm L	1.6805	.4111	.3868	<.001	1.7132	.4054	.4069	<.001
Leg L	4.3219	.4525	.6999	<.001	2.9026	.4106	.5976	<.001
Cormic	-.1420	.0162	-.6681	<.001	-.0983	.0172	-.5150	<.001
Rel Arm	-.0081	.0162	-.0514	ns	.0268	.0200	.1395	ns
Rel Sh	.001	.0127	.0	ns	.0107	.0151	.0744	ns

Table 11-4. *Allometry of Body Dimensions with Height with Standard Errors*

Dimension	Allometry Coefficients			
	Males	SE	Females	SE
Sitting height	.777	.0126	.836	.0192
Shoulder breadth	.671	.0225	.740	.0378
Arm length	.916	.0155	.874	.0268
Leg length	1.239	.0132	1.180	.0209

Note: All coefficients differ from 1.0 at $p < .001$

sus .70 for males and .46 versus .60 for females. This shows that the positive allometry of leg length is responsible for some of the correlation with MAT, but significant correlation remains after adjustment. This justifies further analysis of cormic index.

Regression Analysis

We then performed least-squares linear regression of cormic index on MAT. The regression results of course reflect the correlation results. However, what is of interest is the geographic distribution of the residuals, which identifies geographic regions where MAT does a good job predicting body dimensions and regions where it does a poor job. Figure 11-1 shows the distribution of cormic index residuals for both sexes. The residuals show a distinctive geographical pat-

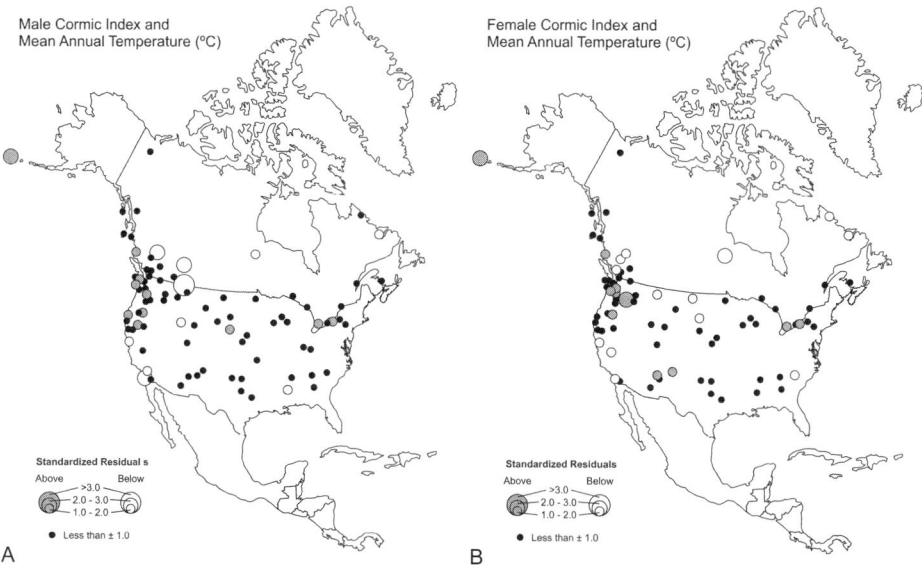

Figure 11-1. *Residuals from regression of cormic index onto mean annual temperature.*

terning, with most of the departures two or more standard deviations from the regression appearing in the Northwest and Subarctic culture areas. The principal aspects of the pattern are that Northwest Coast people tend to have relatively shorter legs than their temperature would predict and that the Subarctic people have relatively longer legs than MAT predicts they should.

Trend Surface Analysis

Additional insight may be obtained by looking at the spatial patterning of cormic index. Table 11-5 shows the standardized regression coefficients of cormic index on x and y coordinates, the Lambert conformal conic projection of latitude and longitude. Both are highly significant, but it can readily be seen that changes in x have a greater effect than changes in y. Figure 11-2 shows the linear trend surfaces, where it can be seen that cormic index increases from the southeast to the northwest. The sexes exhibit a nearly identical pattern.

It is also instructive to look at the spatial patterning of shoulder breadth, given the importance of body breadth in climate adaptation. Figure 11-3 shows the linear trend surfaces of absolute shoulder breadth. The trend direction is oriented more nearly north-south than that of the cormic index. However, the direction of change is the opposite of that predicted by climate models: The widest shoulders are found in the south; the narrowest, in the north. The regression model is much weaker for this than for the cormic index ($R^2 = .084$; $p = .023$ and $R^2 = .049$; $p = .133$, for males and females respectively), especially for females, whose results are not statistically significant.

Table 11-5. *Standardized Regression Coefficient of Cormic Index on* x *and* y *Coordinates*

	Males*			Females**		
	beta	t	p	beta	t	p
x	-.39	-4.30	<.01	-.44	4.92	<.01
y	.28	3.06	<.01	.28	3.09	<.01

Note: Geographic data were transformed using the Lambert conformal conic projection to minimize distance distortion.
* $R^2 = .29$
** $R^2 = .37$

Comparison to World Populations

So far we have considered the climatic and spatial patterning within North America, but for the sake of perspective, it is useful to compare native North Americans to Old World populations. Comparative data were obtained from Roberts (1973) for Europe and East Asia and from Hiernaux (1968) for sub-Saharan Africa. Boas's data from Siberia were used as an additional Old World group and to allow specific comparison to the region from which Native Americans may have come (see Ousley and Jantz 2001 for details of Siberian samples and additional analysis). Data from Roberts are somewhat approximate, since they were obtained from plots.

Table 11-6 presents the mean cormic index, sample size, and the differences between each pair in standard deviation units. Groups are sorted in ascending order. Because sample sizes are for the most part large, most groups are significantly different from each other by the Bonferroni multiple comparison test. Groups tending not to differ significantly are centrally located, as are those with small sample sizes, such as Europe and East Asia.

Table 11-6 allows several interesting observations. Most striking is that the range of means of all four Old World groups is completely contained within the range of native North Americans. The lowest cormic index, and hence the group with the relatively longest legs, is for the Southeast peoples, whose index is even more extreme than that of the Africans. Three New World groups—Arctic, Northwest Coast, and Southwest—have relatively shorter legs than the shortest Old World groups—East Asia and Siberia. The two most extreme groups differ from one another by almost two standard deviations, and both differ from many other groups by more than one standard deviation. These data make clear that native North Americans are extremely variable as far as the cormic index is concerned.

Discussion

There are certain limitations of the data, some of which have been mentioned above, and in the approach taken, which should be addressed before we consider some of the broader implications. Use of MAT as the climate vari-

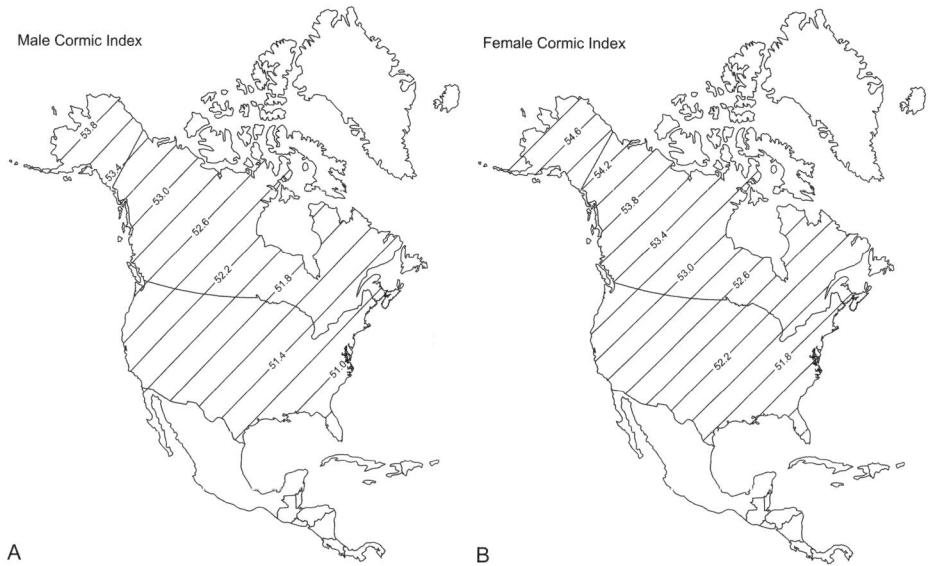

Figure 11-2. *Linear trend surface analysis of cormic index showing NW to SE orientation.*

able is undoubtedly an oversimplification but has the advantage in that it is comparable to earlier work. Newman (1960) suggested temperature of the coldest month would be more relevant to adaptation, but Jantz (2006), also using Boas's data for North America, showed that July temperatures had significantly higher correlations with body measurements than January temperatures did. Climate variables other than temperature may also present significant correlation with body measurements. Hiernaux and Froment (1976) have shown that rainfall and humidity have significant correlations with body measurements, although for height and sitting height correlations with temperature are the most important. Hall and Hall (1995) identified correlations with a variety of climate variables using Boas's data for West Coast populations. Native North American relationships to more general climate variables remain to be investigated.

Our interpretations of anthropometric variation, and especially cormic index, treat them as primarily a reflection of genetic variation. There is evidence that leg length may also exhibit considerable plasticity in response to changing environmental conditions. Tanner and colleagues (1982) have shown that Japanese secular change in height between 1957 and 1977 is due almost entirely to increase in leg length, bringing them closer to European leg length–height proportions. Bogin and associates (2002) found remarkable increase in height, due to gains in both sitting height and leg length, in U.S.–born Maya children compared to those growing up in Guatemala. In addition, leg length increases more than sitting height does, resulting in relatively longer legs in U.S.–born Maya children. Relative increase in leg length has been observed skeletally as a component of the secular increase in height in the general U.S. population (Meadows Jantz and Jantz 1999). This apparent plasticity begs the question of whether the patterns of

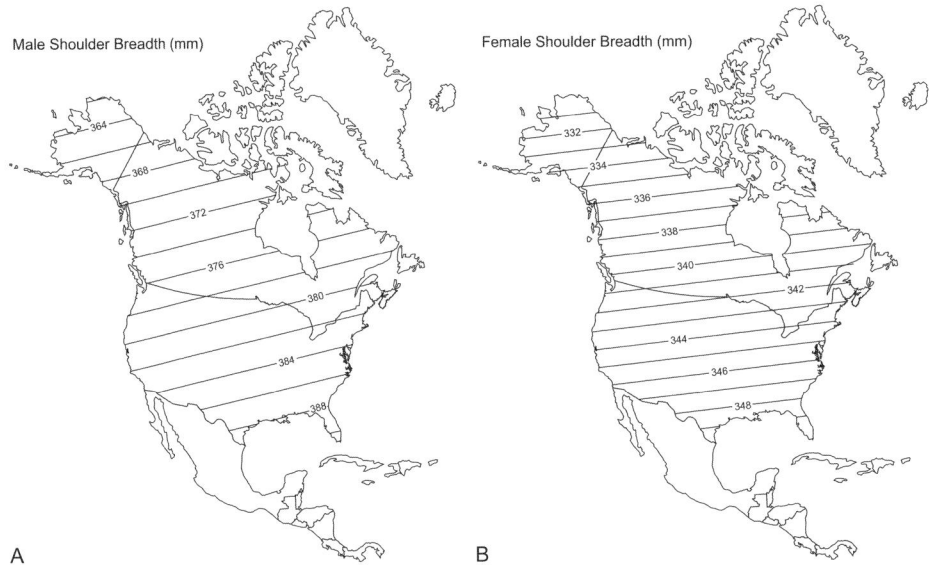

Figure 11-3. *Linear trend surface analysis of shoulder breadth, showing increasing breadth from north to south.*

variation observed could be attributed to environmental variation. We consider that unlikely for the following reasons:

> 1. Some of the change in relative leg length seen in recent groups is allometric. As shown in this study and in Meadows and Jantz (1995), leg length is positively allometric with stature. Secular increases in stature will result in relatively longer legs because of this allometric relationship.
> 2. More important, environmental changes during the twentieth century have resulted in quality of life improvements far exceeding those seen in the preceding centuries (Fogel 2004:Chapter 2).

There was, to be sure, environmental variation among tribes in the Boas samples. For most tribes the nineteenth century, when the adult subjects would have been growing up, was a period of severe cultural disruption accompanied by nutritional and disease stress. In some instances, nutrition may have been adequate during the reservation period, as described by Prince (1995) for the Sioux. However, the health and sanitary conditions on Sioux reservations were deplorable and infant mortality rates were three times as high as those for whites (Prince 1995). It is unlikely any tribe's environment could be described as favorable enough to provide a significant growth advantage.

Ruff (1994) has argued that body breadth is the most relevant in relation to climate adaptation. He obtained a remarkable correlation of .866 of bi-iliac breadth and latitude, used as a proxy for temperature. He further argued that bi-iliac breadth is less responsive to environmental plasticity than are limb proportions. Our use of biacromial breadth cannot be construed as a proxy for bi-iliac breadth because they behave differently. Lintsi and Kaarma (2006) documented greater

Table 11-6. *Sample Sizes, Means, and Pairwise Distances of Culture Areas in Standard Units for Cormic Index*

C area	n	Cormic	SEast	Africa	Calif	NEast	Plains	Subarc	Europe	GBasin	Plateau	Siberia	EAsia	SWest	NWCoast	Arctic
SEast	781	50.35	0													
Africa	153	50.80	.26	0												
Calif	280	51.26	.53*	.27	0											
NEast	1,002	51.49	.66*	.40*	.13	0										
Plains	1,356	51.57	.70*	.45*	.18	.05	0									
Subarc	165	51.62	.73*	.48*	.21	.08	.03	0								
Europe	23	51.70	.78*	.53*	.26	.13	.08	.05	0							
GBasin	209	52.04	.97*	.72*	.45*	.32*	.27*	.24	.19	0						
Plateau	618	52.44	1.21*	.95*	.68*	.55*	.50*	.47*	.42	.23	0					
Siberia	792	52.62	1.31*	1.06*	.79*	.66*	.61*	.58*	.53	.34*	.11	0				
EAsia	25	52.69	1.35*	1.10*	.83*	.70	.65	.62	.57	.38	.15	.04	0			
SWest	286	52.74	1.38*	1.12*	.85*	.72*	.68*	.64*	.60	.40*	.17	.07	.02	0		
NWCoast	410	53.40	1.76*	1.50*	1.23*	1.10*	1.06*	1.03*	.98*	.79*	.55*	.45*	.41	.38*	0	
Arctic	71	53.77	1.97*	1.72*	1.45*	1.32*	1.27*	1.24*	1.19*	1.00*	.77*	.66*	.62	.60*	.21*	0

$^*p < .05$ by Bonferroni multiple comparison

secular change in biacromial than bi-iliac breadths in Estonians. Hiernaux and Froment (1976) showed that biacromial and bi-iliac did not exhibit similar climate correlations, such that body shape changed with climate, rather than becoming uniformly broader or narrower. Biacromial as the only transverse dimensions is one of the primary limitations of Boas's data for research into climate patterning.

One of the questions posed at the outset was whether native North Americans conform to ecogeographical expectations, as has been generally accepted. One way to evaluate the fit to ecogeographical expectations is to use Roberts's (1973) world climate correlations as the expectation and compare it to what is seen in native North Americans. Table 11-7 presents the correlation of several relative and absolute dimensions from Roberts's world sample compared to those from Native Americans. We discuss each of the dimensions in turn:

Height

For world populations height has a negative correlation with MAT, but in native North Americans it is positive. However, height by itself may not have much relevance to climate adaptation. Roberts (1953) has shown that the correlation with MAT disappears when weight is controlled. Lacking weight, height serves as a proxy for body size and indicates that larger bodies are found in warmer climates and smaller bodies in colder climates. The extremely short people of the Northwest coast compared to the tall people of the Southeast are the obvious cases.

Sitting Height/Cormic Index

These two variables reflect the absolute and relative expression of trunk height and the inverse of leg length. The world picture shows that both absolute and relative trunk height are negatively related to mean annual temperature. In native North Americans on the other hand, absolute trunk height is unrelated to MAT, but cormic index is negatively correlated at about the same level as the world sample. That shows that in Native Americans the correlation of cormic index is due entirely to variation in leg length. The world and North American correlations of cormic index with MAT are about the same, which might be taken as support for a climate effect. The strong northwest to southeast orientation may also suggest it is an artifact of colonization history, an idea that will be taken up below.

Relative Shoulder Breadth

Roberts's world data shows that cold-climate people tend to have relatively wider shoulders than populations from warm climates. The relationship is presumably a reflection of the more general association of wide bodies with cold climate (Ruff 1993, 1994). In native North Americans, there is no relationship of relative shoulder breadth to MAT.

Relative Arm Length

In the world picture, warm-climate people have relatively longer arms, which ecogeographic rules predict. In native North Americans, there is no relationship of relative arm length to MAT.

Table 11-7. *World Climate Correlation Compared to America*

Variable	World	America	Fits World Model?
Height	-.351	.570	No
Sitting height	-.438	-.089	No
Cormic index	-.619	-.668	Yes
Relative shoulder height	-.577	.000	No
Relative arm length	.680	-.051	No

Taken as a whole, the body size and shape picture provides little support for a significant climate effect in native North Americans. The only relative dimension to exhibit correlation is cormic index. The value to be attached to this is degraded by the stronger east-west than north-south spatial orientation of the index. One might also expect that if relative leg length responds to climate then relative arm length should also, but that is not the case. At this point, one might ask why so many previous researchers have concluded that Native Americans do conform to Bergmann-Allen predictions (Falsetti 1989; Hall and Hall 1995; Newman 1953). There may be several reasons, the main being that previous analyses were conducted with more limited databases than available to the present analysis. An especially important omission in previous work is subarctic populations, who despite existing in extremely cold conditions (see, e.g., Steegmann et al. 1983) have cormic indices more in line with temperate-climate populations. In addition, including Central and South America, as Newman (1953) did, could alter the picture. The small-bodied people of tropical America are more in-line with what one expects. Others have used a more extensive suite of climate variables (Hall and Hall 1995), which may also bring out aspects of climate response missed by focusing on MAT. Nevertheless, the present findings may be taken as seriously calling into question the commonly held view that native North Americans exhibit ecogeographical patterning in body size and shape distributions.

If body size and shape in native North Americans do not reflect adaptation to climate, what alternative interpretations may be presented? The most interesting finding contained in the present analysis is the exceedingly high among-group variability observed in the cormic index. It includes cormic indices that would be considered tropical adaptations and ones that would be considered Arctic. This high level of variability is in conflict with early ideas of phenotypic uniformity, the so-called American Homotype (Hrdlička 1914; Stewart 1973). This idea has been resurrected recently in order to support an argument for the single origin of Native Americans (Fiedel 2004). However, the high variability is inconsistent with a single origin unless we are to believe that evolutionary processes produced more variability in cormic index in Native Americans than in the entire Old World population, in less time and in a more homogeneous environment.

We are not arguing that the ecogeographical rules lack general validity only that they have not had a significant impact on Native American morphology, presumably because of limited time. Holliday (1997) has shown that early Upper Paleolithic Europeans have tropical limb proportions and required 20,000 years to approach modern proportions. One might, therefore, turn the argument around and ask what limb proportions suggest about the climate in which ancestors of Native Americans might have lived. It has long been assumed that Native American ancestors passed through a "cold filter" as Stewart termed the route from northeast Siberia through Beringia to North America. More recent ideas postulating long-term isolation in Beringia prior to moving south into North and South America (Tamm et al. 2007) imply an even longer period of exposure to cold conditions than envisioned by earlier investigators.

Limb proportions do not support long periods of cold adaptation by Native American ancestors. In that regard they are in conflict with evidence from body breadth. By contrast limb proportions might suggest that Native American ancestors passed through a warm filter. The recent populations of much of North America, including the people of the Plains, Northeast, Southeast, and even the subarctic, are decidedly temperate or tropical in their cormic indices. Temperate, or even tropical, crural indices have a long history in America (Auerbach 2007; Hall et al. 2004). Auerbach (2007) reported crural indices for five early Holocene Americans, yielding an average of 84.7, placing them midway between West Africans and Europeans, compared to data reported by Kurki and colleagues (2008). It is, therefore, extremely interesting to observe that environments in which Clovis technology is found are not in general cold. Backfitting American Clovis environments to Asia corresponds to the mid and southern latitudes of East and Southeast Asia (Gillam et al. 2007).

The evidence from limb proportions is in conflict with evidence from body breadth. Recent Native Americans are wide bodied (Ruff 1994) and this is also true for early Holocene Americans (Auerbach 2007). Body breadth has been interpreted as evidence of cold filtering for Native American ancestors. The position one takes on this issue depends on how one views the evolutionary response of limb proportions and body breadth. On the one hand, rapid changes in limb proportions can be used to argue that plasticity confounds evolutionary response. On the other, long-term stasis in Europe (Holliday 1997, 1999), correlation of relative limb length with climate, and no relationship to mobility (Holliday and Falsetti 1995) supports an evolutionary response.

Compared to limb proportions, body breadth exhibits a stronger correlation with climate and does not seem to be influenced as much by secular trend (Lintsi and Kaarma 2006). At the same time, body breadth does not always track climate (Kurki et al. 2008). It is also apparent that bi-iliac breadth presents a different picture of body-breadth variation than biacromial breadth does. Aleuts are wide bodied on bi-iliac breadth (Laughlin 1951; Ruff 1994) but would be characterized as narrow bodied on biacromial breadth (Laughlin 1951). Finns, on the other hand, have very wide shoulders in relation to bi-iliac breadth (Ruff et al. 2004). The different conclusions arising from body breadth and limb proportions studies remain to be resolved, since both likely reflect responses to many competing factors.

Evidence is mounting that early colonists followed a coastal route to get to America (Erlandson 2002). If the Northwest coast was the entry point of these colonists, then their least cost movement would have taken them primarily in a southeasterly direction (Anderson and Gillam 2000), which could account for the orientation of the cline of relative leg length seen in this analysis. Ironically, the principal place in North America where cold-adapted limb proportions are found, apart from the Arctic, is the Northwest coast. Since early Holocene limb proportions do not suggest cold adaptation, we might infer that these cold adapted populations arrived later. Evidence suggesting interaction between Northwest coast and northeast Siberian populations, dating back to ancient times, has long been known (Black 1983; Gurvich 1988). While the Northwest population has even higher cormic indices than Siberians in general, it is clear that the Northwest Coast people bear a closer relationship to Siberians than to the peoples of the plains, the Southeast, and the Northeast. A more detailed analysis of Siberia may add greater clarity to these relationships.

Conclusion

Body size and shape dimensions of recent native North Americans do not present variation that suggests adaptation to climate as measured by mean annual temperature. Nor is there evidence that Native American ancestors passed through a cold filter. Rather, the data suggest a model deriving Native American ancestors from the mid to lower latitudes of East and Southeast Asia. Limb proportions that can be considered cold adaptations are limited to peoples of the Arctic and the western United States, especially the Northwest coast. People with these limb proportions are likely to have been later arrivals and are responsible for the clinal distribution of cormic index, in particular.

References

Anderson, David G., and J. Christopher Gillam
    2000   Paleoindian Colonization of the Americas: Implications from an Examination of Physiography, Demography, and Artifact Distribution. *American Antiquity* 65:43–66.
Auerbach, Benjamin M.
    2007   Human Skeletal Variation in the New World During the Holocene: Effects of Climate and Subsistence Across Geography and Time. Unpublished Ph.D. dissertation, Center for Functional Anatomy and Evolution, Johns Hopkins University, Baltimore.
Black, Lydia T.
    1983   Some Problems in the Interpretation of Aleut Prehistory. *Arctic Anthropology* 20:49–78.
Bogin, Barry, Patricia Smith, Alicia B. Orden, Maria I. Varell-Silva, and James Loucky
    2002   Rapid Change in Height and Body Proportions of Maya American Children. *American Journal of Human Biology* 14:753–761.

Dillingham, Paul C.
  2005   Geographic Variation in Native American Anthropometrics: A Spatial Analysis of the Boas and Gifford Data Sets. Unpublished Ph.D. dissertation, Department of Anthropology, University of Tennessee, Knoxville.

Erlandson, Jon M.
  2002   Anatomically Modern Humans, Maritime Voyaging and the Pleistocene Colonization of the Americas. In *The First Americans: The Pleistocene Colonization of the New World*, edited by Nina G. Jablonski, pp. 59–92. Memoirs of the California Academy of Sciences, No. 27, San Francisco.

Falsetti, Anthony B.
  1989   Anthropometry of Native North American Indians from the Northwest Coast, Arctic, Subarctic, Great Basin and California: An Examination of Scaling Phenomena. Unpublished Ph.D. dissertation, Department of Anthropology, University of Tennessee, Knoxville.

Fiedel, Stuart J.
  2004   The Kennewick Follies: "New" Theories about the Peopling of the Americas. *Journal of Anthropological Research* 60:75–110.

Fogel, Robert W.
  2004   *The Escape from Hunger and Premature Death, 1700–2100: Europe, America and the Third World*. Cambridge University Press, Cambridge.

Gillam, J. Christopher, David G. Anderson, and A. Townsend Peterson
  2007   A Continental-Scale Perspective on the Peopling of the Americas: Modeling Geographic Distributions and Ecological Niches of Pleistocene Populations. *Current Research in the Pleistocene* 24:86–90.

Gurvich, Il'la S.
  1988   Ethnic Connections Across Bering Strait. In *Crossroads of Continents: Cultures of Siberia and Alaska*, edited by William W. Fitzhugh and Aron Crowell, pp. 17–21. Smithsonian Institution, Washington, D.C.

Hall, Roberta L., and Donald A. Hall
  1995   Geographic Variation of Native People along the Pacific Coast. *Human Biology* 67:407–426.

Hall, Roberta L., Diana Roy, and David Boling
  2004   Pleistocene Migration Routes into the Americas: Human Biological Adaptations and Environmental Constraints. *Evolutionary Anthropology* 13:132–144.

Hiernaux, Jean
  1968   *La diversite humaine en Afrique subsharienne*. Edotopms de l'Institut de Sociologie Universite Libre de Bruxelles, Bruxelles.

Hiernaux, Jean, and Alain Froment
  1976   The Correlations Between Anthropobiological and Climate Variables in Sub-Saharan Africa. Revised Estimates. *Human Biology* 48:757–767.

Hijmans, Robert J., Susan E. Cameron, Juan L. Parra, Peter G. Jones, and Andy Jarvis
  2005   Very High Resolution Interpolated Climate Surfaces for Global Land Areas. *International Journal of Climatology* 25:1965–1978.

Holliday, Trenton W.
  1997   Body Proportions in Late Pleistocene Europe and Modern Human Origins. *Journal of Human Evolution* 32:423–448.
  1999   Brachial and Crural Indices of European Late Upper Paleolithic and Mesolithic Humans. *Journal of Human Evolution* 36:549–566.

Holliday, Trenton W., and Anthony B. Falsetti
    1995   Lower Limb Length of European Early Moderns in Relation to Mobility and Climate. *Journal of Human Evolution* 29:141–153

Holliday, Trenton W., and Christopher B. Ruff
    2001   Relative Variation in Human Proximal and Distal Limb Segment Lengths. *American Journal of Physical Anthropology* 116:26–33.

Hrdlička, Aleš
    1914   Physical Anthropology in America. *American Anthropologist* 16:508–554.

Jantz, Richard L.
    1995   Franz Boas and Native American Biological Variability. *Human Biology* 67:345–353.
    2003   The Anthropometric Legacy of Franz Boas. *Economics and Human Biology* 1:277–284.
    2006   Anthropometry. In *Environment, Origins and Population,* edited by Douglas H. Ubelaker, pp. 777–788. Handbook of North American Indians, Vol. 3, William C. Sturtevant, general editor, Smithsonian Institution, Washington, D.C.

Jantz, Richard L., David R. Hunt, Anthony B. Falsetti, and Patrick J. Key
    1992   Variation among North Amerindians: Analysis of Boas's Anthropometric Data. *Human Biology* 64:435–461.

Jantz, Richard L., and Frank Spencer
    1997   Franz Boas (1858–1942). In *History of Physical Anthropology: An Encyclopedia*, Vol. 1, edited by Frank Spencer, pp. 186–189. Garland, New York.

Katzmarzyk, Peter T., and William R. Leonard
    1998   Climatic Influences on Human Body Size and Proportions: Ecological Adaptations and Secular Trends. *American Journal of Physical Anthropology* 106:483–503.

Kurki, Helen, Jaime Ginter, Jay T. Stock, and Susan Pfeiffer
    2008   Adult Proportionality in Small-Bodied Foragers: A Test of Ecogeographic Expectations. *American Journal of Physical Anthropology* 136:28–38.

Laughlin, William S.
    1951   The Alaska Gateway Viewed from the Aleutian Islands. In *The Physical Anthropology of the American Indian*, edited by William S. Laughlin, pp. 98–126. Viking Fund, New York.

Lintsi, Mart, and Helje Kaarma
    2006   Growth of Estonian Seventeen-Year-Old Boys During the Last Two Centuries. *Economics and Human Biology* 4:89–103.

Meadows, Lee, and Richard L. Jantz
    1995   Allometric Secular Change in the Long Bones from the 1800s to the Present. *Journal of Forensic Sciences* 40:762–767.

Meadows Jantz, Lee, and Richard L. Jantz
    1999   Secular Change in Long Bone Length and Proportion in the United States, 1800–1970. *American Journal of Physical Anthropology* 110:57–67.

Moore, John H., and Janis E. Campbell
    1995   Blood Quantum and Ethnic Intermarriage in the Boas Data Set. *Human Biology* 67:499–516.

Newman, Marshall T.
    1953   The Application of Ecological Rules to the Racial Anthropology of the Aboriginal New World. *American Anthropologist* 55:311–327.
    1960   Adaptations in the Physique of American Aborigines to Nutritional Factors. *Human Biology* 32:288–313.

Ousley, Stephen D., and Richard L. Jantz
  2001   500 Year Old Questions, 100 Year Old Data, Brand New Computers. In *Gateways: Exploring the Legacy of the Jesup North Pacific Expedition, 1897–1902*, edited by Igor Krupnik and William W. Fitzhugh, pp. 257–277. Arctic Studies Center, Contributions to Circumpolar Anthropology 1, National Museum of Natural History, Smithsonian Institution, Washington, D.C.

Prince, Joseph M.
  1995   Intersection of Economics, History and Human Biology: Secular Trends in Stature in Nineteenth-Century Sioux Indians. *Human Biology* 67:387–406.

Roberts, Derek F.
  1953   Body Weight, Race and Climate. *American Journal of Physical Anthropology* 11:533–558.
  1973   *Climate and Human Variability*. Addison Wesley, Reading, Massachusetts.

Ruff, Christopher B.
  1993   Climatic Adaptation and Hominid Evolution: The Thermoregulatory Imperative. *Evolutionary Anthropology* 2:53–59.
  1994   Morphological Adaptation to Climate in Modern and Fossil Hominids. *Yearbook of Physical Anthropology* 37:65–107.

Ruff, Christopher B., Markku Niskanen, Juho-Antil Junno, and Paul Jamison
  2004   Body Mass Prediction from Stature and Bi-iliac Breadth in Two High Latitude Populations, with Application to Earlier Higher Latitude Humans. *Journal of Human Evolution* 48:381–392.

Steegmann, A. Theodore, Marshall G. Hurlich, and Bruce Winterhalder
  1983   Coping with Cold and Other Challenges of the Boreal Forest: An Overview. In *Boreal Forest Adaptations*, edited by A. Theodore Steegmann, pp. 317–351. Plenum Press, New York.

Stewart, T. Dale
  1973   *The People of America*. Charles Scribner's Sons, New York.

Tamm, Erika, Kivisild Toomis, Maere Reidla, Mait Metspalu, David G. Smith,
Connie J. Mulligan, Claudio M. Bravi, Olga Rickards, Cristina Martinez-Labarga,
Elsa K. Khusnutdinova, Sardana A. Fedorova, Maria V. Glolubenko, Vadim A. Stepanov,
Mariana A. Gubina, Sergey I. Zhadanov, Ludmila P. Osipova, Larisa Damba,
Mikhail I. Voevoda, Jose E. Dipierri, Richard Villems, and Ripan S. Malhi
  2007   Beringian Standstill and Spread of Native American Founders. *PLoS ONE* 2(9):e829. doi:10.1371/journal.pone.0000829.

Tanner, James M., Takao Hayashi, Michael A. Preece, and Noel Cameron
  1982   Increase in Length of Leg Relative to Trunk in Japanese Children and Adults from 1957 to 1977: Comparison with British and with Japanese Americans. *Annals of Human Biology* 9:411–423.

Taylor-Weale, Rose, and Lucio Vinicius
  2008   Independent Roles of Climate and Life History in Hunter-Gatherer Anthropometric Variation, *Internet Journal of Biological Anthropology*, Vol. 1, No. 2. Electronic document, http://www.ispub.com/journal/the_internet_journal_of_biological _anthropology/volume_1_number_2_8/article/independent_roles_of_climate_and _life_history_in_hunter_gatherer_anthropometric_variation.html, accessed July 27, 2010.

12. Human Settlement in the New World: Multidisciplinary Approaches, the "Beringian" Standstill, and the Shape of Things to Come

David G. Anderson

As we have seen from the papers in this volume, understanding the colonization and subsequent occupational history of the Americas is a daunting task. The events that unfolded and the reasons for them appear to be far more complex and diversified than the stories or models we as practitioners of individual disciplines have typically been producing. Fortunately, our understanding is growing all the time, and while our explanations are becoming ever more complex and qualified, they also appear to be closer and closer approximations to what was likely actually happening, at least in some times and some places and for some processes. What may seem on first inspection to be a confusing picture is actually a marked improvement on the state of our understanding compared to even a few years and certainly a few decades ago. We continue to explore the same questions raised by Fewkes, Hrdlička, and their colleagues about the origin of and variation among New World peoples (Fewkes et al. 1912; Auerbach, Chapter 1), but as the papers in this volume have demonstrated, the data, methods, and theoretical approaches we now bring to bear on these questions are immense, well-grounded, and sophisticated, and our understanding is improving daily.

Perhaps the most obvious common thread among the papers in this volume is their multidisciplinary nature, drawing on the findings of a number of disciplines, most notably physical anthropology and archaeology, but also commonly involving research by linguists, geneticists, paleoenvironmental scientists, and other special-

Human Variation in the Americas: The Integration of Archaeology and Biological Anthropology, edited by Benjamin M. Auerbach. Center for Archaeological Investigations, Occasional Paper No. 38. © 2010 by the Board of Trustees, Southern Illinois University. All rights reserved. ISBN 978-0-88104-095-1.

ists. Because of this, the need for synthetic and synergistic theoretically informed analyses capable of interrelating evidence from numerous sources is arguably the greatest challenge facing us if we are to arrive at satisfying descriptions of and explanations for the colonization and postcolonization settlement history of the Americas. Specialized analyses by scholars working on their own or as part of collaborative ventures will remain critical and comprise the vast majority of research undertaken, but the results of this work will need to be considered, integrated, and evaluated from broader and multiple geographic, temporal, theoretical, and comparative analytical perspectives (see also commentary by Sassaman, Chapter 13).

Studying the peopling of the Americas, however, is also a classic example of a scientific research endeavor in which evidence from diverse subdisciplines—and even from various investigators within a subdiscipline—has sometimes yielded divergent or overtly contradictory results. Examples of this—such as the results of morphometric and genetic analyses suggesting that the location of New World source populations may have been in southeastern, eastern, or northeastern Asia or that one or more populations or migrations are implicated in the colonization (cf. Chapters 2, 11)—are readily apparent in the papers in this volume. When archaeology is added to the mix, it is evident that we don't currently know where classic late Pleistocene stone tool technologies like Clovis or Nenana originated (Goebel 2004), and homelands much farther afield than eastern Asia for colonizing populations have been advanced and heatedly debated, such as western Europe (cf. Meltzer 2009 and Strauss et al. 2005; Stanford and Bradley 2002 and Bradley and Stanford 2004). Linguistic arguments have been raised in favor of both comparatively recent and much more ancient dates for initial human entry into the Americas (cf. Greenberg et al. 1986; Nettle 1999; Nichols 1990, 2002, 2008), and while both the archaeological and genetic records are also somewhat ambiguous on this question, a later rather than an earlier initial entry, after circa 20,000 cal yr B.P. rather than upward of 25,000 or more years B.P. appears to be gaining ground (Goebel et al. 2008; Meltzer 2009; see Chapters 2, 11; cf. Madsen 2004a; Stanford et al. 2005).

What are we to make of results that are incompatible and hence in apparent disagreement, and how are we to proceed when they occur? The papers in this volume, including the introductory and concluding commentaries, offer examples of how multidisciplinary scientific research endeavors are undertaken and how their results can indeed be integrated into a better overall understanding of the past. Some of the major themes explored in this manner, as well as strengths and weaknesses of these approaches, are discussed in what follows. I conclude with some thoughts on sources of New World founding populations and, specifically, whether, where, and how the so-called Beringian Standstill may have occurred.

Tracing Population Movement in the Americas

Several papers in this volume illustrate how multiple lines of evidence can be combined in the reconstruction of the rates and routes of past population movements or migrations, often generating new insights in the process. Kemp and Schurr (Chapter 2) summarize how different kinds of genetic evidence are employed, often in a complementary and mutually reinforcing manner in such studies, including nuclear, mitochondrial, and Y chromosomal DNA. Genetic data, several papers in this volume demonstrate, are being used to explore a wide array of big picture type questions, such as (1) how the settlement of the Americas may have proceeded, both initially and in subsequent, post-initial colonization population movements (Kemp and Schurr, Chapter 2); (2) how agriculture may have spread into the Southwest (Watson, Chapter 6); and (3) how the highland empires in western South America expanded into coastal regions (Shinoda et al., Chapter 9). Explicit correlates or genetic signatures for rapid versus slower patterns of movement as well as for greater or lesser affinity are employed in these efforts, illustrating the potential of genetic information to inform on much more than ancestral descendant relationships. If haplogroups and subhaplogroups occur widely, for example, it implies fairly rapid movement (at least faster than the mutation rate) as opposed to the occurrence of "nested sets of variation" produced by slower patterns of movement (Kemp and Schurr, Chapter 2). The same distributions can also be used to explore interaction between different groups, complementing archaeologically based analyses of spatiality in mating network, interaction, or political relationships, that is, how and why human populations position themselves as they do on landscapes (e.g., Kelly 1995; Wobst 1974, 1976).

The widespread occurrence of mtDNA subhaplogroups across the Americas, for example, suggests that population dispersal occurred very quickly, although the distinctive variation found within specific regions also implies that what some authors have called "tribalization" or the formation of more-or-less endogamous cultures or populations occurred quite early on (after Kemp and Schurr, Chapter 2; Malhi et al. 2002; Torroni et al. 1993; Watson, Chapter 6). Similar patterning is observed in the archaeological record (Figure 12-1), where restricted distributions of particular projectile point types are observed soon after 12,900 cal yr B.P., following the demise of the widespread Clovis culture, patterning interpreted as evidence for the emergence of subregional cultural traditions (Anderson 1990, 1995; Meltzer 2003, 2004, 2009). Watson (Chapter 6) suggests the emergence of distinctive cultures in the Americas is related to geographic isolation and the relative proximity of groups to each other.

In something of a contradiction, the widespread occurrence of mtDNA haplogroups and subhaplogroups also appears due, at least in part, to interaction among New World populations over fairly large areas, regardless of how isolated the archaeological assemblages might appear (Kemp and Schurr, Chapter 2; Chatters, Chapter 3). Determining the extent and directions over which interaction occurred, several papers in this volume demonstrate, is clearly an area where archaeology and physical anthropology can productively work together. Indeed, the other discussant for this volume, my friend and colleague Ken Sassaman, has argued that prehistoric population movement should be considered commonplace in prehistoric North America, the rule rather than the exception, and that traditional models positing that the cultures in a given area reflect descent and diversification from a local Pleistocene age founding population are likely to be wildly unlikely (Sassaman 2010; Sassaman, Chapter 13). I agree, as have scholars in many parts of North America, as migration, population replacement, and the

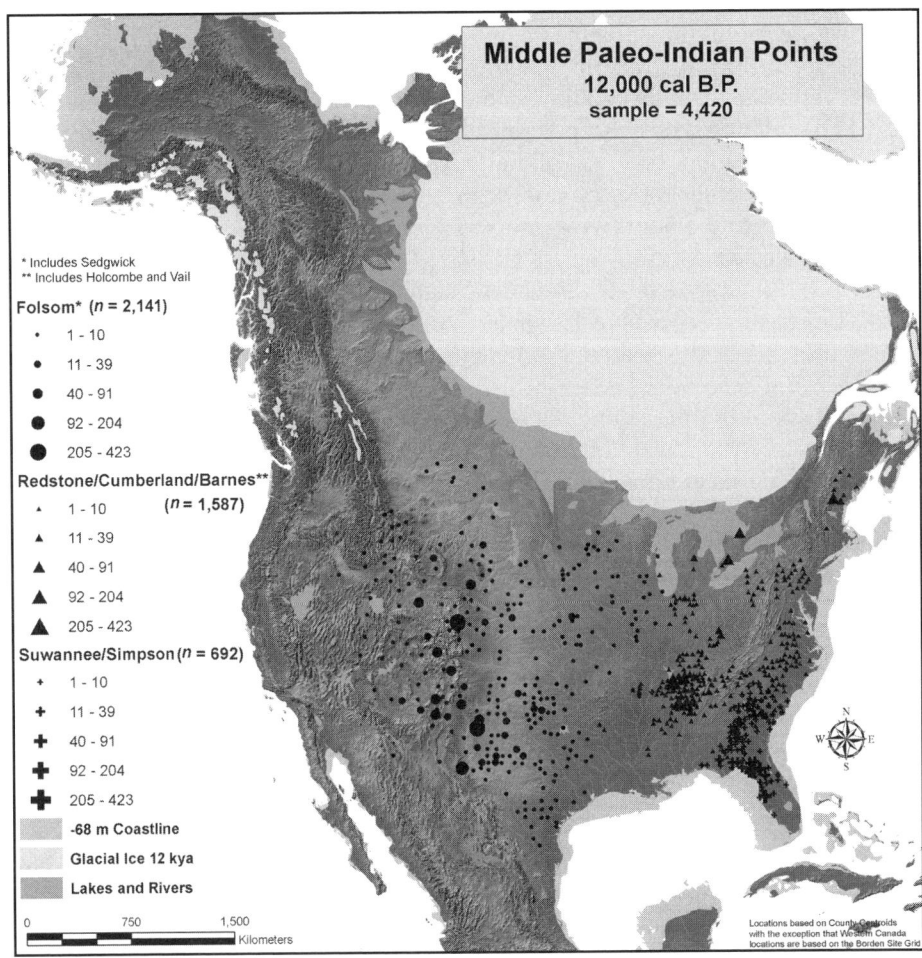

Figure 12-1. *Subregional archaeological cultural traditions in the immediate post-Clovis period in North America, circa 12,850–12,000 cal yr B.P. (Image courtesy PIDBA [Paleoindian Database of the Americas, http://pidba. utk.edu/].)*

abandonment and reoccupation of regions have again become acceptable phenomena for study (e.g., Anderson 1996, 1999; Anthony 1990; Cameron 1995; Lekson 1999, 2008; Pauketat 2007; Snow 1995, 2009; to cite a few of many such works). Too many archaeological analyses still never look beyond the site or locality, even though analyses of hunter-gatherer mobility and interaction indicate we should be thinking at much larger scales, encompassing large numbers of people over truly vast areas, and with population movement and replacement as likely as stasis and continuity (e.g., Kelly 1995:111–203; Wobst 1974, 1976). How regular and widespread gene flow was among culturally divergent populations would appear to be an ideal question to explore with genetic data in the Americas, as several papers demonstrate (e.g., Auerbach, Chapter 7; Shinoda et al., Chapter 9).

Other bioanthropological procedures used to explore important big-picture questions and employing archaeological, linguistic, or other lines of evidence include (1) the use of craniometric data to examine the possible spread of the Western Cordilleran cultural tradition into the Columbia Plateau during the Early Holocene (Chatters, Chapter 3); (2) the use of morphometric evidence to explore the emergence of distinct historical groups on the plains (Auerbach, Chapter 7); (3) the use of discrete dental traits to explore migration in the Southwest, in a test of one aspect of Lekson's (1999) "Chaco Meridian" argument that peoples from Chaco Canyon moved north to Aztec Ruins (Durand et al., Chapter 5); and (4) the use of genetic markers, discrete dental traits, molar wear angles, caries incidence, and other measures of skeletal biology and health to examine the spread of agriculture into the Southwest, and the changes that occurred in local populations following its adoption (Watson, Chapter 6). All of these studies made use of multiple lines of evidence, most commonly archaeological and physical anthropological data sets. When the results from individual analyses were not in agreement, decisions about which was more likely were typically based on the quality and preponderance of evidence (i.e., sample size, the number of different corroborating lines of evidence, how robust the analytical results were, and so on). Like Sassaman (Chapter 13), I accept that some kinds of evidence are more useful than others for resolving certain questions, but I remain sanguine, indeed unabashedly optimistic, about what we can learn using multiple approaches to the study of the past.

Physical anthropology, the chapters herein demonstrate, can sometimes provide direct evidence of population relationships that complement archaeological interpretations or clarify ambiguous or contentious archaeological results. The reaction of archaeologists in the Southwest to Lekson's (1999) arguments proposed in *The Chaco Meridian*, for example, has been decidedly mixed, with some scholars arguing that the meridianal alignment of Chaco, Aztec, Paquimé, and other sites is due to chance and that no movements or relocations of peoples occurred between these centers. Lekson (1999, 2008:337–341), in contrast, thinks quite differently, arguing that the alignment of centers reflects direct interaction and population movements between these centers, albeit from one to another over time. In this regard, the bioanthropological evidence indicating that a direct movement of people apparently did occur between Chaco and Aztec Ruins (Durand et al., Chapter 5) shows that Lekson's argument has merit and that further aspects of it, such as the hypothesized movements of peoples to and from Aztec Ruin and Paquimé, or between other centers on the same meridian, are eminently testable.

Watson's (Chapter 6) examination of the spread of agriculture into the Southwest likewise serves as an excellent example of a well-integrated multidisciplinary approach that scholars in other regions should emulate. Archaeologists in many parts of the world have argued for generations about whether agriculture was adopted by indigenous groups, spread by the migration of peoples, or some combination thereof. A good way to begin to resolve the question, Watson notes, is to examine haplogroup frequencies and other bioanthropologically derived measures in pre-farming and initial farming populations and in any contemporary nonfarming neighboring groups (in addition, of course, to using other lines of evidence). Such genetically based analyses should also prove useful to exploring language spread in various parts of the world. Diamond (1997, 2002), for example, has argued that demographic pressure led agriculturalists to expand, which is why the languages they spoke spread widely (see also Cavalli-Sforza 2003). Indeed, the distribution of dominant (in terms of numbers of speakers) languages or language families in the world reflects, in part, the spread of agricultural populations. Computer simulations can also prove useful in predicting and interpreting genetic data associated with possible population movements or differences to help take into account the effect of drift or low levels of gene flow (Cabana et al. 2008), among other things.

Of course, not all movement is unidirectional and long-distance. People move over the landscape all the time in the course of their normal subsistence pursuits and to exchange information, maintain mating networks, and familiarize themselves with resources in different parts of their normal ranges. Care must be taken to avoid confusing signatures of movement obtained from range mobility with outright migration or at least to recognize the possibility that hunter-gatherers can move over large areas and different habitats over the course of a year, a decade, or a lifetime (e.g., Binford 1983; Kelly 1995:111–160). Isotopic data can be used to determine homelands of origin, as in the famous case of the Amesbury archer found near Stonehenge, who originally appears to have come from central Europe (Fitzpatrick 2003; Richards 2007). Such analyses and interpretations, of course, can be complicated by many factors, including use of food sources that may themselves have come from long distances, such as anadromous fish (Cybulski, Chapter 4; see below).

Several of the studies in this volume, in fact, explore paleosubsistence practices and impacts using archaeological and bioanthropological data (e.g., Chatters, Chapter 3; Cybulski, Chapter 4; Watson, Chapter 6). Bioanthropological data, of course, has long been used to examine the impact of the adoption of agriculture on human populations. Conventional wisdom on this subject has it that only rarely have conditions improved for most people when this has occurred (e.g., Cohen and Armelagos 1984; Diamond 1987), although I personally think there is a lot to be said for the impact on human health of modern medicine, dentistry, or even clean water, some of the other things that resulted from agricultural food production and the generation of surpluses capable of supporting a wide range of specialists. The role foraging played in the transition— that is, how important wild resources continued to be—and how the adoption of agriculture influenced such things as gender and labor relations are considered less often but can be explored using a number of approaches, such as stable isotope analyses, molar wear angle, incidence of caries, robusticity, and other measures of relative health and diet (Watson, Chapter 6). That increased mechanical processing of agricultural foods can lead to changes in tooth wear angles and that wild plant foods with similar carbohydrate or isotopic signatures can mask the effects of the transition to agriculture are also things that researchers should routinely consider (Watson, Chapter 6). Finally, while one does not necessarily lead to the other, I find it interesting that relative skeletal health declined over the course of the Hohokam until the final collapse of this culture occurred

(Watson, Chapter 6). The success of complex organizational systems appears to some degree linked to the health of their constituent populations (e.g., Cohen and Armelagos 1984; Diamond 1987), a lesson we are not grasping very well in our own country at present.

Ecogeographic Variation in New World Populations

Several papers in this volume use morphometric analyses of skeletal or body proportions in conjunction with ecogeographic rules (Mayr 1956) to help determine the source populations of specific groups (Auerbach, Chapter 7) and New World populations in general (Jantz, Chapter 11; King, Chapter 10). That is, the settlement of the Americas would have likely resulted, over time, in clinal variation in body size, shape, or other characteristics (e.g., Auerbach 2007; Holliday 1997, 1999; Ruff 2002; Weinstein 2005:569), following classic ecogeographic rules like those of Bergmann (1847) and Allen (1877). Using the assumption that human morphology covarys with environment and length of settlement in that environment, several investigators have examined body shape and proportions to evaluate longevity of occupation in given regions and whether migration from elsewhere might have occurred (Auerbach 2007; Auerbach, Chapter 7; Jantz et al., Chapter 11; King, Chapter 10; Weinstein 2005). These kinds of analyses have generated useful insights about sources and rates of human movements in the Old World. Trinkaus (1981), for example, has demonstrated that the earliest Upper Paleolithic anatomically modern humans in western Europe exhibited morphological characteristics indicative of derivation from a warmer climate, probably Africa, and that over time these populations became increasingly cold adapted (see also Holliday 1997, 1999). Analyses of New World populations should, in a comparable fashion, provide clues about where their ancestral homeland(s) were located, provided they spent appreciable time there.

Recognizing ecogeographical patterning can prove difficult, however, for a number of reasons. What environmental variables should be considered, and how have these changed over the last few tens of thousands of years, globally and in specific regions? How does culture (i.e., things like clothing, use of fire, and type of shelter) influence thermoregulatory response? How long does it take human populations to respond morphologically to climate conditions and to climate change? Are skeletal samples dating thousands of years after colonization likely to exhibit the same patterning as those of initial immigrants? Analyses to date based in part on comparison with changes in Old World populations suggest that some characteristics, such as intralimb proportions, take a long time to change significantly, from many thousands of years to upwards of ten or twenty thousand years (Auerbach 2007; Auerbach, Chapter 7; Holliday 1997; Jantz et al., Chapter 11; Ruff 2002; Trinkaus 1981). Franz Boas's (1910, 1912) classic research with immigrants, in contrast, showed that some things, such as human cranial morphology, can change rapidly in response to new conditions, particularly diet (see discussion below). Human body size may also be linked to extent of population duration within specific physiographic regions, such as mountain ranges or low-lying plains, and to warmer and colder or wetter and dryer climates, particularly if studies from other species where such patterns have been observed can be applied to our own (James 1970; Trinkaus 1981:210).

However, the record from the New World on the subject of ecogeographic variation in human populations, as the papers in this volume indicate, is somewhat ambiguous and nowhere near as pronounced as observed in portions of the Old World (Auerbach 2007; Ruff 2002; Trinkaus 1981). This may be due to the much shorter time frame over which adaptation within our own species has played out in the New World as opposed to in the Old World, on the order of 20,000 as opposed to perhaps 150,000 or more years. Some patterns are evident, however. Auerbach (2007, Chapter 7) has argued that the New World colonizing population was "cold filtered" with wide bodies being one result. If so, this would suggest that the source populations for New World peoples spent appreciable time in a cold climate. Such an inference is plausible if their homeland was northeastern Asia and if the Beringian Incubation or Standstill (BIM/Standstill) described by Kemp and Schurr (Chapter 2; see discussion below) is correct, and the Standstill occurred in an area with a cold climate like Beringia during the last glacial period. King's (Chapter 10) examination of postcranial variation in later Holocene North American populations was able to demonstrate fairly strong linkages between climate and morphology, indicating that New World/North American populations have also apparently undergone considerable adaptation to local climate conditions since colonization.

Jantz and his colleagues' (Chapter 11) analysis of Boas's modern body measurement data from native North American populations (Jantz 1995, 2003), in contrast, concluded that the sample provided little evidence for climate-linked patterning but instead reflected the colonization history of the continent. While regional clines in morphology were documented, these were mixed with regard to expectations based on ecogeographic rules, with the latitudinal occurrence of cormic indices[1] conforming to and mean shoulder breadth the opposite of expectations, with a further longitudinal component of trends from roughly northwest to southeast in these measurements (Jantz et al., Chapter 11; see also Auerbach and Ruff 2010). Ecogeographic factors, they argue, "have not had a significant impact on North American morphology, presumably because of limited time" and, more interestingly, that "limb proportions do not support long periods of cold adaptation in Native American ancestors" (Jantz et al., Chapter 11). Their analyses suggest instead that New World colonizing populations may have derived from more temperate climates or passed through a "warm filter" rather than developed over an extended period in a colder climate, as the BIM/Standstill and "cold filter" models suggest (although it should be noted that the BIM/Standstill model is actually neutral, save only in its name, as to where the colonizing populations actually "incubated"). The clinal distributions in morphological characteristics were used to suggest initial entry in the Pacific Northwest, followed by movement to the south and east (referencing Anderson and Gillam's [2000] least-coast pathway analyses that suggest that dispersal in these directions is facilitated by physiographic features on the continental scale). The ultimate origins for New World populations were suggested to lie in mid- to lower-latitude Southeast

Asia, based on earlier craniometric studies (Jantz and Owsley 2001, 2005) and in part on recent analyses back plotting the ecological associations of Clovis sites to temperate portions of eastern Asia, suggesting early Paleo-Indian populations were better adapted to temperate than to arctic conditions (Gillam and Tabarev 2006; Gillam et al. 2007). If Jantz and colleagues' (Chapter 11) interpretations are correct, the "Beringian" part of the BIM/Standstill model may need to renamed or at least acknowledge a different starting point for the populations that subsequently diversified (see discussion below).

Interestingly, Jantz and colleagues' (Chapter 11) analyses also examined modern Siberian body measurement data for comparative purposes and concluded that these populations exhibited more temperate body proportions, suggesting they hadn't been in Northeast Asia long enough to more fully adapt to the cold climate. This suggests that they, as well as the New World source populations presumed to have derived from them, likely came from somewhere else in the not-too-distant past, perhaps from farther south. Of course, the relationship of analyses based on measurements taken from living individuals as opposed to analyses of skeletal samples needs to be worked out (Auerbach and Ruff 2010). Ecogeographic patterning in the New World, it would appear, has been complicated by the comparatively short time since initial colonization and by subsequent population movements throughout prehistory (Auerbach 2007; Auerbach, Chapter 7; Jantz et al., Chapter 11). The research does highlight the critical importance of finding and examining early skeletal remains from both the New World and northeastern and eastern Asia.

Diet, Nutrition, and Warfare

Several case studies reported herein also show how other aspects of life—such as subsistence and warfare—are yielding new insights through the use of multiple analytical approaches, especially in times and places remote from ethnohistoric/ethnographic analogs (e.g., Chatters, Chapter 3; Cybulski, Chapter 4; Durand et al., Chapter 5; Watson, Chapter 6). Using analyses encompassing morphometrics, stable isotopes, and dental pathology, in conjunction with archaeological data, Cybulski (Chapter 4), for example, examines adaptations between different culture areas, the Northwest Coast and the Plateau, as well as within a culture area, the Plateau. Skeletal indications for watercraft use are more apparent in Northwest Coast populations, with greater use of plant foods relative to fish and animal protein inferred from higher caries rates among Plateau groups. In an important cautionary finding, anadromous sea fish consumption was shown to yield stable carbon isotope signatures suggesting appreciable consumption of marine resources by peoples located well inland; decreasing signatures for marine fish use with distance from the coast was also indicated (Cybulski, Chapter 4). Care must thus be taken to avoid inferring a coastal origin for peoples found in the interior if they consumed anadromous fish, as apparently initially happened with regard to the Kennewick remains; one way to do this, Cybulski shows, is to compare ancient skeletal samples with historically documented or modern samples of people from the same area whose dietary choices are well documented.

Multidisciplinary research can sometimes lead to surprising, indeed counterintuitive findings, as demonstrated by Schmidt and colleagues' (Chapter 8) examination of burial practices and remains in the later Archaic period American Midwest. Experimental studies showed there was a "right" way that was consistently employed to take body part trophies among local Archaic populations. These procedures, which were apparently different from butchering strategies employed with game animals, were in use for perhaps two thousand or more years, suggesting warfare was not intermittent or infrequent, but routinized, with specialized associated behaviors passed down from generation to generation. Our understanding of Archaic period warfare in the East has emphasized its presence, but evidence for its frequency has been more ambiguous (e.g., Dye 2009:61–67; Mensforth 2007:256ff; Milner 1999, 2004:46–47; Smith 1996). Schmidt and colleagues' (Chapter 8) analyses also showed that there were lengthy traditions as well as changes in the treatment of victims by aggressors and survivors alike over time. During the Archaic, at least in this part of the lower Midwest, trophy taking was fairly standardized, with most people, including victims, buried the same way. Victims of conflict in the Late Prehistoric era, in contrast, were subject to more diversified and seemingly more haphazard postmortem indignities and were sometimes placed in mass graves and given "less stylized treatment" than the burial treatment accorded individuals dying in times of peace. The treatment of victims over time, the example demonstrates, can be profitably evaluated and cannot be assumed to be consistent.

Craniometric Differences Between Earlier and Later New World Populations

A critical question touched on in several papers is why the earliest Americans were apparently morphologically quite different from later Holocene and modern Native American populations, particularly in craniofacial characteristics (see also Jantz and Owsley 2001, 2005). Small colonizing populations, founder effects, and genetic drift, as well as subsequent population movements all appear to have played a role (Auerbach, Chapters 1 and 7; Chatters, Chapter 3; Jantz et al., Chapter 11; King, Chapter 10). Chatters (Chapter 3), for example, argues that the hypothesized early Holocene Old Cordilleran expansion was one of many migrations from the north that resulted in the distinctive morphologies of later Holocene Native Americans and part of the movements that Sassaman (Chapter 13) argues had a major impact on cultural developments in the Eastern Woodlands, such as the rise of the Shell Mound Archaic.

The papers in this volume also highlight the fact that great care must be taken in interpreting craniofacial data. Human crania can be somewhat plastic in response to changes in diet and environment, as Boas (1912) first demonstrated. How "plastic" human crania are and the rates by which changes in morphology occur, in fact, remain a subject of some debate in physical anthropology (cf.

Gravlee et al. 2003; Powell 2005:232–236; Relethford 2004; Sparks and Jantz 2002, 2003; van Vark et al. 2003). While cranial plasticity clearly exists, however, no one today would argue that it completely erases the population structure and history information contained in cranial morphology. Softer diets and lower protein intake, brought about in part by increased consumption of small game and plant foods, may explain some of the differences between earlier and later New World populations (e.g., Chatters, Chapter 3; Sardi et al. 2006). Changes in craniofacial morphology were also likely accelerated when food-processing technologies like milling stones and ceramics came into widespread use and when domesticated plants appeared, reducing the need for more massive masticatory features. Likewise, the development of specialized stone tools reduced the need for the use of teeth as tools, producing a similar reduction (e.g., Brace 1962).

Even within our relatively small sample of early New World crania, however, appreciable variability is evident (González-José et al. 2008; Jantz and Owsley 2001, 2005). How much this reflects sampling variability, and perhaps the emergence of isolated populations in minimal interaction with one another, is unknown. Many of these specimens, including some of the best known like Kennewick or Spirit Cave, actually date up to several millennia after widespread archaeological evidence for settlement circa 13,000 cal yr B.P. (González-José et al. 2008; Meltzer 2009:175–181). Even given a small and fairly uniform founding population (itself something that is not too likely given the genetic variation evident in descendant populations), perhaps such morphological variability should not be too surprising, since it had thousands of years to develop. To effectively resolve questions about the affinities and morphological and genetic characteristics of the earliest Americans, including their relationship with later Americans, the analyses herein demonstrate that we need many more human skeletal remains, ideally with accompanying well-preserved genetic material, from the late Pleistocene of the Americas and Northeast Asia.

Implications of the Beringian Incubation/Standstill Model

Genetic evidence appears to be narrowing toward a robust and well-grounded consensus about when and from where the Americas were colonized: It took place sometime after 20,000 cal yr B.P., from a single source population located somewhere in eastern or northeastern Asia (e.g., Kemp and Schurr, Chapter 2; Tamm et al. 2007). The evidence from genetics, however, is at odds with current interpretations based on archaeological research, which has shown that multiple stone-tool industries were present in Northeast Asia, Alaska, and North America in the late Pleistocene and, hence, presumably reproductively more or less distinctive human populations as well (e.g., Goebel 1999, 2004; Goebel and Slobodin 1999; Goebel et al. 2008; Hamilton and Goebel 1999). Morphological analyses, in turn, particularly craniometric analyses have, as discussed above, somewhat controversially suggested that the earliest human populations in North America, so-called Paleoamericans were distinct from later Holocene and recent American

Indians (e.g., Jantz and Owsley 2001, 2005; Jantz et al., Chapter 11). Consideration of the Beringian incubation model, or BIM/Standstill, offers the opportunity to reconcile some of these and other apparent contradictions, as well as indicates how genetic and morphometric data can guide archaeological research.

What is the modern genetic consensus on the colonization of the Americas, and how does the BIM/Standstill hypothesis fit into it? Analyses of mtDNA mutation rates, as summarized by Kemp and Schurr (Chapter 2), indicate that New World populations apparently split from a single Old World source population comparatively recently, probably after circa 25,000 cal yr B.P., and whose modern descendants are located in the Lake Baikal/Altai mountain area of eastern Asia. Entry into the Americas occurred sometime afterward, probably between around 20,000 to 15,000 cal yr B.P. The difference between time of divergence and time of entry, on the order of 5,000–10,000 or more years, is because the populations that ultimately settled the New World were apparently genetically isolated from their Old World source population for a long period of time, sufficient for distinctive haplotype/lineage mutations to occur. This period of isolation has been called the Beringian Incubation or Beringian Standstill model, here shortened to BIM/Standstill (Kemp and Schurr, Chapter 2; Tamm et al. 2007). That isolation of the proto–Native American population *must have occurred* is apparently not in question among geneticists at present. A contrasting scenario, the "Direct Colonization Model," no longer considered viable, held that New World populations moved into Beringia and beyond into the Americas with minimal genetic, and hence temporal, separation from their Asian source populations (Kemp and Schurr, Chapter 2).

What is remarkable about the BIM/Standstill model is how quickly it was replicated and essentially confirmed, with four major studies appearing within a few months of one another, encompassing mtDNA, Y chromosomal DNA, and autosomal marker studies (Achilli et al. 2008; Fagundes et al. 2008a, 2008b; Kemp and Schurr, Chapter 2; Kitchen et al. 2008; Tamm et al. 2007). This consensus among geneticists, of course, conflicts with arguments from linguistics, skeletal biology, and archaeology favoring multiple migration events, as well as linguistic and archaeological evidence and arguments, albeit some quite controversial, for a much greater antiquity for the peopling of the Americas. When evidence from differing disciplines or at least scholars is in significant disagreement, as it is with the peopling of the Americas, then we must either find a way to reconcile these differences or admit that something is not right and re-examine our fundamental assumptions (i.e., mtDNA mutation rates, language diversification rates, and archaeological evidence). Archaeological remains from earlier failed migrations (*sensu* Meltzer 1989) may, of course, still be found from peoples who entered the New World much earlier yet left no surviving genetic signatures. Unlike artifacts, though, people had to survive to pass on their language, suggesting that linguistic models used to advocate a very early entry, on the order of 30,000–40,000 years B.P. (e.g., Nichols 1990, 2002, 2008), need some rethinking. Specifically, if the genetic evidence holds up, and no archaeological evidence is found for occupations earlier than circa 15–20 k cal yr B.P., then New World languages must have diversified at a much faster rate than assumed in some models (e.g., Nettle 1999).

There are details to consider when using genetic evidence, of course. Are the mtDNA mutation rates well established and constant, or might they have been different or varied in the past? How long have the presumed source populations whose descendants are currently living in the Lake Baikal source area actually resided there?[2] Were they there since the Last Glacial Maximum (LGM) or earlier, or did they arrive from somewhere else in more recent times (i.e., from farther south in temperate Asia, perhaps pushed there in recent millennia by expanding agricultural populations)? Wherever the source populations were located, how and why did a subset of these people come to be isolated from the parent group for several thousand years? Why is there so little evidence for reverse migration, a backflow of people and mtDNA and Y chromosomal lineages into Asia early on, especially when it is clear that both cultural and genetic exchanges occurred later, including by modern Eskimos (Forsyth 1992; Karafet et al. 1997:307–309)? That is, why was the initial movement of people that resulted in the widespread settlement in the New World apparently only in one direction? Or are we simply not recognizing or missing the evidence that it was not? While the presence of fluted points in Alaska has been taken by some to be possible evidence for a "back-migration" of people, most likely Clovis and immediate post-Clovis hunter-gatherers living farther south, these technologies and hence presumably the people who made them did not get beyond the Seward Peninsula nor into Northeast Asia (Goebel and Slobodin 1999; Hamilton and Goebel 1999).[3]

At present there is no strong archaeological evidence for people living in or near Beringia, at least not in extreme Northeast Asia for many thousands of years prior to circa 14,000 or 15,000 cal yr B.P. (e.g., Goebel 2004; Goebel and Slobodin 1999), even though the genetic evidence indicates that a standstill or incubation interval did indeed occur.[4] Assuming no flaws in the BIM/Standstill model develop (cf. Meltzer 2009:367, who remains properly cautious until more time has passed), *the challenge facing archaeologists is thus to determine where, when, how, and why the "standstill" or "incubation" took place.* Furthermore, if New World source populations were isolated prior to circa 15,000 cal yr B.P., it means the standstill also took place prior to the opening of the ice-free corridor in Canada, which was not traversable from circa 34,000 until sometime after circa 15,000 cal yr B.P., although movement along the intermontane valleys of the western Cordillera may have been possible somewhat longer, to perhaps 24,000 cal yr B.P. or slightly later, when this region too was closed by expanding ice sheets (Dyke 2004; Madsen 2004b:12; Mandryk et al. 2001). Glacial conditions thus all but ensure that human settlement of the Americas south of the North American ice sheets, if it initially occurred between circa 24,000 and 15,000 cal yr B.P., was almost certainly via a Pacific coastal route, which in turn means sophisticated watercraft had to have been used. This latter inference, fortunately, is not at all problematic, since extended maritime voyaging by our species dates well back into the last glaciation, and support for coastal entry using this technology has been mounting (Dixon 1999; Erlandson 2002; Fedje et al. 2004:120–123; Fladmark 1979; Madsen 2004b).

But where would be a favorable place for incubation or standstill to occur, and where would the use of boats have been likely? Several possibilities come to mind, including late glacial era Japan, the Russian Far East, and southern Beringian coastal regions and archipelagoes.[5] Human use of the first two regions, apparently with a well-developed maritime adaptation, dates well back into the last glaciation, upward of 15,000 cal yr B.P., and they have indeed been suggested as possible source areas of New World populations (Dixon 1999; Fedje et al. 2004:135; Goebel 1999, 2004; Goebel and Slobodin 1999; Ikawa-Smith 2004; Madsen 2004b:6). The problem with areas to the south in the Russian Far East and Japan is that it is difficult to conceive how a long period of isolation from other human populations could have occurred, something essential for the BIM/Standstill to be viable. Coastal and indeed much of the now-submerged portions of Beringia, in contrast, offered a vast and potentially resource-rich area that may well have been sufficiently distant and difficult to access, providing the necessary isolation. Central Beringia certainly had to be negotiated to reach the Americas, either by land in the interior or by boat along the coast, barring movement exclusively along the Aleutian island chain. While the latter is a possibility, there were formidable water gaps between some of the islands in the Aleutians, even at the LGM, such as from western Kamchatka in the Gulf of Kamchatka/Ozernoi Gulf area and Ostrov Beringa and Ostrov Medneyy/the Medneyy Seamount (>75 km), and particularly between there and the seamount defined by the modern islands of Attu, Agattu, Alaid, Niski, and Shemya (>400 km), although beyond this part of the late Pleistocene Aleutians, islands—most appreciably larger and with more area available for colonization than at present—would have been fairly closely spaced all the way to the Alaska Peninsula and the continental mainland.

An easier and potentially environmentally far richer passage to the New World was available, however, a comparatively short distance to the north. The continental-scale Beringian landmass was exposed and accessible, and along its southern margins were a remarkable series of archipelagoes (Figures 12-2 through 12-7) that persisted for some ten thousand years following the LGM, with new islands and island chains appearing and disappearing as sea levels rose (Brigham-Grette et al. 2004:59; Erlandson et al. 2007, 2008:2234; Manley 2002). The existence of these archipelagoes was dramatically illustrated in a video by Manley (2002), showing the flooding of Beringia from circa 21,000 cal yr B.P. to the present in 1,000-year intervals, based on bathymetric data available at the time. The potential of southern Beringia for human settlement was evaluated in a subsequent paper by Brigham-Grette and her colleagues (2004:36–40, 57–61), who presented a figure created from Manley's video that included highlighted outlines of several of the larger islands present at six moments in time during the late Pleistocene and after, when sea levels were -120, -88, -77, -64, -54, and 0 meters below the modern stand (Brigham-Grette et al. 2004:38). Manley's video was developed from the 2001 version of the ETOPO2 database with elevation and bathymetric data from a two minute latitude-longitude grid (i.e., with cells ca. 3.7 km on a side) and approximately 1 meter vertical precision (U.S. Department of Commerce 2006; this is the most recent version of the database available online).

It is now possible to evaluate changes in topography, including the extent of the Beringian/Aleutian archipelagoes at higher resolution, employing the newly available ETOPO1 database employing elevation and bathymetric data from a 1-minute grid, with cells approximately 1.85 km on a side (Amante and Eakins

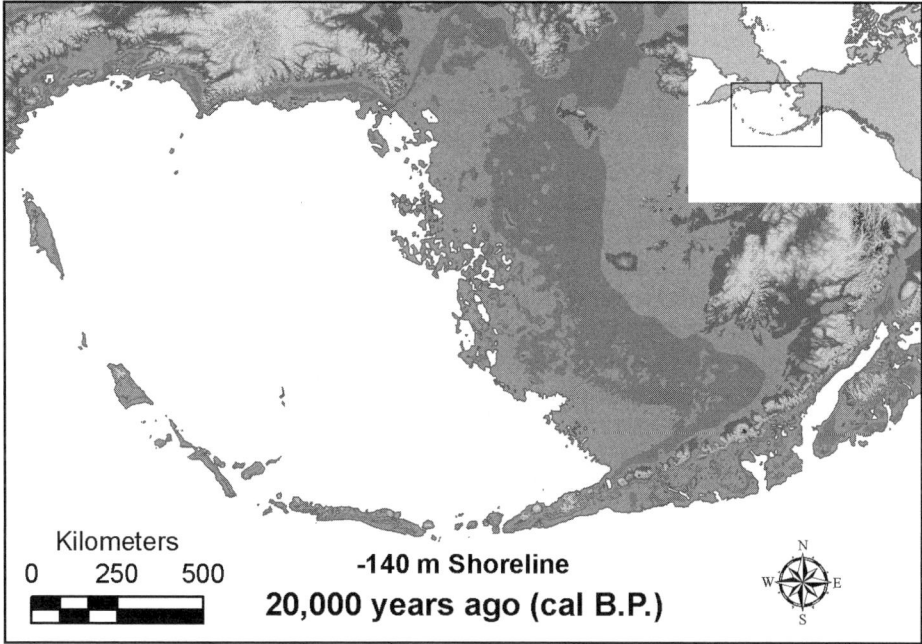

Figure 12-2. *The island archipelago of southern Beringia and the Aleutians at circa 20,000 cal yr B.P.[6]*

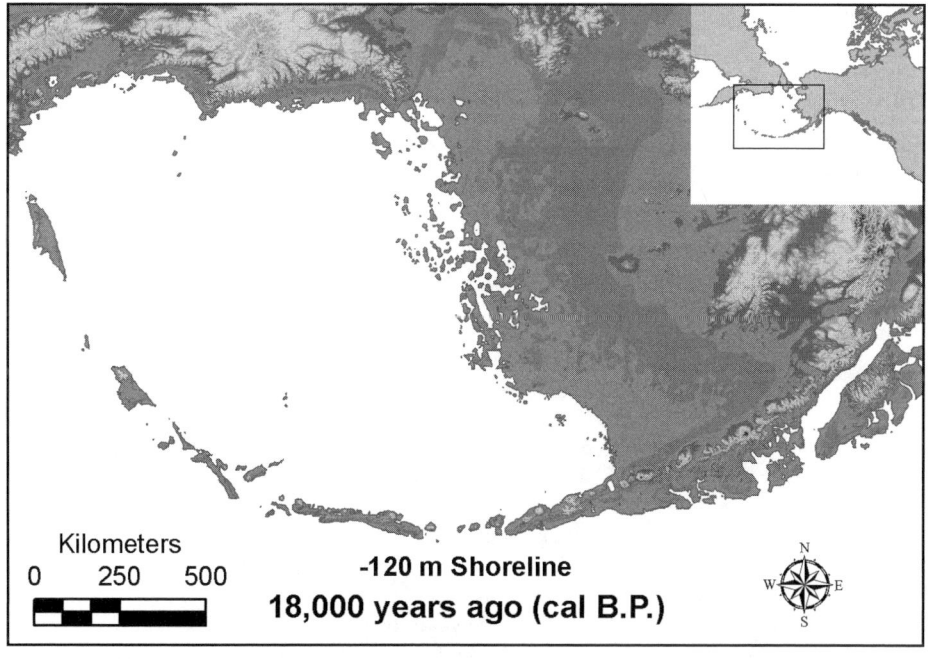

Figure 12-3. *The island archipelago of southern Beringia and the Aleutians at circa 18,000 cal yr B.P. [6]*

2009). Using Manley's (2002) depth values for global sea-level change for the Late Pleistocene, which correspond to reconstructions derived from multiple locations (Lambeck et al. 2002), it proved possible to use the ETOPO1 data to produce a series of maps (Figures 12-2 through 12-7) showing shorelines and land areas at circa 2,000 year intervals, from the LGM at circa 20,000 cal yr B.P. to the early Holocene at circa 10,000 cal yr B.P. (Anderson et al. 2010). Whether sea levels were at the exact elevations indicated on these maps at the specific moments indicated is debatable. As discussed by Brigham-Grette and colleagues (2004:36–40), well-dated sea-level curves need to be worked out for different parts of Beringia to control for isostatic and tectonic factors. Fortunately, corrected sea-level values for specific areas and times, once determined, can be fairly easily substituted into these maps. Regardless of precisely when sea levels were at the stands illustrated, there is no question that they were at these stands at some point in the Late Pleistocene, as sea levels rose from a low of perhaps -140 m at the LGM (Lambeck et al. 2002:358). What is apparent from inspection of these maps is that a remarkable number of islands were present along the southern margins of Beringia throughout the Late Pleistocene. These islands were closely spaced, furthermore, allowing for movement between them without likely losing sight of land. Movement could have proceeded to the southeast to the Alaska Peninsula, with people either looping around the peninsula and then heading back to the east and south along the Pacific Northwest coast or moving to the west out into the Aleutians. Both routes were likely taken once people reached the Alaska Peninsula. A route through the archipelagoes of southern Beringia and out into the Aleutians from the east likely would have been far less dangerous than crossing the Aleutian chain from west to east from Kamchatka, even during periods of greatly reduced sea level.

In the absence of much direct physical evidence, considerable uncertainty exists as to how productive those portions of the now-submerged Beringian landmass and coastal zones actually may have been to human populations. Up to several months of open water, free of sea-ice cover and a rich habitat for marine life, however, has been inferred to have been present on the southern Beringian coast during much of the last late glacial era, including during the LGM (Brigham-Grette 2004:59-61; Clague et al. 2004:82; Sancetta et al. 1985). As Erlandson and colleagues (2008:2234) note:

> Once portrayed as a harsh and relatively unproductive area for human habitation (e.g., Hopkins et al., 1982), recent research suggests that the south coast of Beringia may have been "geomorphically complex during the late glacial, with hundreds of islands located just off a coast riddled with bays and inlets" (Brigham-Grette et al., 2004, p. 59). During the summer months, such convoluted coastlines—when combined with the low gradient of the Beringian platform—may have offered broad expanses of productive intertidal and nearshore habitats for early maritime peoples to hunt, forage, and gather in. Even covered with sea ice much of the year, the south coast of Beringia would have provided rich habitat for seals, walrus, and a variety of other marine organisms. Erlandson et al. (2007) have argued that much of Beringia's south coast may have supported productive kelp forests after the end of the LGM.

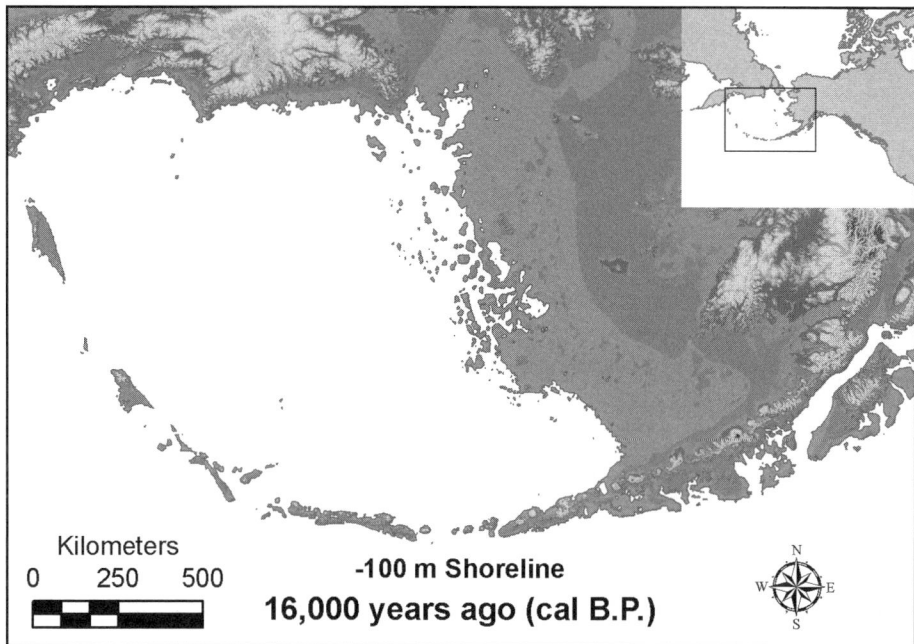

Figure 12-4. *The island archipelago of southern Beringia and the Aleutians at circa 16,000 cal yr B.P.*[6]

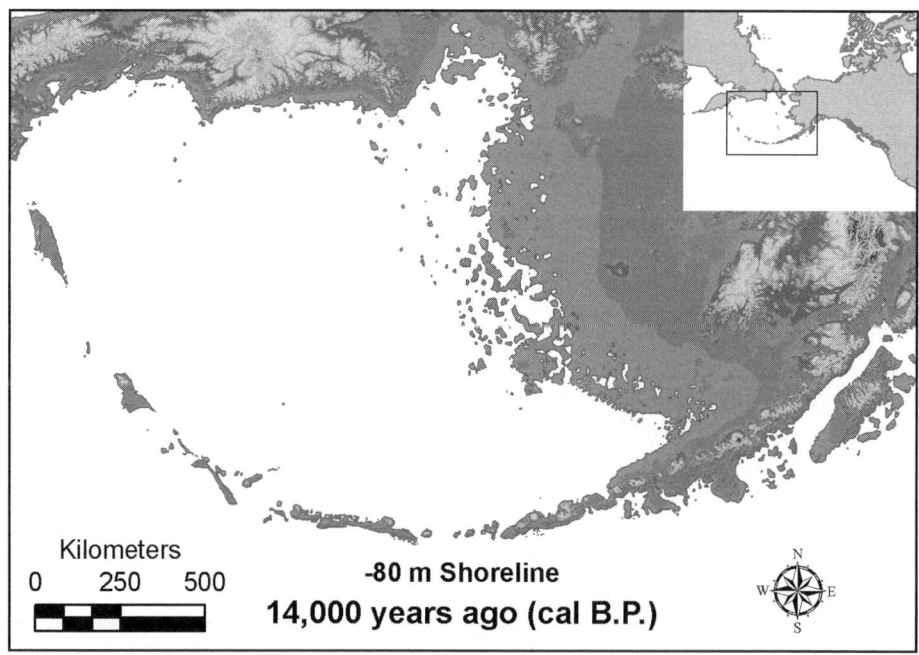

Figure 12-5. *The island archipelago of southern Beringia and the Aleutians at circa 14,000 cal yr B.P.*[6]

If southern Beringia's coasts and offshore islands were habitable, given the existence of precursor maritime populations in Japan and quite likely in the Russian Far East prior to 20,000 cal yr B.P., the island archipelagoes that existed on the southern coast of Beringia in the millennia prior to the flooding of the land bridge might well have been settled early on and could have served as an ideal habitat for human populations to exist, "incubate" and, indeed, thrive. The area, minimally, would have offered a navigable route into the Americas with numerous islands and bays along the way, obviating the need to make much if any use of the terrestrial resources of the Beringian continental landmass to the north and east.

Living in the potentially rich and diversified environments of the Beringian archipelago would have facilitated coastal migration and maritime adaptations, since watercraft would have been essential for survival in such a setting. As sea levels rose following the LGM, the location and extent of the islands in the archipelago and the shoreline of the larger Beringian landmass itself shifted dramatically over time (Figures 12-2 through 12-7), with new islands appearing as old ones were submerging (Brigham-Grette et al. 2004; Manley 2002). Living in such a habitat would have predisposed coastal migration, especially as sea levels rose and people were forced to move to other islands within the Beringian archipelago, and would have made feasible further movement to the south and west into the Aleutians and, ultimately, to the east and southeast along the northwest Pacific Coast.[7] Given the numerous closely spaced islands revealed by the bathymetric data, furthermore, early populations could have island hopped much of the way from northern Kamchatka to the eastern Aleutians and only rarely been out of sight of land or another island. Movement very far inland on the Beringian landmass proper may not have been necessary or even attempted, if the numerous coastal bays and offshore islands proved to be sufficiently attractive habitats to sustain a maritime fisher-forager way of life.

Unfortunately, likely locations along the southern Beringian coast and coastal archipelagoes where hypothetical early Amerind precursor populations could have incubated are now submerged beneath some of the most treacherous waters on the planet. Archaeological verification of human settlement in this area, while feasible, would be challenging (e.g., Fedje et al. 2004; Josenhans et al. 1997). Deepwater surveys for LGM and later Pleistocene archaeological sites and shorelines have been considered or are under way in several locations, such as the eastern Gulf of Mexico, off Baja California, along the Pacific Northwest coast, and on the Atlantic continental shelf of eastern North America, but even in these locations the logistical challenges are daunting (e.g., Adovasio and Hemmings 2008; Faught 2004; Josenhans et al. 1997; Stright 1986).[8]

Alternate or indeed several incubator locations may exist, of course, such as somewhere in eastern Asia, in Alaska–western Beringia, in the Pacific Northwest, or even perhaps south of the North American ice sheets.[9] LGM climate conditions appear to have been instrumental in creating the conditions necessary for population isolation and incubation. Subarctic eastern Siberia from around 55° to 65° N latitude was apparently colonized only after circa 30,000 cal yr B.P., and perhaps as late as circa 26,000 cal yr B.P. (Goebel 2004:319) and was largely depopulated after circa 22,000 cal yr B.P., during the LGM, "except perhaps in

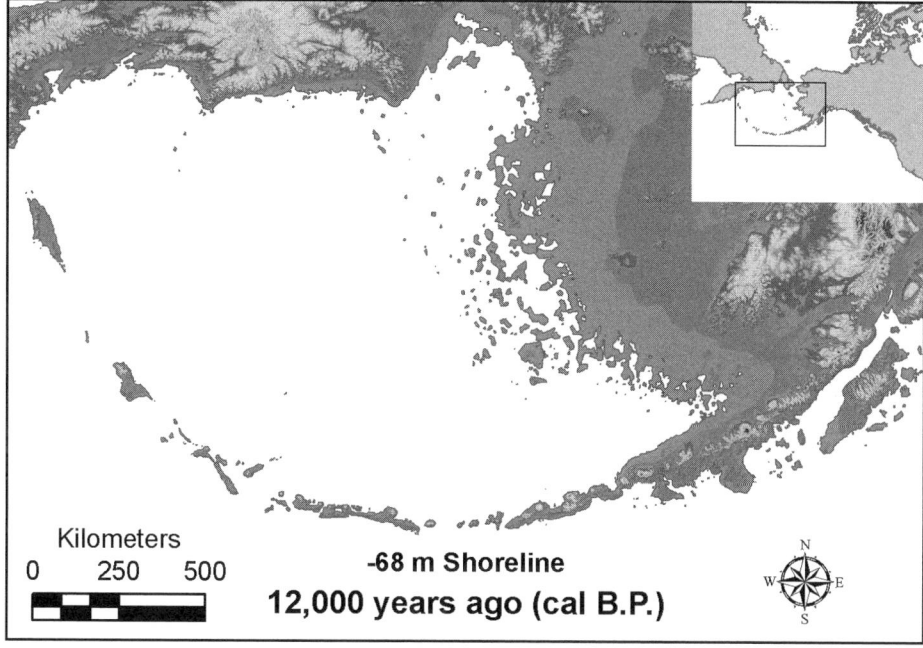

Figure 12-6. *The island archipelago of southern Beringia and the Aleutians at circa 12,000 cal yr B.P.*[6]

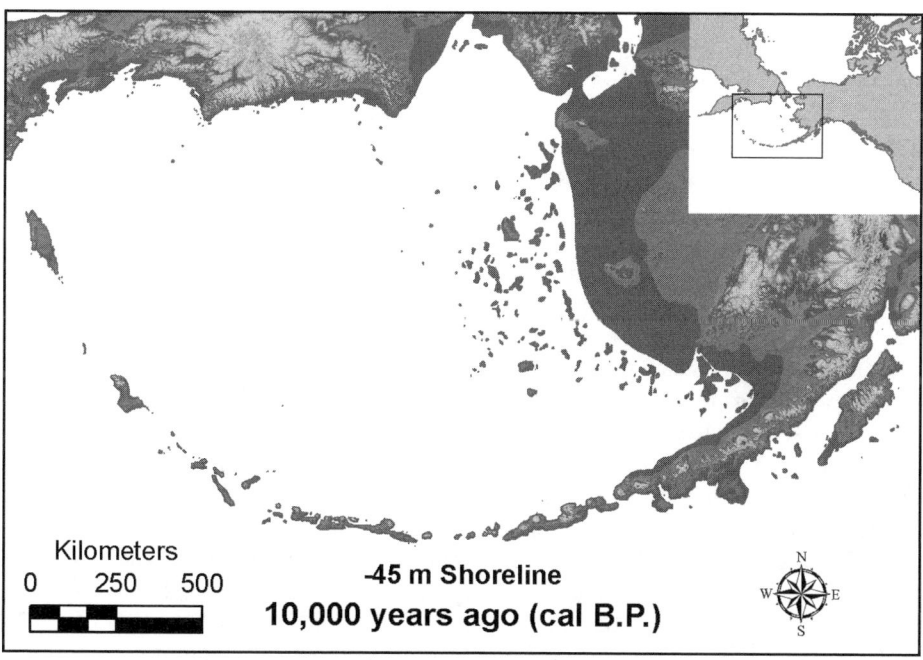

Figure 12-7. *The island archipelago of southern Beringia and the Aleutians at circa 10,000 cal yr B.P.* [6]

small refuges like the southern Yenisei or Transbaikal region" (Goebel 1999:222). The extreme cold not only likely resulted in Northeast Asian populations retreating south and eastward (Goebel 1999, 2004; Goebel and Slobodin 1999; Madsen 2004b:11) but would have also lowered sea levels (opening up new coastal habitats) as well as creating cold and ice-sheet barriers to movement. Whether human populations occupied coastal regions or the hypothesized interior refugia of Northeast Asia (and perhaps Alaska) during the millennia on either side of the LGM is unknown, but these are areas that, if settled, could have provided the time required of the BIM/Standstill model. The recolonization of northwestern Siberia after the LGM occurred fairly quickly, within a few thousand years, with human populations inferred to have reached Beringia by circa 14,000 cal yr B.P. (Goebel 1999:224; 2004:344; Goebel et al. 2003). Whether human populations reached or at least passed through Beringia earlier than this remains controversial, but movement along southern Beringia would appear likely given archaeological sites near the west coast of the Americas at this time, at Monte Verde, Chile, (Dillehay 1997, 2000) and possibly at Paisley Caves, Oregon, (cf. Gilbert et al. 2008; Goldberg et al. 2009; Poinar et al. 2009).

If the now-submerged coastal habitats in eastern Asia and southern Beringia were not where population isolation occurred, what would be the likely characteristics of other possible locations? First, they would have to be at some distance or separated by some barrier (i.e., ice sheets, mountains, ocean crossings) from the source populations, since the closer the incubating populations were, the greater the likelihood of gene flow back and forth, something that does not appear to have occurred (and why incubation in the Russian Far East or Japan appears improbable). Second, incubating groups likely stayed in environments they were familiar with, which if located in coastal settings may have precluded much likelihood of gene flow with their source populations if these were in the interior. But what about completely different possibilities? Could human populations have reached eastern Beringia or areas to the south prior to the LGM or been pushed there during the LGM as climate deteriorated? That is, could population refugia have been established earlier than currently thought in Alaska or in the Pacific Northwest, perhaps during the LGM or slightly earlier? Or could people have even made it south of the ice sheets prior to the LGM, as some archaeologists have controversially suggested (e.g., see summaries in Meltzer 2009; Stanford et al. 2005)? Admittedly, there is no convincing archaeological evidence in support of such inferences (Fedje et al. 2004; Goebel 2004; Goebel et al. 2008; Hamilton and Goebel 1999; Madsen 2004b), nor is it very likely that human populations could have existed for as long as 10,000 years in Alaska, in the Pacific Northwest, or south of the ice sheets without leaving a significant archaeological signature, given likely human population growth rates (Anderson and Gillam 2000; Bettinger and Young 2004; Fedje et al. 2004:132-135; Goebel 2004:354; Madsen 2004a:392–395). Resolving whether such scenarios are plausible will require far more research, and it is clear that archaeology—both on land in Northeast Asia and Alaska, as well as underwater along the submerged coasts and archipelagoes of eastern Asia, southern Beringia, and northwestern North America—will be essential to substantiating the argument for the BIM/Standstill model currently based on genetics.

The BIM/Standstill model, fortunately, does help reconcile some major problems and contradictions in existing arguments. For one, it provides more time for the linguistic diversity observed in the Americas to have developed. If Nichols (1990, 2002, 2008) is correct about the time it takes languages to diversify, than genes may not have been the only thing incubating: Languages might have been as well. The standstill also allows time for the so-called high-latitude cold or "germ filter" to develop, that is, to allow populations to shed at least some of their Old World disease burdens, albeit with tragic consequences millennia later once contact between the hemispheres was reestablished (Dillehay 1991; Stewart 1973:19–20). It also helps explain why archaeologists have so much trouble identifying New World progenitor technologies in eastern Asia and why such diversity in stone-tool industries exists across Beringia from circa 14,000 to 12,000 cal yr B.P., as is evident at the Denali, Nenana, and Dyuktai complexes and in the assemblages at Uski-1 and Ushki-5 in Kamchatka (Goebel 2004:353–356; Goebel and Slobodin 1999; Goebel et al. 2003). Extended isolation from the presumed eastern Asian source populations likely had an influence on other aspects of culture as well. Jomon pottery, for example, was being made in the Japanese archipelago and the Russian Far East at the same time that Clovis was radiating in North America (Kuzmin 2006; Lutaenko et al. 2007:362–364; Odai Yamamoto I Site Excavation Team 1999), but no evidence for ceramics has been found on New World Paleo-Indian sites. Finally the BIM/Standstill model gives ecogeographically linked human body characteristics more time to incubate, such as those presumably related to cold adaptation (e.g., Auerbach 2007; King, Chapter 10). Given what it helps explain, if the BIM/Standstill model had not been posited by geneticists, archaeologists might well have had to invent it themselves.

Conclusions: The Shape of Things to Come

Beyond valuable insights into particular research questions, the papers in this volume offer important lessons for researchers contemplating working with a wide array of data types. As Kemp and Schurr (Chapter 2) noted in their review of the history of mtDNA research, for example, a conscious attempt to simplify analyses led to a reduction in interpretive ability. Eliminating high-resolution sequencing precluded resolving variation within haplogroups, which was later shown to be critically important. Haplogroup X found in North American populations was initially assumed by some researchers to represent ancient European contacts with the Americas (e.g., Bradley and Stanford 2004; Stanford and Bradley 2002:265) yet, upon detailed analysis, was shown to be different from that found in western Europe (Derenko et al. 2001; Kemp and Schurr, Chapter 2). A similar situation was noted in skeletal studies (Durand et al., Chapter 5), where the analysis or consideration of more rather than fewer discrete dental traits was found to yield more satisfying analytical results. Data can always be removed from consideration, but it is better to have collected it in the first place whenever possible. The genetics example also shows that we don't always know what kinds of information might be important down the line. For the future, doing more research and data collection would appear to be far better than doing less.

Another important lesson from this volume is that bioanthropological analyses can tell us incredibly valuable things about the past and answer major questions of broad interest to scholars and the general public alike, but only if we have the samples to work with. The amount of well-preserved Late Pleistocene human DNA and skeletal material available for examination from the Americas and from eastern Asia is small, which is unfortunate given how much can be learned from these remains. We need to become better at finding early human remains and at working with descendant populations to ensure access to those that are found. With regard to the first point, two examples from the southeastern United States indicate human remains might not always be where we expect to find them and that we need to cast our searches wider. The Late Paleo-Indian Dalton culture Sloan site in Arkansas, dating to circa 12,000 cal yr B.P., was a cemetery located on a sand dune well away from known occupation sites (Morse 1997). While the associated artifacts—hypertrophic Dalton points and other tools—are fairly well known in the central Mississippi Valley, no other burial areas like Sloan have been found, probably because neither archaeologists nor collectors spend much time looking in places where they don't expect to find things.[10] The second unusual location in the Southeast where human remains were found was at the Windover site in Florida, a "subaqueous cemetery" with burials placed and in some cases staked down in a pond (Doran 2002). Fortunately, Windover is not unique, as several other similar submerged burial sites have been found in Florida, most dating from circa 10,000 to 7000 cal yr B.P. Of course, cultural practices in the past and present, such as the widespread use of cremation or scaffold burial, or the locations where archaeological research tends to occur can also skew sample availability. Well-preserved remains may just not exist from certain cultures, nor do archaeologists routinely look in some parts of the landscape, most notably in swampy terrain or water-saturated soils.

Finding ancient human remains is only part of the challenge; we must also be able to excavate and analyze them, which means we must work with and be considerate of the concerns of locally resident and presumed descendant populations (e.g., Fine-Dare 2002; Thomas 2000).[11] The papers in this volume provide excellent examples of what can be learned from the responsible study of human remains and can provide guidance in making the case for their examination when found. In the United States, collection and analysis protocols for presumed Native American human remains are shaped by the wording, implementation regulations, and legal decisions associated with the Native American Graves Protection and Repatriation Act (NAGPRA), as well as by state and local laws and regulations. If we are ever to understand the settlement of the Americas, cooperation with descendant populations must always guide our actions, and they must be the first peoples we inform when such remains are found. But we must also, as a last resort and when all else has failed, be willing to raise legal challenges when appropriate (cf. Kintigh 2004; Schneider 2004; Schneider and Bonnichsen 2005; Watkins 2004). In my opinion, human remains of Pleistocene age found in the Americas clearly meet such a threshold, that is, where both full consultation as well as thorough analysis should occur. Given the likelihood that Pleistocene age human remains uncovered in the Americas may have no living descendants or descendants near where the remains were found, genetic and other bioanthropologically informed analyses would appear to be the logical and appropriate first step. While analyses directed to recovering DNA or isotopic data require destructive analysis, the samples involved are small compared with the information gained, to the point of becoming truly microscopic in some cases. Furthermore, many analyses, such as of skeletal morphology or discrete dental traits, are entirely nondestructive, an important consideration in the event that DNA and other analyses are precluded (Durand et al., Chapter 5).

Above all, the papers in this volume have shown that bioanthropological analyses can provide effective evidence where traditional archaeological signatures of relationship (i.e., basketry, pottery, architecture, specific stone-tool forms) are rare, lacking, ambiguous, or contentious, as in the case of population movements associated with the hypothesized "Chaco Meridian" (Durand et al., Chapter 5; Lekson 1999, 2008). The best analyses, we have seen, focus on important questions, bring together information and analyses from a range of sources, and produce results that help advance the theoretical foundations and interpretations of the subject matter under investigation. Of course, as Auerbach (Chapter 7) also notes, the "correlation of biological variation with cultural identity, let alone history, is problematic" and our inferences and models still all too often suffer from an oversimplification that does not accurately reflect the complexity of past behavior. We are outgrowing these limitations, however, as we come to view different kinds of evidence as complementary sources of information and insight. While the integration of different disciplinary theoretical perspectives, data sets, and analyses is far from perfect (see commentaries by Auerbach, Chapter 1, and Sassaman, Chapter 13), the papers in this volume demonstrate how consideration of skeletal biology, genetics, and archaeology can synergistically arrive at more complex and compelling models of the colonization and postcolonization history of the Americas. Let us all continue to work together to make it so.

Acknowledgments

I wish to thank Benjamin M. Auerbach for asking me to participate in the commentary on the papers in this volume. While I would have liked to attend the SIU Visiting Scholar Conference this volume is based on, that weekend I was attending "Early Paleoindian Colonization of the North American Midcontinent," a workshop at the University of Illinois Urbana-Champaign. Thanks also to my fellow discussant, Kenneth E. Sassaman, for his thoughts on these subjects, as well as to Ben Auerbach, Ted Goebel, Richard Jantz, and two anonymous reviewers for their comments on the draft manuscript. The maps of topographic change in the Beringian area in the late Pleistocene employing the ETOPO1 database were produced by informatics/GIS guru Stephen R. Yerka of the Archaeological Research Laboratory at the University of Tennessee. His help is deeply appreciated. Mary Lou Wilshaw-Watts did an excellent job editing the manuscript, and I appreciate the time and effort she put into improving my prose. Any errors or omissions are, of course, the responsibility of the author.

Notes

1. The cormic index, or sitting height to total height (SH/H), is a measure of truck to leg length. The ratio is expected to be larger in cold-adapted and smaller in warm-adapted populations, following expectations for ecogeographic rules (see Jantz et al., Chapter 11).

2. Multidisciplinary research by the Baikal Archaeology Project, directed to documenting the Mid-Holocene record of hunter-gatherers of the Lake Baikal region, should help answer this question (e.g., Schurr et al. 2008; Weber et al. 2008).

3. While large bifaces are fairly common in northeastern Siberia, fluted points have not been found to date. The only example of such an artifact, a biface from the Uptar site in Siberia, has a "flute" that appears caused by impact damage, from the tip to the base, representing an accidental rather than an intentional production (King and Slobodin 1996; Meltzer 2009:189).

4. The Yana RHS site (ca. 28–25 rc yr B.P.) on the Yana River in Siberia near the Laptev Sea/Arctic Ocean—at 71° N latitude located well above the Arctic Circle and east of the Verkhoyansk Mountains in Siberia—is a possible exception (Meltzer 2009:189–190; Pitulko et al. 2004). Although located well to the west of extreme northeastern Asia, its presence demonstrates that people were able to live north of the Arctic Circle prior to the Last Glacial Maximum. The genetic differentiation postulated by the BIM/Standstill could have occurred among such northern populations, assuming they were indeed isolated from other groups, as might have happened during the LGM, and assuming they remained in the north instead of retreating southward. Where in the north they may have stayed is currently unknown, although herein I suggest the Beringian archipelago may be one possibility. Another, of course, would be Alaska, while a third possibility could be elsewhere in eastern Asia. A final possibility could be in the Americas themselves, south of the ice sheets, assuming a rapid movement from eastern Asia; there is little evidence to support this at present.

5. The inspiration for this idea comes from many sources that I have tried to acknowledge, but it is perhaps most explicitly stated in a paper by Fedje and his colleagues: "[W]e suggest that areas that were near-shore coastal archipelagoes during and immediately after the Last Glacial Maximum (LGM) might be key for early maritime-adapted peoples and therefore worthy of the most intensive investigation" (Fedje et al. 2004:135).

6. The island archipelagoes of southern Beringia and the Aleutians at various periods in the late Pleistocene and initial Holocene (based on sea-level data in Manley [2002] and Lambeck and colleagues [2002], bathymetric data from http://ngdc.noaa.gov/mgg/global/global.html [Amante and Eakins 2009], and a

mapping approach adapted from Manley [2002] and Brigham-Grette and associates [2004:38]).

7. Some human populations, of course, may have moved into the interior from these coastal habitats, colonizing central and northern Beringia, including interior Alaska. Such movement into the interior may have been prompted by fluctuating or rising sea levels, although if watercraft were present it may have been viewed as a last resort for these coastally adapted peoples.

8. As global sea levels rise in the decades and centuries to come, our profession should get better and better at examining underwater archaeological resources, assuming archaeology remains a priority in a civilization likely to be severely challenged by the flooding of so much densely occupied or farmed terrain (Anderson et al. 2007:15–18).

9. In his comments on this paper, Ted Goebel noted that central Alaska still remains the only area where late-glacial pre-Clovis sites have been found in eastern Beringia, like Swan Point and Broken Mammoth. He further suggested that, based on evidence for a surprisingly early opening of the Alaska Range passes into southern Alaska (Dyke 2004; England et al. 2006)—especially the pass connecting the upper Tanana and Copper River valleys—that the opening of an ice-free corridor along the Copper River drainage could have occurred prior to the opening of the western Canadian interior corridor. This would have given people a way out of the Alaskan interior to the coast at a very early date. That is, it is possible that the BIM/Standstill may have occurred in interior central Alaska, with people reaching the northern northwest coastal archipelagoes in the Gulf of Alaska region well before Clovis times, perhaps meeting groups moving into the area from the western Beringian and Aleutian archipelagoes or perhaps, if the first to reach the area, adopting their own maritime technology and moving onward on their own. Of course, and as Goebel also noted, a lot of archaeological and geological research will be needed to evaluate these arguments. Given the increasing numbers of apparent pre-Clovis sites in the Americas, it is even possible that a more southerly incubator may ultimately be considered feasible, rather than or in addition to one in central Alaska.

10. Unfortunately, "collectors" finding early cemetery sites like Sloan may be unlikely to report them, given the market value of the associated artifacts. Indeed, another unfortunate criterion influencing whether burials are likely to be preserved is whether they possess associated funerary objects of value to looters and traffickers in antiquities.

11. Determining who the descendants of a given sample or population are remains a major challenge that can often only be resolved with any degree of certainty by using physical anthropological evidence, such as discrete dental traits or genetic evidence. The 10,300 cal yr B.P. human remains found at On Your Knees Cave on Prince of Wales Island, Alaska, for example, were shown to represent a distinct and previously unknown haplotype of haplogroup D, whose closest genetic match in the Americas were the Cayapa of Ecuador (Kemp et al. 2007:616–617). The On Your Knees Cave analysis, coupled with the discovery of two circa 5,000-year-old individuals at the China Lake site in British Columbia possessing mtDNA haplogroup M, not found in modern New World populations

(Malhi et al. 2008), indicates that early settlers in the Americas were more genetically heterogeneous than once thought and that some of this early variation has been lost through the extinction of local groups at some time in the past. Some early populations in the New World, quite simply, may not have modern descendants, while others may have moved great distances from where their ancestors' remains occur.

References

Achilli, Alessandro, Ugo A. Perego, Claudio M. Bravi, Michael D. Coble, Qing-Peng Kong, Scott R. Woodward, Antonio Salas, Antonio Torroni, and Hans-Jürgen Bandelt
    2008   The Phylogeny of the Four Pan-American mtDNA Haplogroups: Implications for Evolutionary and Disease Studies. *PLoS ONE* 3(3):e1764. doi:10.1371/journal.pone.0001764.

Adovasio, James M., and C. Andrew Hemmings
    2008   The First Snowbirds: The Archaeology of Inundated Late Pleistocene Landscapes in the Northeastern Gulf of Mexico, NOAA Ocean Explorer. Paper available at http://oceanexplorer.noaa.gov/explorations/08negmexico/welcome.html.

Allen, Joel Asaph
    1877   The Influence of Physical Conditions in the Genesis of Species. *Radical Review* 1:108–140.

Amante, Christopher, and Barry W. Eakins
    2009   ETOPO1 1 Arc-Minute Global Relief Model: Procedures, Data Sources and Analysis. *NOAA Technical Memorandum NESDIS NGDC-24*, National Geophysical Data Center, Boulder, Colorado.

Anderson, David G.
    1990   The Paleoindian Colonization of Eastern North America: A View from the Southeastern United States. In *Early Paleoindian Economies of Eastern North America*, edited by Kenneth B. Tankersley and Barry L. Isaac, pp. 163–216. Research in Economic Anthropology, Supplement 5. JAI Press, Greenwich, Connecticut.
    1995   Paleoindian Interaction Networks in the Eastern Woodlands. In *Native American Interaction: Multiscalar Analyses and Interpretations in the Eastern Woodlands*, edited by Michael S. Nassaney and Kenneth E. Sassaman, pp. 1–26. University of Tennessee Press, Knoxville.
    1996   Chiefly Cycling and Large-Scale Abandonments as Viewed from the Savannah River Basin. In *Political Structure and Change in the Prehistoric Southeastern United States*, edited by John F. Scarry, pp. 150–191. Florida Museum of Natural History: Ripley P. Bullen Series, University Press of Florida, Gainesville.
    1999   Examining Chiefdoms in the Southeast: An Application of Multiscalar Analysis. In *Great Towns and Regional Polities in the Prehistoric American Southwest and Southeast*, edited by Jill E. Neitzel, pp. 215–241. Amerind Foundation New World Study Series 3, University of New Mexico Press, Albuquerque.

Anderson, David G., and J. Christopher Gillam
    2000   Paleoindian Colonization of the Americas: Implications from an Examination of Physiography, Demography, and Artifact Distribution. *American Antiquity* 65:43–66.

Anderson, David G., Kirk A. Maasch, Daniel H. Sandweiss, and Paul A. Mayewski
    2007   Climate and Culture Change: Exploring Holocene Transitions. In *Climate*

Change and Cultural Dynamics: A Global Perspective on Mid-Holocene Transitions, edited by David G. Anderson, Kirk A. Maasch, and Daniel H. Sandweiss, pp. 1–23. Academic Press, Amsterdam.

Anderson, David G., Stephen J. Yerka, and J. Christopher Gillam
    2010   Employing High Resolution Bathymetric Data to Infer Possible Migration Pathways of Late Pleistocene Groups. Manuscript under review. On file at Department of Anthropology, University of Tennessee, Knoxville.

Anthony, David W.
    1990   Migration in Archeology: The Baby and the Bathwater. *American Anthropologist* 92:895–914.

Auerbach, Benjamin M.
    2007   Human Skeletal Variation in the New World During the Holocene: Effects of Climate and Subsistence Across Geography and Time. Unpublished Ph.D. dissertation, Center for Functional Anatomy and Evolution, Johns Hopkins University, Baltimore.

Auerbach, Benjamin M., and Christopher B. Ruff
    2010   Stature Estimation Formulae for Indigenous North American Populations. *American Journal of Physical Anthropology* 141:190–207.

Bergmann, Karl
    1847   Ueber die verhaltnisse der warmeokonomie der thiere zu ihrer grosse. *Gottinger Stud* 3:595–708.

Bettinger, Robert L., and David A. Young
    2004   Hunter-Gatherer Population Expansion in North Asia and in the New World. In *Entering America: Northeast Asia and Beringia Before the Last Glacial Maximum*, edited by David B. Madsen, pp. 239–251. University of Utah Press, Salt Lake City.

Binford, Lewis R.
    1983   Long-Term Land Use Patterns: Some Implications for Archaeology. In *Lulu Linear Punctated: Essays in Honor of George Irving Quimby*, edited by Robert C. Dunnell and Donald K. Grayson, pp. 27–53. Anthropological Paper 72, University of Michigan, Museum of Anthropology, Ann Arbor.

Boas, Franz
    1910   *Report Presented to the 61st Congress on Changes in Bodily Form of Descendants of Immigrants*. U.S. Government Printing Office, Washington, D.C.
    1912   Changes in the Bodily Form of Descendants of Immigrants. *American Anthropologist* 14:530–562.

Brace, C. Loring
    1962   Cultural Factors in the Evolution of the Human Dentition. In *Culture and the Evolution of Man*, edited by M. F. Ashley Montagu, pp. 343–354. Oxford University Press, New York.

Bradley, Bruce A., and Dennis Stanford
    2004   The North Atlantic Ice-Edge Corridor: A Possible Palaeolithic Route to the New World. *World Archaeology* 36:459–478.

Brigham-Grette, Julie, Anatoly V. Lozhkin, Patricia M. Anderson, and Oyu Y. Glushkova
    2004   Paleoenvironmental Conditions in Western Beringia Before and During the Last Glacial Maximum. In *Entering America: Northeast Asia and Beringia Before the Last Glacial Maximum*, edited by David B. Madsen, pp. 29–61. University of Utah Press, Salt Lake City.

Cabana, Graciela S., Keith Hunley, and Frederika A. Kaestle
    2008   Population Continuity or Replacement? A Novel Computer Simulation Approach and Its Application to the Numic Expansion (Western Great Basin, USA). *American Journal of Physical Anthropology* 135:438–447.

Cameron, Catherine M.

    1995   Migration and the Movement of Southwestern Peoples. *Journal of Anthropological Archaeology* 14:104–124.

Cavalli-Sforza, L. Luca

    2003   Demic Diffusion as the Basic Process of Human Expansions. In *Examining the Farming/Language Dispersal Hypothesis*, edited by Peter Bellwood and Colin Renfrew, pp. 79–88. McDonald Institute for Archaeological Research, Cambridge, United Kingdom.

Clague, John J., Rolf W. Mathewes, and Thomas A. Ager

    2004   Environments of Northwestern North America Before the Last Glacial Maximum. In *Entering America: Northeast Asia and Beringia Before the Last Glacial Maximum*, edited by David B. Madsen, pp. 63–94. University of Utah Press, Salt Lake City.

Cohen, Mark N., and George J. Armelagos (editors)

    1984   *Paleopathology at the Origins of Agriculture*. Academic Press, Orlando, Florida.

Derenko, Miroslava V., Tomasz Grzybowski, Boris A. Malyarchuk, Jakub Czarny, Danuta Miscicka-Sliwka, and Ilia A. Zakharov

    2001   The Presence of Mitochondrial Haplogroup X in Altaians from South Siberia. *American Journal of Human Genetics* 69:237–241.

Diamond, Jared

    1987   The Worst Mistake in the History of the Human Race. *Discover* 8(5):64–66.

    1997   *Guns, Germs, and Steel: The Fates of Human Societies*. W. W. Norton, New York.

    2002   Evolution, Consequences and Future of Plant and Animal Domestication. *Nature* 418:703–707.

Dillehay, Tom D.

    1991   Disease Ecology and Initial Human Migration. In *The First Americans: Search and Research*, edited by Tom D. Dillehay and David J. Meltzer, pp. 231–264. CRC Press, Boca Raton, Florida.

    1997   *Monte Verde: A Late Pleistocene Settlement in Chile, The Archaeological Context and Interpretation*, Vol. 2. Smithsonian Institution, Washington, D.C.

    2000   *The Settlement of the Americas: A New Prehistory*. Basic Books, New York.

Dixon, E. James

    1999   *Boats, Bones & Bison: Archaeology and the First Colonization of Western North America*. University of New Mexico Press, Albuquerque.

Doran, Glen H. (editor)

    2002   *Windover: Multidisciplinary Investigations of an Early Archaic Florida Cemetery*. University Press of Florida, Gainesville.

Dye, David H.

    2009   *War Paths, Peace Paths: An Archaeology of Cooperation and Conflict in Native Eastern North America*. AltaMira Press, Lanham, Maryland.

Dyke, Arthur S.

    2004   An Outline of North American Deglaciation with Emphasis on Central and Northern Canada. In *Quaternary Glaciations-Extant and Chronology*, Pt. II, 2b, edited by Jürgen Ehlers and Philip L. Gibbard, pp. 373–424. Elsevier Science and Technology Books, Amsterdam.

England, John H., Nigel Atkinson, Jan Bednarski, Arthur S. Dyke, Douglas A. Hodgson, and Colm Ó Cofaigh

    2006   The Innuitian Ice Sheet: Configuration, Dynamics and Chronology. *Quaternary Science Reviews* 25:689–703.

Erlandson, Jon M.

    2002   Anatomically Modern Humans, Maritime Voyaging, and the Pleistocene Colonization of the Americas. In *The First Americans: The Pleistocene Colonization of the New World*, edited by Nina G. Jablonski, pp. 59–92. Memoirs of the California Academy of Sciences No. 27, Wattis Symposium Series in Anthropology, University of California Press, San Francisco.

Erlandson, Jon M., Michael H. Graham, Bruce J. Bourque, Debra Corbett, James A. Estes, and Robert S. Steneck

    2007   The Kelp Highway Hypothesis: Marine Ecology, the Coastal Migration Theory, and the Peopling of the Americas. *Journal of Island and Coastal Archaeology* 2:161–174.

Erlandson, Jon M., Madonna L. Moss, and Matthew Des Lauriers

    2008   Life on the Edge: Early Maritime Cultures of the Pacific Coast of North America. *Quaternary Science Review* 27:2232–2245.

Fagundes, Nelson J., Ricardo Kanitz, and Sandro L. Bonatto

    2008a   A Reevaluation of the Native American mtDNA Genome Diversity and Its Bearing on the Models of Early Colonization of Beringia. *PLoS ONE* 3(9):e3157. doi:10.1371/journal.pone.0003157.

Fagundes, Nelson J. R., Ricardo Kanitz, Roberta Eckert, Ana C. S. Valls,
Mauricio R. Bogo, Francisco M. Salzano, David Glenn Smith, Wilson A. Silva Jr.,
Marco A. Zago, Andrea K. Ribeiro-dos-Santos, Sidney E. B. Santos,
Maria Luiza Petzl-Erler, and Sandro L. Bonatto

    2008b   Mitochondrial Population Genomics Supports a Single Pre-Clovis Origin with a Coastal Route for the Peopling of the Americas. *American Journal of Human Genetics* 82:583–592.

Faught, Michael K.

    2004   The Underwater Archaeology of Paleolandscapes, Apalachee Bay, Florida. *American Antiquity* 69:235–249.

Fedje, Daryl W., Quentin Mackie, E. James Dixon, and Timothy H. Heaton

    2004   Late Wisconsin Environments and Archaeological Visibility on the Northern Northwest Coast. In *Entering America: Northeast Asia and Beringia Before the Last Glacial Maximum*, edited by David B. Madsen, pp. 97–138. University of Utah Press, Salt Lake City.

Fewkes, J. Walter, Aleš Hrdlička, William H. Dall, James W. Gidley, Austin Hobart Clark,
William H. Holmes, Alice C. Fletcher, Walter Hough, Stansbury Hagar, Paul Bartsch,
Alexander F. Chamberlain, and Roland B. Dixon

    1912   The Problems of the Unity or Plurality and the Probable Place of Origin of the American Aborigines. *American Anthropologist* 14:1–59.

Fine-Dare, Kathleen S.

    2002   *Grave Injustice: The American Indian Repatriation Movement and NAGPRA*. University of Nebraska Press, Lincoln.

Fitzpatrick, Andrew P.

    2003   The Amesbury Archer. *Current Archaeology* 184:146–152.

Fladmark, Knut

    1979   Routes: Alternate Migration Corridors for Early Man in North America. *American Antiquity* 44:55–69.

Forsyth, James

    1992   *A History of the Peoples of Siberia*. Cambridge: Cambridge University Press.

Gilbert, M. Thomas P., Dennis L. Jenkins, Anders Götherstrom, Nuria Naveran, Juan
J. Sanchez, Michael Hofreiter, Philip Francis Thomsen, Jonas Binladen, Thomas F. G.
Higham, Robert M. Yohe II, Robert Parr, Linda Scott Cummings, and Eska Willerslev

    2008   DNA from Pre-Clovis Human Coprolites in Oregon, North America. *Science* 320:786–789.

Gillam, J. Christopher, David G. Anderson, and A. Town Peterson
  2007   A Continental-Scale Perspective on the Peopling of the Americas: Modeling Geographic Distributions and Ecological Niches of Pleistocene Populations. *Current Research in the Pleistocene* 24:86–90.

Gillam, J. Christopher, and Andrei V. Tabarev
  2006   Geographic Information Systems and Predictive Modeling: Prospects for Far East Archaeology. In *Archaeological Elucidation of the Japanese Fundamental Culture in East Asia*, pp. 63–76. Kokugakuin University 21st Century COE Program, Archaeology Series, Vol. 7, Tokyo.

Goebel, Ted
  1999   Pleistocene Human Colonization of Siberia and Peopling of the Americas: An Ecological Approach. *Evolutionary Anthropology* 8:208–227.
  2004   The Search for a Clovis Progenitor in Subarctic Siberia. In *Entering America: Northeast Asia and Beringia Before the Last Glacial Maximum*, edited by David B. Madsen, pp. 311–356. University of Utah Press, Salt Lake City.

Goebel, Ted, and Sergei B. Slobodin
  1999   The Colonization of Western Beringia: Technology, Ecology, and Adaptations. In *Ice Age Peoples of North America: Environments, Origins, and Adaptations of the First Americans*, edited by Robson Bonnichsen and Karen L. Turnmire, pp. 104–155. Center for the Study of the First Americans, Corvallis, Oregon.

Goebel, Ted, Michael R. Waters, and Margarita Dikova
  2003   The Archaeology of Ushki Lake, Kamchatka, and the Pleistocene Peopling of the Americas. *Science* 301:501–505.

Goebel, Ted, Michael R. Waters, and Dennis H. O'Rourke
  2008   The Late Pleistocene Dispersal of Modern Humans in the Americas. *Science* 319:1497–1502.

Goldberg, Paul, Francesco Berna, and Richard I. Macphail
  2009   Comment on "DNA from Pre-Clovis Human Coprolites in Oregon, North America." *Science* 325:148.

González-José, Rolando, Maria C. Bortolini, Fabricio R. Santos, and Sandro L. Bonatto
  2008   The Peopling of America: Craniofacial Shape Variation on a Continental Scale and Its Interpretation from an Interdisciplinary View. *American Journal of Physical Anthropology* 137:175–187.

Gravlee, Clarence C., H. Russell Bernard, and William R. Leonard
  2003   Boas's Changes in Bodily Form: The Immigrant Study, Cranial Plasticity, and Boas's Physical Anthropology. *American Anthropologist* 105:326–332.

Greenberg, Joseph H., Christy G. Turner II, and Stephen L. Zegura
  1986   The Settlement of the Americas: A Comparison of the Linguistic, Dental, and Genetic Evidence. *Current Anthropology* 27:477–497.

Hamilton, Thomas D., and Ted Goebel
  1999   Late Pleistocene Peopling of Alaska. In *Ice Age Peoples of North America: Environments, Origins, and Adaptations of the First Americans*, edited by Robson Bonnichsen and Karen L. Turnmire, pp. 156–199. Center for the Study of the First Americans, Corvallis, Oregon.

Holliday, Trenton W.
  1997   Body Proportions in Late Pleistocene Europe and Modern Human Origins. *Journal of Human Evolution* 32:423–447.
  1999   Brachial and Crural Indices of European Late Upper Paleolithic and Mesolithic Humans. *Journal of Human Evolution* 36:549–566.

Hopkins, David M., John V. Matthews, Charles E. Schweger, and Steven B. Young
    1982   *Paleoecology of Beringia*. Academic Press, New York.
Ikawa-Smith, Fumiko
    2004   Humans Along the Pacific Margin of Northeast Asia Before the Last Glacial Maximum. In *Entering America: Northeast Asia and Beringia Before the Last Glacial Maximum*, edited by David B. Madsen, pp. 285–309. University of Utah Press, Salt Lake City.
James, Frances C.
    1970   Geographic Size Variation in Birds and Its Relationship to Climate. *Ecology* 51:365–390.
Jantz, Richard L.
    1995   Franz Boas and Native American Biological Variability. *Human Biology* 67:345–353.
    2003   The Anthropometric Legacy of Franz Boas. *Economics and Human Biology* 1:277–284.
Jantz, Richard L., and Douglas W. Owsley
    2001   Variation Among Early North American Crania. *American Journal of Physical Anthropology* 114:146–155.
    2005   Circumpacific Populations and the Peopling of the New World: Evidence from Cranial Morphometrics. In *Paleoamerican Origins: Beyond Clovis*, edited by Robson Bonnichsen, Bradley T. Lepper, Dennis Stanford, and Michael R. Waters, pp. 185–193. Center for the Study of the First Americans, Texas A&M University Press, College Station.
Josenhans, Heiner, Daryl Fedje, Reinhard Plenitz, and John Southon
    1997   Early Humans and Rapidly Changing Holocene Sea Levels in the Queen Charlotte Islands–Hecate Strait, British Columbia, Canada. *Science* 277:71–74.
Karafet, Tatiana, Stephen L. Zegura, Jennifer Vuturo-Brady, Olga Posukh, Ludmila Osipova, Victor Wiebe, Francine Romero, Jeffrey C. Long, Shinji Harihara, Feng Jin, Bumbein Dashyam, Tudevdagva Gerelsaikhan, Keiichi Omoto, and Michael F. Hammer
    1997   Y Chromosome Markers and Trans-Bering Strait Dispersals. *American Journal of Physical Anthropology* 102:301–314.
Kelly, Robert L.
    1995   *The Foraging Spectrum: Diversity in Hunter-Gatherer Lifeways*. Smithsonian Institution, Washington, D.C.
Kemp, Brian M., Ripan S. Malhi, John McDonough, Deborah A. Bolnick, Jason A. Eshleman, Olga Rickards, Cristina Martinez-Labarga, John R. Johnson, Joesph G. Lorenz, E. James Dixon, Terence E. Fifield, Timothy H. Heaton, Rosita Worl, and David Glenn Smith
    2007   Genetic Analysis of Early Holocene Skeletal Remains from Alaska and Its Implication for the Timing of the Peopling of the Americas. *American Journal of Physical Anthropology* 132:605–621.
King, Maureen L., and Sergei B. Slobodin
    1996   A Fluted Point from the Uptar Site, Northeastern Siberia. *Science* 273:634–636.
Kintigh, Keith W.
    2004   A Delicate Balance: The Society for American Archaeology and the Development of a National Repatriation Policy. In *New Perspectives on the First Americans*, edited by Bradley T. Lepper and Robson Bonnichsen, pp. 193–196. Center for the Study of the First Americans, Texas A&M University Press, College Station.

Kitchen, Andrew, Michael M. Miyamoto, and Connie J. Mulligan
  2008   A Three-Stage Colonization Model for the Peopling of the Americas. *PLoS ONE* 3(2):e1596. doi:10.1371/journal.pone.0001596.
Kuzmin, Yaroslav V.
  2006   Chronology of the Earliest Pottery in East Asia: Progress and Pitfalls. *Antiquity* 80:362–371.
Lambeck, Kurt, Yusuke Yokoyama, and Tony Purcell
  2002   Into and Out of the Last Glacial Maximum: Sea-Level Change During Oxygen Isotope Stages 3 and 2. *Quaternary Science Review* 21:343–360.
Lekson, Stephen H.
  1999   *The Chaco Meridian: Centers of Political Power in the Ancient Southwest*. AltaMira Press, Walnut Creek, California.
  2008   *A History of the Ancient Southwest*. School for Advanced Research Press, Santa Fe, New Mexico.
Lutaenko, Konstantin A., Irina S. Zhushchikhovskaya, Yuri A. Mikishin, and
Alexander N. Popov
  2007   Mid-Holocene Climatic Changes and Cultural Dynamics in the Basin of the Sea of Japan and Adjacent Areas. In *Climate Change and Cultural Dynamics: A Global Perspective on Mid-Holocene Transitions*, edited by David G. Anderson, Kirk A. Maasch, and Daniel H. Sandweiss, pp. 331–406. Academic Press/Elsevier, Amsterdam.
Madsen, David B.
  2004a   Recapitulation: The Relative Probabilities of Late Pre-LGM or Early Post-LGM Ages for the Initial Occupation of the Americas. In *Entering America: Northeast Asia and Beringia Before the Last Glacial Maximum*, edited by David B. Madsen, pp. 389–396. University of Utah Press, Salt Lake City.
  2004b   Colonization of the Americas Before the Last Glacial Maximum: Issues and Problems. In *Entering America: Northeast Asia and Beringia Before the Last Glacial Maximum*, edited by David B. Madsen, pp. 1–26. University of Utah Press, Salt Lake City.
Malhi, Ripan S., Jason A. Eshleman, Jonathan A. Greenberg, Deborah A. Weiss,
Beth A. Schultz Shook, Frederika A. Kaestle, Joseph G. Lorenz, Brian M. Kemp,
John R. Johnson, and David G. Smith
  2002   The Structure of Diversity Within the New World Mitochondrial DNA Haplogroups: Implications for the Prehistory of North America. *American Journal of Human Genetics* 70:905–919.
Malhi, Ripan S., Brian M. Kemp, Jason A. Eshleman, Jerome Cybulski,
David Glenn Smith, Scott Cousins, and Harold Harry
  2008   Haplogroup M Discovered in Prehistoric North America. *Journal of Archaeological Science* 34:642–648.
Mandryk, Carole A. S., Heiner Josenhans, Daryl W. Fedje, and Rolf W. Mathewes
  2001   Late Quaternary Paleoenvironments of Northwestern North America: Implications for Inland Versus Coastal Migration Routes. *Quaternary Science Review* 20:301–314.
Manley, William F.
  2002   Postglacial Flooding of the Bering Land Bridge: A Geospatial Animation. IN-STAAR, University of Colorado, http://instaar.colorado.edu/qgisl/bering_land_bridge/, accessed August 2, 2010.
Mayr, Ernst
  1956   Geographical Character Gradients and Climactic Adaptation. *Evolution* 10:105–108.

Meltzer, David J.

1989   Why Don't We Know When the First People Came to North America? *American Antiquity* 54:471–490.

2003   Peopling of North America. *Development in Quaternary Science* 1:539–563.

2004   Modeling the Initial Colonization of the Americas Issues of Scale, Demography, and Landscape Learning. In *The Settlement of the American Continents: A Multidisciplinary Approach to Human Biogeography*, edited by C. Michael Barton, Geoffrey A. Clark, David R. Yesner, and Georges A. Pearson, pp. 123–137. University of Arizona Press, Tucson.

2009   *First Peoples in a New World: Colonizing Ice Age America*. University of California Press, Berkeley.

Mensforth, Robert P.

2007   Human Trophy Taking During the Archaic Period: The Relationship to Warfare and Cultural Complexity. In *The Taking and Displaying of Body Parts by Amerindians*, edited by Richard J. Chacon and David H. Dye, pp. 222–277. Springer Science and Business Media, New York.

Milner, George R.

1999   Warfare in Prehistoric and Early Historic Eastern North America. *Journal of Archaeological Research* 7:105–151.

2004   *The Moundbuilders: Ancient Peoples of Eastern North America*. Thames and Hudson, London.

Morse, Dan F.

1997   *Sloan: A Paleoindian Dalton Cemetery in Arkansas*. Smithsonian Institution, Washington, D.C.

Nettle, Daniel

1999   Linguistic Diversity of the Americas Can Be Reconciled with a Recent Colonization. *Proceedings of the National Academy of Sciences of the United States of America* 96:3325–3329.

Nichols, Johanna

1990   Linguistic Diversity and the First Settlement of the New World. *Language* 66:475–521.

2002   First American Languages. In *The First Americans: The Pleistocene Colonization of the New World*, edited by Nina G. Jablonski, pp. 273–293. Memoirs of the California Academy of Sciences, No. 27, Wattis Symposium Series in Anthroplogy, University of California Press, San Francisco.

2008   Language Spread Rates and Prehistoric American Migration Rates. *Current Anthropology* 49:1109–1118.

Odai Yamamoto I Site Excavation Team (editors)

1999   *Archaeological Research at the Odai Yamamoto I Site*. Kokugakuin University. Tokyo.

Owsley, Douglas W., and Richard L. Jantz

2005   Nearsightedness in Paleoamerican Research: Historical Perspective and Contemporary Analysis. In *Paleoamerican Origins: Beyond Clovis*, edited by Robson Bonnichsen, Bradley T. Lepper, Dennis Stanford, and Michael R. Waters, pp. 289–294. Center for the Study of the First Americans, Texas A&M University Press, College Station.

Pauketat, Timothy R.

2007   *Chiefdoms and Other Archaeological Delusions*. AltaMira Press, Lanham, Maryland.

Pitulko, Vladimir V., Pavel A. Nikolsky, Evgeny Yu. Girya, Alexander E. Basilyan,
Vladimir E. Tumskoy, S. A. Koulakov, S. N. Astakhov, Elena Yu. Pavlova, and
Mikhail A. Anisimov
    2004   The Yana RHS Site: Humans in the Arctic Before the Last Glacial Maximum. *Science* 303:52–56.

Poinar, Hendrik, Stuart Fiedel, Christine E. King, Alison M. Devault, Kirsti Bos,
Melanie Kuch, and Regis Debruyne
    2009   Comment on "DNA from Pre-Clovis Human Coprolites in Oregon, North America." *Science* 325:148.

Powell, Joseph F.
    2005   *The First Americans: Race, Evolution and the Origins of Native Americans*. University of New Mexico Press, Albuquerque.

Relethford, John H.
    2004   Boas and Beyond: Migration and Craniometric Variation. *American Journal of Human Biology* 16:379–386.

Richards, Julian
    2007   *Stonehenge, the Story so Far*. English Heritage, Swindon.

Ruff, Christopher B.
    2002   Variation in Human Body Size and Shape. *Annual Review of Anthropology* 31:211–232.

Sancetta, Constance, Linda Heusser, Laurent Labeyrie, A. Sathy-Naidu, and
Stephen W. Robinson
    1985   Wisconsin-Holocene Paleoenvironment of the Bering Sea: Evidence from Diatoms, Pollen, Oxygen Isotopes, and Clay Minerals. *Marine Geology* 62:55–68.

Sardi, Marina L., Paula S. Novellino, and Héctor M. Pucciarelli
    2006   Craniofacial Morphology in the Argentine Center-West: Consequences of the Transition to Food Production. *American Journal of Physical Anthropology* 130:333–343.

Sassaman, Kenneth E.
    2010   *The Eastern Archaic, Historicized*. AltaMira Press, Lanham, Maryland.

Schneider, Alan L.
    2004   Public Policy and Prehistory. In *New Perspectives on the First Americans*, edited by Bradley T. Lepper and Robson Bonnichsen, pp. 197–202. Center for the Study of the First Americans, Texas A&M University Press, College Station.

Schneider, Alan L., and Robson Bonnichsen
    2005   Where Are We Going? Public Policy and Science. In *Paleoamerican Origins: Beyond Clovis*, edited by Robson Bonnichsen, Bradley T. Lepper, Dennis Stanford, and Michael R. Waters, pp. 297–312. Center for the Study of the First Americans, Texas A&M University Press, College Station.

Schurr, Theodore, Ludmilla P. Osipova, Sergey I. Zhadanov, Matthew C. Dulik,
Omer Gokcumen, Athma Pai, and Damian Labuda
    2008   Genetic Diversity in Contemporary Indigenous Populations of Siberia and the Americas: Implications for the Prehistoric Settlement of the Cis-Baikal Region. In *Prehistoric Hunter-Gatherers of the Baikal Region, Siberia: Bioarchaeological Studies of Past Lifeways*, edited by Andrzej Weber, M. Anne Katzenberg, and Theodore G. Schurr. University of Pennsylvania Press, Philadelphia.

Smith, Maria O.
    1996   Bioarchaeological Inquiry into Archaic Period Populations of the Southeast: Trauma and Occupational Stress. In *Archaeology of the Mid-Holocene Southeast*, edited by Kenneth E. Sassaman and David G. Anderson, pp. 134–154. University of Florida Press, Gainesville.

Snow, Dean R.

1995   Migration in Prehistory: The Northern Iroquoian Case. *American Antiquity* 60:59–79.

2009   The Multidisciplinary Study of Human Migration: Problems and Principles. In *Ancient Human Migrations: A Multidisciplinary Approach*, edited by Peter N. Peregrine, Ilia Peiros, and Marcus Feldman, pp. 6–20. Santa Fe Institute. University of Utah Press, Salt Lake City.

Sparks, Corey S., and Richard L. Jantz

2002   A Reassessment of Human Cranial Plasticity: Boas Revisited. *Proceedings of the National Academy of Sciences of the United States of America* 99:14636–14639.

2003   Changing Times, Changing Faces: Franz Boas' Immigrant Study in Modern Perspective. *American Anthropologist* 105:333–337.

Stanford, Dennis, Robson Bonnichsen, Betty J. Meggers, and D. Gentry Steele

2005   Paleoamerican Origins: Models, Evidence, and Future Directions. In *Paleoamerican Origins: Beyond Clovis*, edited by Robson Bonnichsen, Bradley T. Lepper, Dennis Stanford, and Michael R. Waters, pp. 313–353. Center for the Study of the First Americans, Texas A&M University Press, College Station.

Stanford, Dennis, and Bruce Bradley

2002   Ocean Trails and Prairie Paths? Thoughts about Clovis Origins. In *The First Americans: The Pleistocene Colonization of the New World*, edited by Nina G. Jablonski, pp. 255–271. Memoirs of the California Academy of Sciences No. 27, Wattis Symposium Series in Anthropology, University of California, San Francisco.

Stewart, T. Dale

1973   *The People of America*. Scribner, New York.

Straus, Lawrence G., David J. Meltzer, and Ted Goebel

2005   Ice Age Atlantis? Exploring the Solutrean-Clovis 'Connection.' *World Archaeology* 37:507–532.

Stright, Melanie J.

1986   Human Occupation of the Continental Shelf During the Late Pleistocene/Early Holocene: Methods for Site Location. *Geoarchaeology* 1:347–364.

Tamm, Erika, Toomas Kivisild, Maere Reidla, Mait Metspalu, David Glenn Smith, Connie J. Mulligan, Claudio M. Bravi, Olga Rickards, Cristina Martinez-Labarga, Elsa K. Khusnutdinova, Sardana A. Fedorova, Maria V. Golubenko, Vadim A. Stepanov, Marina A. Gubina, Sergey I. Zhadanov, Ludmila P. Osipova, Larisa Damba, Mikhail I. Voevoda, Jose E. Dipierri, Richard Villems, and Ripan S. Malhi

2007   Beringian Standstill and Spread of Native American Founders. *PLoS ONE* 2(9):e829. doi:10.1371/journal.pone.0000829.

Thomas, David H.

2000   *Skull Wars: Kennewick Man, Archaeology, and the Battle for Native American Identity*. Basic Books, New York.

Torroni, Antonio, Theodore G. Schurr, Margaret F. Cabell, Michael D. Brown, James V. Neel, Merethe Larsen, David G. Smith, Carlos M. Vullo, and Douglas C. Wallace

1993   Asian Affinities and Continental Radiation of the Four Founding Native American mtDNAs. *American Journal of Human Genetics* 53:563–590.

Trinkaus, Erik

1981   Neandertal Limb Proportions and Cold Adaptation. In *Aspects of Human Evolution*, Society for the Study of Human Biology, Vol. 21, edited by Christopher B. Stringer, pp. 187–224. Taylor and Francis, London.

U.S. Department of Commerce

2006   National Oceanic and Atmospheric Administration, National Geophysical

Data Center. *2-minute Gridded Global Relief Data (ETOPO2v2),* http://www.ngdc.noaa.gov/mgg/fliers/06mgg01.html, accessed August 2, 2010.

van Vark, Gerrit N., Don Kuizenga, and Frank L'Engle Williams
    2003   Kennewick and Luzia: Lessons from the European Upper Paleolithic. *American Journal of Physical Anthropology* 121:181–184.

Watkins, Joe
    2004   Public Policy and Native Americans: Where Do We Go from Here? In *New Perspectives on the First Americans,* edited by Bradley T. Lepper and Robson Bonnichsen, pp. 203–207. Center for the Study of the First Americans, Texas A&M University Press, College Station.

Weber, Andrzej, M. Anne Katzenberg, and Theodore G. Schurr (editors)
    2008   *Prehistoric Hunter-Gatherers of the Baikal Region, Siberia: Bioarchaeological Studies of Past Lifeways.* University of Pennsylvania Press, Philadelphia.

Weinstein, Karen J.
    2005   Body Proportions in Ancient Andeans from High and Low Altitudes. *American Journal of Physical Anthropology* 128:569–585.

Wobst, H. Martin
    1974   Boundary Conditions for Paleolithic Social Systems: A Simulation Approach. *American Antiquity* 39:147–178.
    1976   Locational Relationships in Paleolithic Societies. *Journal of Human Evolution* 5:49–58.

13. Bridging the Empirical Divide of Human Variation Research in the New World

Kenneth E. Sassaman

The empirical bases of research in genetics, skeletal morphology, and archaeology inform alternative readings of human diversity in the New World. The recent burgeoning of data in molecular genetics has been especially influential in revealing contradictions between the biological and cultural dimensions of New World variation. Predictably, new data have eroded the credibility of extant knowledge claims, causing some to rally the troops against new thinking—that is, remain dutifully skeptical—and others to simply resign orthodoxy to the wave of new science. While no specialist, molecular or otherwise, truly believes that his or her data are unassailable, and thus more "truthful" than that of others, efforts to compare and contrast, possibly even to reconcile divergent data sets are not terribly common.

This volume arose from an honest attempt to address this shortcoming. In this commentary, I assess the fruits of this effort from the vantage point of Americanist archaeology. My empirical basis for this assessment is the material record of human activity in the New World since its inception at least 14,000 years ago. This record bears only passing attention in some of the papers of this volume, those whose defining data sets consist primarily of biological attributes. Importantly, archaeological data have potential for reconciling some of the contradictions arising from research on genotypic and phenotypic human variation, arguably the signature feature of this volume. Still, like my colleagues in genetics and skeletal biology, I cannot privilege archaeological data over other data sets for at least two reasons: (1) genes, biology, and culture interact in ways so pervasive and complex as to preclude the unassailability of reductionist claims; and (2) despite the inter-

Human Variation in the Americas: The Integration of Archaeology and Biological Anthropology, edited by Benjamin M. Auerbach. Center for Archaeological Investigations, Occasional Paper No. 38. © 2010 by the Board of Trustees, Southern Illinois University. All rights reserved. ISBN 978-0-88104-095-1.

dependent qualities of genes, biology, and culture, they also vary independently to produce overlapping, nonlinear patterns of variation. Thus, my commentary on the chapters of this volume is intended to encourage continued dialogue among diverse investigators. Reaching consensus is not the point unless we all agree at the outset that the goal is not to seek "truth," however that is imagined, but rather to allow our respective data sets to be drawn into comparisons for the purpose of critically evaluating underlying logics and founding assumptions. This "dialogic" process ensures that extant and emerging new data sets of human variation in the New World remain situated in a network of alternative interpretations.

To anticipate my overall assessment of this volume, I must conclude that the dominant knowledge claims of geneticists and morphologists are a bit at odds (see also Goebel et al. 2008). Examining the DNA of modern Native Americans, geneticists advance the argument that founding populations of the New World came from a single source in Asia, which "incubated" in Beringia or elsewhere for several millennia, and then made their way into the Western Hemisphere to beget most, if not all, descendant indigenous populations. As Kemp and Schurr note in their contribution to this volume (Chapter 2), at least one subsequent immigration of people into the New World is evident, but apparently from the same source population(s). Genetic variation among extant populations is attributed to a history of genetic drift against an overall backdrop of marked homozygosity. In contrast, some analysts of skeletal morphology see at least two distinct populations in ancient America: Paleoamericans and Amerinds. Differences between the two are attributed by some analysts to microevolution since colonization, while others envision multiple founding migrations from distinct source populations (see Anderson's commentary, Chapter 12, for a review of alternative perspectives). Alternative readings of genetic or morphological data turn on the temporal and spatial scales in question, as several chapters of this volume exemplify.[1] On balance, and at multiple scales, I would have to conclude that archaeological data map onto morphological data better than onto genetic data. It is my intent in this assessment to bring some of these observations to bear in the context of a dialogue that is generally lacking archaeological enunciation.

Causes, Units, and Scales of Variation

Before delving into the specific contributions of this volume, it may be useful to consider aspects of epistemology that help us differentiate among the analytical logics of genetics, morphology, and archaeology. As implied by the double-ended arrows of Figure 13-1, I do not consider these realms of inquiry to be categorically different, for indeed they coexist within Western ontologies of empiricism, rationality, and objectivity. Likewise, genotypic, phenotypic, and cultural dimensions of human existence interact in ways that are nonlinear, nonisomorphic, and nondeterministic, but nonetheless mutually constitutive, as in the panarchical qualities of complex adaptive systems (Gunderson and Holling 2001). Thus, at the risk of oversimplifying the analytical logics of genetics, morphology, and archaeology, I aim to summarize below aspects of theory and

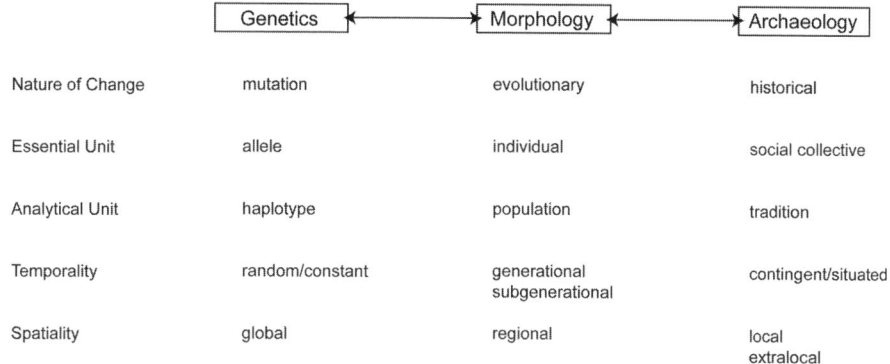

	Genetics	Morphology	Archaeology
Nature of Change	mutation	evolutionary	historical
Essential Unit	allele	individual	social collective
Analytical Unit	haplotype	population	tradition
Temporality	random/constant	generational subgenerational	contingent/situated
Spatiality	global	regional	local extralocal

Figure 13-1. *Epistemological contrasts in the causes, units, and scales of genetics, morphology, and archaeology.*

method that inform units and scales of analysis involved. The intent in Figure 13-1 and the paragraphs to follow is to draw contrasts among these areas of inquiry in an effort to understand how they lead to divergent interpretations of ancient human variation. I admit that some of these contrasts are artifacts of earlier modes of reason and have since been eroded, or at least softened, by more recent research. Nonetheless, these epistemological differences represent the historical context for specialties that continue to produce alternative readings of data on human variation.

Molecular genetics is a relatively recent approach to investigating human variation and, despite its technical complexity, it is informed by logic that is straightforward and elegant. No body of data on human variation is as structured by independent "natural" forces as is genetics. The same rules governing the genesis and transfer of molecular matter that is the focus of "peopling" research (i.e., mtDNA, Y chromosome) apply to all sexually reproducing organisms. In contrast, not all organisms express the same level of phenotypic plasticity that leads to different morphology (and function) under different conditions (both microevolutionary and developmental), and no nonhuman organism, as far as we know, is a full-fledged "cultural" being (the essence of which is self-awareness vis-à-vis others, hence consciousness that is patently social and semiotic). For reasons that are hard to dispute, geneticists focus on units of analysis—alleles and haplotypes—that are indicative of biological descent but are not subject to forces of evolution that inflect patterns of survivorship and successful mating. In other words, the genetic markers most commonly used in research on "peopling" and other modes of population restructuring are chosen by design to provide a definitive means to reconstruct biological ancestry. The ultimate causes of new molecular variation in these cases are point mutations, distinct markers of emergent new genetic matter that register descent in a neutral (nonselective) fashion. Of course, variation in the frequency of alleles is subject to all manner of demographic and historical processes that determine the distribution of people (breeding populations, in this case) across space. Ultimately, this is where genetics becomes archaeologically relevant: at the level of population history.

Patterns of descent from comparisons of genetic markers among living people are both mapable and datable, given certain assumptions. Among the more fundamental assumptions is that the rate of mutation has been more-or-less constant over time or at least can be treated as an average rate of random events over protracted time. The average rate of mutation then serves as a basis for inferring the amount of time necessary to account for divergences in the allele frequencies between two units of comparison. The "molecular clock," as it has come to be known, gives tempo to biological process, enabling inferences about the absolute duration of events, such as mass migration, that account for major restructuring of human populations.

With time comes space for a species that has a penchant for exploring and colonizing. Indeed, the human story is truly global and molecular genetics provides a powerful account of origins, migrations, coalescence, and divergence. At a global scale of observation, the relative degree of genetic similarity maps onto a chronological scale that has been largely verified with paleontological and archaeological data. At lesser scales of observation, the contours of genetic variation begin to reveal smaller movements of people and barriers to interbreeding that do not always agree with the timing and content of morphological and archaeological data. This is not to say that genetics data have no utility beyond the global only that it is best expressed, perhaps uniquely efficacious, at large scales of time and space.

The biggest difference between the logics of genetics and those of morphological and archaeological inquiry are that the latter two involve relationships between organisms and their environments. With this added dimension we have to consider realms of variation that are at once independent and mutually constitutive. It stands to reason that patterned variation in morphological and archaeological data is not reducible to single-cause phenomena nor can it be interpreted without contextual detail. The processes of interaction include both evolutionary and historical factors, but in general those of morphology refer to evolutionary forces, while those of culture are uniquely historical, as I elaborate below.

The essential unit of analysis in morphology is the individual organism, and its analytical unit of comparison is the population (Figure 13-1). Studies of morphology have a long, somewhat checkered past, with a period of gestation that contributed directly to the scientific definition of "race." In reality, morphology is a complex mix of biology and environment that, unlike alleles, cannot be classified as discrete units. Instead, as phenotypic variation, attributes commonly studied by morphologists embody continuous variation, much of it a product of and a precedent for microevolution, the stuff of adaptation. Tempo-wise, microevolutionary forces such as natural selection operate at the scale of reproductive generations. But also at play are subgenerational events, such as migration (gene flow, conditions of drift) and, of course, the complications of plasticity, much of it precipitated by "cultural" choices that vary free of physical environment (e.g., cradle-boarding, foot-binding).

The spatiality of morphology reflects the contours of natural and cultural processes that occur at scales no smaller than the region. Latitudinal and elevational gradients are good measures of the natural units, while polities, macrobands, and interaction networks exemplify some of the historical dimensions of relevance. Notably, these two spatial realms are not necessarily isomorphic (cf. the Kulturkreis school of the early twentieth century); indeed, a sociohistorical structure, such as a regional exchange (and mating) network, may crosscut a range of environmental variation and emerge, whether intended or not, in a "system" for buffering the risk of local survival, for example.

Aside from variations attending microevolutionary processes, morphology in humans responds developmentally to conditions of actual experience. Of particular interest are those that are imposed by the will of cultural beings. Schmidt and colleagues (Chapter 8) show us how this plays out in perimortem fashion in the Archaic Midwest, and one can cite examples of other cultural practices, even institutions, that contributed to secular and developmental change in the human form. This is where archaeology contributes something other than a means to date and locate sites of human activity or to avail sites to biological data mining.

To think of archaeology as the study of historical process (through materiality) is more than saying it is concerned with long-term change. Arguably, evolution is a historical process when situated in context (Vayda 1995a, 1995b), and the patterns of molecular variation showcased in this volume entailed specific events of migration and gene flow over long periods of time. Thus, it is not merely time or chronicle that makes something historical in a cultural sense but rather the ongoing process of reproducing culture through social practice. This process is historical in that it entails a consciousness about the past that informs and motivates action in the present. Simply put, to say that human action or practice is constrained by structure is to say that it is constrained by history, by what came before. Importantly, however, what humans actually experience and how they regard it in the cultural construction of a "past" have room to vary independently. In other words, humans invent or make tradition (Hobsbawm and Ranger 1983; Pauketat 2001), even if, as Marx (1963 [1852]) suggested, in terms they themselves do not decide. Culture is no more self-generative than human biology is gene determinant.

It may be a bit disingenuous to suggest that culture only weakly tracks biology. Granted, it does not strain any sense of reason to suggest that in certain times and places, notably in the peopling of an unoccupied land, that a given group of closely related humans carried only a limited range of biological variation and an even smaller range of cultural variation attending shared recent history. Such is the case for the envisioned colonists of the late Pleistocene. Small in size, close through alliance, and like-minded, these first people are not expected to have carried any more cultural variation than they carried genetic variation, which is projected to be small. However, their entry into the New World was no less historical than the European one to follow 20 millennia later insofar as these early colonists did not appear out of thin air but were situated in lineages of biology and culture that entailed many more people of diverse makeup, experience, and disposition. In other words, they came with cultural baggage. Even if there were a one-to-one correspondence between genetic variation and cultural variation (with our colonists, again, bearing little of each), the tempo and rhythm of change would vary independently, especially if migrating populations experienced subgenerational change in the spatial and social scale of interacting with "others." Put another way, the pace of cultural change (including that of linguistics) is contingent on the complexity and scale of the networks of interaction that links people together in mutually constitutive relationships of alterity. We simply do not know what culture is like in the sort of vacuum implied by the Beringian Incubation Model (see Anderson, Chapter 12; Kemp and Schurr, Chapter 2), especially as it may have played out over several millennia. Rather, the modern understanding of culture, as historical process, is that it is a patently *relational* property of social consciousness, manifest in demographic terms by the collectives we formalize as bands, or tribes, or a "people." This begs the questions about the history of the first colonists of the New World: What was their cultural affiliation in donor populations, both kin and nonkin, and how did these historical predispositions affect the decisions they made to move eastward and southward and, once they did, to assert practices that structured the social contours of affiliation and interaction? No known human society is without these sorts of cultural constructs, and they are not determined simply by the biological and environmental forces in play.

Despite some important differences in the logics of practice, morphological and archaeological inquiry have more in common than either does with most genetics research. The results of investigations into the molecular genetics of early migrants offer a uniquely elegant record of biological descent, which, when situated in space, provides insight on events of mass migration. However, the findings of geneticists do not provide all the answers, and some of the operational assumptions of genetics (e.g., rate of mutation) are not beyond reproach. Indeed, orienting the relationship of genetics research to our understanding of the layers of variation that result from "being in the world," that is, from experience, is the challenge facing contributors to this volume.

The Volume's Contributions

Every paper in this volume makes a significant contribution to efforts to explain and interpret human variation in the Americas. The volume editor, Ben Auerbach, seized on an opportune moment in history to juxtapose the results of a relatively new line of inquiry (genetics) with those of two lines of inquiry (morphology and archaeology) that have been around for a good long while. With longevity, as Auerbach alludes in the introduction, the practice of these other two fields has witnessed big changes in ways of thinking, most brought on by critical analysis from alternative viewpoints and many precipitated by technical advances, such as in radiocarbon dating and computers. As an admirably precise science, genetics persuades us to direct archaeological and morphological inquiry to its service or, more accurately perhaps, to lend such data to independent tests of hypotheses generated from genetics data. Drawing on genetics data to test hypotheses derived from archaeological and morphological observations is likewise persuasive.

As this volume shows, bridging the empirical divide of these diverse viewpoints is a daunting task but unquestionably one of great promise. In his commentary on this volume, David Anderson (Chapter 12) highlights some of the cross-disciplinary bridges assembled by the authors, and editor Auerbach makes a compelling argument for advancing this effort. I remain somewhat unconvinced that archaeology has achieved equal billing in most of the chapters that showcase the findings of genetics or skeletal morphology. Perhaps this is an unfair assessment, for very few sciences have found much success in comparisons of data sets structured by significantly different scales of abstraction. As it stands now, inquiry into ancient human variation might be likened to three blindfolded analysts trying to describe an elephant. One has the tail, one has the trunk, and the geneticists have the gonads. We are thus in great need of methods of "cross-scale" analysis, a challenge facing much of science today and the expected outcome of inquiry that has become increasingly cross-disciplinary. This, then, is the real work ahead. Contributions to this volume offer a strong basis for moving the project along.

Kicking off the case studies, Kemp and Schurr (Chapter 2) provide a first-rate review and analysis of genetics data on modern and ancient human variation in the Americas. This chapter is noteworthy, not only for its cogent synthesis, but also for providing the requisite background for interpreting genetic variation. We have to understand one another's specialties if we hope to achieve any sort of cross-scale analysis, and Kemp and Schurr do an excellent job making genetics accessible to the nonspecialist.

I am in no position to evaluate most of the substantive findings outlined by Kemp and Schurr, and I have no reason to doubt the authority of their views. I am more interested in how they regard the fit between the findings of genetics and other lines of inquiry. In the opening paragraph of their chapter, Kemp and Schurr herald the value of molecular data as an independent assessment of the pace and direction of human evolution. This assessment as it pertains to the peopling of the Americas has been bolstered by agreement among multiple analysts looking at the problem from several molecular and demographic angles, a point elaborated on by Anderson (Chapter 12). Unfortunately, the fit with archaeological and morphological data is not so good. The big picture in the Americas does not enjoy the concatenation of data seen in the Old World, perhaps for lack of greater time depth. Instead we have a more interesting situation in the Americas, where data sets tend to contradict one another, pointing to more complex scenarios than we have yet imagined.

Cross-scale comparisons of archaeological and genetic data in the Americas require that either genetics be downscaled to the level of experience (currently not possible) or that archaeology be upscaled to the level of subglobal variations (a scale not unusual to much of archaeology). Imagine, for instance an upscaling of archaeological observations in which the appropriate spatial scale would be the total extent of each of the five major haplogroups of Amerindians. Sum together the total cultural variation of the spatial extent at any point in time, as evident in archaeological data, and you have the total potential reservoir of cultural history to consider. A critical point of observation would be the sum of cultural variations attending the geographical epicenter from which founding Amerind populations are believed to have originated. Importantly, this is not simply an image capture but rather a time-lapse process that enfolds all forward-moving traditions of experience. Admittedly slippery stuff, but nonetheless this is relevant to the bigger picture.

Geneticists primarily attribute variations in the relative proportions of haplogroups from the point of human entry into the New World forward to drift. This implies that the specific instances of drift operated to differentially distribute existing variations in the founding population, as well as any that arose in the interim. Multiple "splintering" events are implicated, each entailing a subset of the full range of variation or at least different proportions of existing haplogroups among individuals of immigrant populations. I suggest it may be fruitful to consider that each such event was not random but instead elapsed along cleavage planes of cultural affiliation/differentiation. Again, there is no reason to assume, a priori, that such cultural variations will map onto distinct haplogroups or subhaplogroups.

The application of modern genetics to finer-scale movements of people in the past holds great promise, as Kemp and Schurr detail, but ultimately it may take ancient DNA (aDNA) to substantiate specific claims about the timing, direction, magnitude, and cause of human migrations. Shinoda and colleagues (Chapter 9) provide a good example of this approach in their analysis of tooth enamel from the Moquegua Valley of Peru. Noting the lack of concordance among linguistic, genetic, and archaeological data, Shinoda and colleagues test competing models of interaction between highland and coastal populations of the Formative era. Linking the results of aDNA to paleoenvironmental reconstruction of droughts and El Niño events, these researchers are able to document the movement of highland people into the Moquegua Valley after the onset of the Middle Horizon (A.D. 550–600). The ethnic and linguistic identity of these immigrants remains a matter of contention among investigators. Modern population distributions, specifically those of the Aymara, are likely an artifact of European colonialism, despite archaeological claims to the contrary. That linguistic markers crosscut or even subvert other evidence of cultural heritage may explain why archaeological and linguistic data do not agree, bolstering the utility of aDNA data as an independent arbitrator. With regard to scale, however, the linguistic evidence for Aymara origins expands inquiry well beyond the scale Shinoda and colleagues address.

Returning again to the "peopling" issue, craniometric analyses of ancient skeletons in recent years shows that at least two distinct populations inhabited the Americas by the early Holocene. The earlier of the two consist of individuals with long, narrow crania and prognathous faces. Referred to generally as "Paleoamericans," individuals of this group have posed something of a dilemma for archaeologists because they bear little resemblance to either modern Amerinds or the northeast Asiatic populations of presumed common ancestry.

Most morphological studies emphasize either microevolution or migration as a source of variation, but in his contribution to this volume, Chatters (Chapter 3) sees both forces at work. His example is from the Columbia Plateau of the American Northwest, where, during the ninth radiocarbon millennium B.P., a group of Amerindians known to archaeologists as members of the Old Cordilleran tradition moved south from Alaska onto the Plateau. Indigenous people of Paleoamerican ancestry occupied this region since the Ice Age but were eventually displaced by immigrants from the north. With this assertion, Chatters opens up a line of inquiry I will address more fully below in discussion of my own archaeological perspective.

The interpretive line between microevolutionary and historical factors in accounting for morphological variations is nicely drawn in the chapters by King and by Jantz and colleagues. King's contribution (Chapter 10) outlines a classic example of adaptive radiation in a quickly colonizing population or multiple founding populations who settled virtually every habitat in the Americas by the early Holocene. King focuses on environmental factors, notably climate, that would have shaped, through microevolution, the postcranial variation of ancient Americans. The pace of evolutionary change implicated here was quick, suggesting further that selective pressures attending climate were severe. In broad outline, the patterns King illustrates between skeletal robusticity and climate make logical sense and are empirically valid, to a degree. Missing in the equation are the time-transgressive trends of population movement during an era of rapid postglacial climate change. Attempting to maintain a given ecological niche may have been the engine of rapid cultural, if not biological change (cf. Romer's Rule).

Jantz and colleagues (Chapter 11) read the evidence for microevolution that King discusses as an indication of migration history. Theirs is an important paper because it underscores the empirical reality that people tended to move around a lot, under all sorts of sociocultural circumstances, and thus one of the most defining features of the biological history of humans has been, and continues to be, gene flow. Jantz and colleagues describe morphological variations that do not square with a straightforward ecogeographic model in which all humans in the New World passed through the "cold filter" of Beringia. Rather, their analysis reveals trends for cormic indices indicative of both arctic- and tropical-adapted bodies, supporting the case for multiple migrations early on, at least one following a maritime route that delivered humans to South America no later than the late Pleistocene. Archaeological evidence in support of an early Pacific route grows stronger every year (Erlandson et al. 2008).

Chatters (Chapter 3) provides perhaps the best effort to integrate the morphological, archaeological, and genetics within a model that accounts for at least two very different types of people on the landscape at the close of the Pleistocene. Accepting the Beringian Incubator Model, Chatters envisions multiple immigrations from a "displaced" Asian core, which then introduced multiple instances of drift into the Americas as the remaining portion of the founding population was "molded" in the harsh climate of Beringia to become the root stock of modern Amerinds. When the "evolved" descendants of this Beringian core eventually moved into the Northwest Plateau area, they encountered descendants of earlier immigrations, with whom they shared common ancestry. However, aside from variations in genetic markers not affected by selection, we simply do not find genetic markers that are commensurate with a biosocial history that elapsed over several millennia to beget vastly different-looking people with vastly different ways of life.

Other detailed case studies in morphology and archaeology introduce additional complexity to the big picture. In his contribution to this volume Cybulski (Chapter 4) takes a detailed look at postcranial metric variations to explore microevolutionary and developmental outcomes of adaptations to diverse niches of the Northwest Coast and Plateau. Combining isotopic signatures with morphological data, Cybulski is able to discriminate between coastal and Plateau groups in their differential reliance on marine resources. Among the oldest skeletons, Cybulski notes examples of two young males (China Lake site) who do not fit the usual range of postcranial variation, notably in their gracile long bones. Both tested as haplogroup M (Malhi et al. 2007), a type not seen in modern Native Americans, and they both have dental and isotopic characteristics that fit comfortably in the Plateau. These findings remind us of the likelihood that lineages of founding populations did not necessarily survive, that is biologically, but nevertheless had a decided impact on the direction and pace of sociocultural histories.

Migrations in the American Southwest were also pivotal in the direction and pace of cultural change. Among the more celebrated events were the abandonment of Chaco Canyon in the early twelfth century (Lekson 1999) and the subsequent southward migration of communities from the Four Corners as early as the late twelfth and early thirteenth centuries (Duff and Wilhusen 2000). Durand and colleagues (Chapter 5) bring to bear on these events data on discrete dental traits as proxies for biological affinity. They rightfully note that artifact similarities can occur without migration and that direct aDNA studies are not always possible because they are destructive. They find close affinity between individuals from Chaco and Aztec Ruin, supporting Lekson's hypothesis, but evidence for immigration of Mesa Verde groups into the San Juan basin is ambiguous. The evidence clearly shows immigrants arriving in the region, but the source of these newcomers remains uncertain.

A similar problem bedevils archaeological reconstructions of emergent agriculture in the Southwest. As reviewed by Watson (Chapter 6), genetics does not find strong evidence for migration of farmers from Mexico (cf. as in the Neolithic frontier of Europe), despite the seeming veracity of other data sets. Watson tests alternative hypotheses about emergent farming, noting that different methods of inquiry should not be pitted against one another but rather integrated to inform on synergistic properties that affect genetic outcomes. For instance, Watson argues that major changes in lifeways (namely the transition to "settled" village life and "tribalization") restructured human relationships in ways that affected genetic distributions. He concludes that the best genetic test of migration hypothesis will be to compare the aDNA of preagricultural and early agricultural people, but unfortunately, the former is not available at this time.

In another deeply contextualized study, Auerbach (Chapter 7) uses body morphology to compare patterns of human distributions against linguistic-cultural groups on the Great Plains over the past 1,000 years. His results highlight the discrepancy of sociodemographic patterns inferred from archaeological and morphological data, again underscoring some of the independent historical factors that dislodge cultural affiliation from biological affinity. This work also reminds us that the dichotomy between population replacement and continuity is facile. The political implications of Auerbach's work for Native American repatriation are acute.

Origins and Diversity
from an Eastern Woodlands Perspective

Auerbach reminds us in the introduction to this volume (Chapter 1) that anthropologists have been struggling to integrate the data sets of archaeology and biology since the nascent days of the profession a century ago. He also reminds us that variation in biology is what we make of it and that with recent advances in science, such as those of genetics, cross-disciplinary integration grows increasingly challenging. Only geneticists have truly essential units of observation with which to work (alleles), and in the analytical elegance this enables, other data sets are too often subordinated. Those trying to explain other dimensions of biological diversity have less discrete traits to consider and outcomes that are subject to equifinality. Morphologists, of course, have nothing over archaeologists confronting the seemingly intractable variations of "culture." It should be evident by now, as Auerbach notes in the introduction, that biology and culture may not always operate in ways that satisfy the goals of holistic inquiry. But before we give up on efforts at synthesis and integration, we should remind ourselves of the epistemological challenges of integrating inquiry using different units and scales of observation. This leads us in a slightly different direction: one of developing a framework linking different scales and knowledge systems, not for synthesis and consensus, but for critical dialogue. Ultimately, the operative issue is the uses to which alternative knowledge claims are to be put.

Before closing this commentary with some thoughts on bridging scales for continuing dialogue, I want to situate myself in the matrix of claims and counterclaims by briefly outlining my perspective on diversity in the Eastern Woodlands of North America.

My specialty is the Archaic period of the Eastern Woodlands, with particular emphasis on the Archaic of the American Southeast. Not that many practitioners focus on the archaeology of this 8,000+ year period, but most North American archaeologists know the standard interpretation given in introductory textbooks. As the story goes, Archaic populations of the Eastern Woodlands were the direct descendants of Paleoindian settlers who successfully colonized most of the subcontinent by the end of the Ice Age. Against a backdrop of post-Pleistocene climate change, Archaic groups diversified and grew to beget a range of regional, subregional, and local traditions. The subtext to this process, following on the influential work of Joseph Caldwell (1958), is that Archaic people became increasingly efficient at exploiting the environment, adopting innovations along the way (e.g., groundstone, pottery) to enhance their economic returns. With success came greater numbers, and with greater numbers came greater social and economic challenges. Agriculture, settled village life, and sociopolitical structures arose from this increasingly intensified lifestyle, but by then the Archaic had ended and the Woodland era had begun, as the Mesolithic era of Europe similarly gave way to the Neolithic Revolution. Although the anachronistic elements of this line of argumentation are obvious from a modern perspective (e.g., see

Gamble 2007), its general premises continue to shape our perception of Archaic diversity. It is a narrative that mirrors the macroevolutionary model of adaptive radiation, with one common ancestor, geographic isolationism, and speciation being the operative metaphors.

I view the diversity of the Eastern Archaic from an alternative perspective (see Sassaman 2010 for the full account). I start with the premise that at least two very different lineages begot the Archaic cultures of the Eastern Woodland: one that descended culturally (and perhaps biologically) from the Eastern Clovis baseline of the Paleoindian era, and a second, western line that relocated to land east of the Mississippi River after about 9,000 years ago. I have no clear sense of the common ancestry of these two distinct lines, except to note that it does not appear to have existed in the Eastern Woodlands, or if it did, it existed within a pre-Clovis tradition yet to be identified. The common root of what I refer to as Ancestry I and II can be accommodated by models that involve either one founding population at a standstill in Beringia during the late Pleistocene or multiple episodes of immigration from East Asia during the late Pleistocene and/or early Holocene. My model does not implicate an Atlantic point of entry for Paleoamericans, as proposed by Bradley and Stanford (2004), but it does allow for a northern expansion of South American groups who colonized the New World via the Pacific Coast (e.g., Faught 2008; cf. Anderson and Gillam 2000).

Descendant Archaic populations aside, I must agree with Faught (2008) that the diversity of Paleoindians is too great to be accounted for by a single wave of migration from a single ancestral source. Biology notwithstanding, the rate of cultural diversification is marked. One might jump to the conclusion that this diversity simply reflects an advanced state of adaptive radiation, as in so many species of Galapagos finches. Certainly this sort of process was likely at play, but to problematize diversity is to demand better data on the relationships between technological strategies and ecological niches, and the time-transgressive trends of both. We simply do not have very good analogs for the sorts of processes unfolding at the close of the Pleistocene in the New World; considering the rate of environmental change at this time (including rates of sea-level rise that have never been matched since), there may be none.

Irrespective of the diversity of Paleoindian/American populations at the close of the Pleistocene, by the second millennium of the Holocene, an entirely new wave of cultural tradition appears in the Eastern Woodlands in the guise of the Shell Mound Archaic. Centered on the Interior Plateau between the Ohio and Tennessee rivers, the Shell Mound Archaic was a series of closely related cultural traditions dating from roughly 8,000 to 3,000 years ago. As the name implies, these traditions involved the collection and accumulation of shellfish from freshwater rivers in the Midsouth and lower Midwest, as well as a suite of artifact forms, such as bannerstones, that set them apart from neighbors in adjacent regions. To my way of thinking, the Shell Mound Archaic, as a whole, was a diasporic tradition with not only a major displaced core that dwelled inland between the Ohio and Tennessee rivers but also with several enclaves farther afield, including those of the Northeast, the lower Southeast, and the upper Midwest.

Separating evidence of the onset of the Shell Mound Archaic from that of the descendants of Paleoindian populations is a circa 500-to-800-year hiatus in occupation across many areas of the Eastern Woodlands. One hypothesis to account for such widespread abandonment is that late Paleoindian communities responded to post-Pleistocene warming by incrementally moving farther northward and into zones of higher elevation (i.e., the Appalachians). The record of human settlement in higher latitudes and altitudes of the East indeed reflects greater continuity with the Clovis horizon than in locations of lower latitude and elevation. Elsewhere, in subregions that were apparently abandoned by descendants of Clovis (Ancestry I), reoccupation several centuries later entailed vastly different cultural traditions, notably those of the Shell Mound Archaic. The basal levels of Shell Mound Archaic sites in the Midsouth contain assemblages of objects and features with no precedent in the region. While one can argue that the emergence of these novel riverine adaptations arose from a prolonged spike in the magnitude of postglacial warming (e.g., Dye 1996), I suggest that prior abandonment of large tracts of the east by Paleoindian descendants afforded opportunities for nonlocal, nondescendant populations to colonize newly open niches.

From where did these alleged interlopers come? The answer eludes us for now, and vast portions of the early Holocene landscape remain unexplored beneath the waters of the Gulf of Mexico, the Atlantic coast, and the Great Lakes (but see Faught 2004; Lovis 2009; O'Shea and Meadows 2009). I can imagine all sorts of possible scenarios, but the one that makes the most sense from examination of current data is a migration of Archaic groups from the west. In contemplating the origins of the Morrow Mountain tradition—artifacts of which are found in the basal strata of Shell Mound Archaic sites (Dye 1996) —Joffre Coe (1964) imagined a western origin. Archaeologists have since either dismissed out of hand such long-distance connections or simply focused their attentions on local-scale phenomena (such as subsistence or lithic raw material use). Despite such preoccupations or outright denial about foreign influences, a local genesis for the Shell Mound Archaic is not to be found and we certainly have no basis for restricting possibilities to local precedents.

Chatters's contribution to this volume (Chapter 3) provides a clue to one plausible scenario (see also Chatters et al. 2010). Having argued for a migratory history for the Amerind ancestors known from the Old Cordilleran tradition, Chatters envisions a multicultural, or at least a bicultural landscape of potential conflict and contest. Encounters between communities of distinct identity and heritage had as much potential to be hospitable as to be hostile, but no matter the relations between indigenous and nonlocal people, encounters likely stimulated even further displacement and realignment. It is but a short (if rugged) trip over the Rockies from the Columbia basin of Washington to the headwaters of the Missouri River drainage. A continuing, unimpeded sojourn eastbound would have delivered immigrants to the American Bottom, the gateway to the Ohio and Tennessee river drainages.

Parallels between Old Cordilleran and a nascent Shell Mound Archaic are appreciable (see Sassaman 2010), and the timing between Plateau encounters (ca.

8500 B.P. or 9500 cal B.P.) and reoccupation of the interior Midsouth follows the expected sequence. I do not have any strong sense of the biological outcomes of these postulated events. Apparently, migrations such as this did not result in dramatic differences in the interpopulational distribution of mtDNA or Y-chromosome markers, given the history of relatively recent common ancestry and the bottleneck of Beringia, if truly *the* place of entry. Still, cultural differences were marked, and in places where such differences precluded much intermixing, the phenotypic divergences may have likewise been marked. I would imagine that in many contexts of encounter and interaction, admixture was actually relatively advanced. Culture-wise, of course, dimensions of variation were free to behave independently of strict biology or the admixture of interactions, even to operate with relational qualities that preserved or enhanced differences. Here we find the emergent structure of cultural affiliation and differentiation—a product of intense interactions, not isolationism, and thus the antithesis of a speciation-like process inasmuch as gene flow and other forms of biological and cultural interaction were rampant.

Finally, the nature of interactions must have varied greatly, from hostile to peaceful, from transformative to inconsequential. In their contribution to this volume, Schmidt and colleagues (Chapter 8) give one glimpse into the physical and presumably violent confrontations at one of the more active "border zones" between communities of Ancestry I and II, the Ohio River Valley. Other examples of interactions abound (see Sassaman 2010), all contributing to diversity, not homogeneity, and setting the stage for the sorts of institutional structures that would determine the contours of demographic continuity and change through the remainder of the pre-Columbian era.

Conclusion

Throw culture into the mix and we confront a kaleidoscope of ever-changing relations and emergent structures. Hold culture constant at the inter-specific scale of biological comparison, and all humans are the same. Articulating a limited (intraspecific) range of biological diversity with an advanced range of cultural diversity would seem to be futile, one might even argue irrelevant, were it not for the fact that the two impinge on each other. If we truly hope to be able to understand how such impingements play out over long stretches of time and vast geographies, we will need to develop means to compare data and scales of analysis that are nonisomorphic and asynchronous. The contributions to this volume expose the challenges such an effort might entail.

In the spirit of continuing dialogue, I suggest in closing that we may find guideposts for progress among our colleagues who deal with other complex, multiscalar phenomena. Those charged with ecosystem assessment, for instance, strive to look at a complex problem from lots of different angles, to bridge or integrate those angles into a coherent "story" (e.g., Millennium Ecosystem Assessment 2003). Consider the multiscalar nature of climate change and then consider further the varying epistemologies involved in deciding its causes and consequences. The processes that account for global climate change are truly cosmic and millennial in scope, but they are realized in real human time by the sorts of calamities attending drowned coastlines, salinized water sources, and prolonged drought. Human perceptions of climate change vary as widely as do relevant scales, as do the knowledge systems that inform mitigative action (or lack thereof). And, like all things modern, climate-change research is highly politicized, with stakeholders who alternatively see no change where there is plenty and plenty of change where there is none.

Bridging the scales and knowledge systems of climate-change research is requisite to problem solving (Reid et al. 2006). Approaches taken by those charged with assessing change and recommending mitigative action vary, but they are grounded in the assumption that complex phenomena, like climate change, are structured by cross-scale interactions, that is, conditions whereby phenomena at one scale influence phenomena at another scale (Wilbanks 2006). Multiscalar approaches to complex phenomena are diverse and include methods to integrate data at a single scale (i.e., upscaling, downscaling), multiscalar synthesis (usually nested scales), and modeling cross-scale linkages (focus on relational properties or networks). Often the methods and models are highly quantitative, but the point of cross-scaling is to draw in diverse approaches. This then leads to a multivocality with no bounds other than those of self-imposed peer evaluation (which varies from one knowledge system to the next). Particularly challenging, but extremely fruitful, has been the effort to integrate the work of science with indigenous non-Western epistemologies. Although the varying epistemologies of genetics, morphology, and archaeology may not be as divergent as, say, the approaches of biomedicine and shamanistic religion, the empirical differences among these three fields are considerable and seemingly incompatible.

Methods for bridging scales and integrating multiple knowledge systems are neither intuitive nor enabled by the usual disciplinary models. In ecosystem assessment and related fields involving forecasting, methods to draw diverse perspectives and data sets into critical dialogue require a certain detachment from standard scientific practice. One alternative is to construct *scenarios*, essentially stories or narratives that attempt to integrate two or more points of view in simple language. The Millennium Ecosystem Assessment describes scenarios as plausible alternative futures (Bennett and Zurek 2006), but this same method applies to plausible alternative pasts. The proposal, debate, and revision of scenarios draws out the unspoken, taken-for-granted assumptions of each paradigm involved, forcing adherents to justify scales and units of analysis, explanations for change, and ultimately, how they might integrate with other epistemologies.

Scenarios also help bring to light key uncertainties. Many of us have more in common by what we still don't know, and drawing our diverse perspective into comparison, as this volume does so well, is the best means we have for learning more.

Acknowledgments

My thanks to Ben Auerbach for inviting me to participate in the 25th Annual Visiting Scholar Conference at Southern Illinois University Carbondale and in the book project that followed. Before these experiences, I had little occasion to compare the work of geneticists, skeletal biologists, and archaeologists over problems of common interest. I am glad to have had the chance to do so, and I trust I have not misrepresented the work of my fellow participants in my effort to understand how we articulate. Improvements to an earlier draft of this chapter were enabled by the salient comments of Ben Auerbach and an anonymous reviewer. My thanks as well to fellow commentator, David Anderson, for sharing his thoughts and an earlier draft of his excellent contribution, and lastly to all the authors of the stimulating research reported in this book.

Note

1. In his introduction to this volume, for example, Auerbach (Chapter 1) notes that the fundamental difference between the genetic and the morphological reading of early human variation in the Americas is that the former deals with models for peopling events that predate the skeletal populations of the latter by millennia.

References

Anderson, David G., and J. Christopher Gillam
  2000   Paleoindian Colonization of the Americas: Implications from an Examination of Physiography, Demography, and Artifact Distribution. *American Antiquity* 65:43–66.
Bennett, Elena, and Minika Zurek
  2006   Integrating Epistemologies through Scenarios. In *Bridging Scales and Knowledge Systems: Concepts and Applications in Ecosystem Assessment*, edited by Walter V. Reid, Fikret Berkes, Thomas Wilbanks, and Doris Capistrano, pp. 275–294. Island Press, Washington, D.C.
Bradley, Bruce A., and Dennis Stanford
  2004   The North Atlantic Ice-Edge Corridor: A Possible Palaeolithic Route to the New World. *World Archaeology* 36:459–478.
Caldwell, Joseph R.
  1958   *Trend and Tradition in the Prehistory of the Eastern United States*. American Anthropological Association Memoir 88. Menasha, Wisconsin.
Chatters, James C., Steven Hackenberger, Brett Lenz, Anna M. Prentiss, and Jayne-Leigh Thomas
  2010   The Paleoindian to Archaic Transition in the Pacific Northwest: In Situ Development or Ethnic Replacement? In *On the Brink: Transformations in Human Organization and Adaptation at the Pleistocene-Holocene Boundary in North America,* edited by C. Britt Bousman and Bradley J. Vierra. Texas A&M Press, College Station, in press.

Coe, Joffre L.
1964   *The Formative Cultures of the Carolina Piedmont.* Transactions of the American Philosophical Society Vol. 54, Pt. 5. Philadelphia.
Duff, Andrew I., and Richard H. Wilshusen
2000   Prehistoric Population Dynamics in the Northern San Juan Region, A.D. 950–1300. *Kiva* 66:167–190.
Dye, David H.
1996   Riverine Adaptation in the Midsouth. In *Of Caves and Shellmounds*, edited by Kenneth C. Carstens and Patty Jo Watson, pp. 140–158. University of Alabama Press, Tuscaloosa.
Erlandson, Jon M., Madonna L. Moss, and Matthew Des Lauriers
2008   Life on the Edge: Early Maritime Cultures of the Pacific Coast of North America. *Quaternary Science Reviews* 27:2232–2245.
Faught, Michael K.
2004   The Underwater Archaeology of Paleolandscapes, Apalachee Bay, Florida. *American Antiquity* 69:275–289.
2008   Archaeological Roots of Human Diversity in the New World: A Compilation of Accurate and Precise Radiocarbon Ages from Earliest Sites. *American Antiquity* 73:670–698.
Gamble, Clive
2007   *Origins and Revolutions: Human Identity in Earliest Prehistory.* Cambridge University Press, Cambridge.
Goebel, Ted, Michael R. Waters, and Dennis H. O'Rouke
2008   The Late Pleistocene Dispersal of Modern Humans in the Americas. *Science* 319:1497–1502.
Gunderson, Lance, and C. S. Holling
2001   *Panarchy: Understanding Transformations in Systems of Humans and Nature.* Island Press, Washington, D.C.
Hobsbawm, Eric, and Terence Ranger (editors)
1983   *The Invention of Tradition.* Cambridge University Press, Cambridge.
Lekson, Stephen H.
1999   *The Chaco Meridian: Centers of Political Power in the American Southwest.* AltaMira, Walnut Creek, California.
Lovis, William A.
2009   Hunter-Gatherer Adaptations and Alternative Perspectives on the Michigan Archaic: Research Problems in Context. In *Archaic Societies: Diversity and Complexity across the Midcontinent,* edited by Thomas E. Emerson, Dale L. McElrath, and Andrew C. Fortier, pp. 725–754. State University of New York Press, Albany.
Malhi, Ripan S., Brian M. Kemp, Jason A. Eshleman, Jerome S. Cybulski,
David Glenn Smith, Scott Cousins, and Harold Harry
2007   Mitochondrial Haplogroup M Discovered in Prehistoric North Americans. *Journal of Archaeological Science* 34:642–648.
Marx, Karl
1963[1852] *The Eighteenth Brumaire of Louis Bonaparte.* International Publishers, New York.
Millennium Ecosystem Assessment
2003   *Ecosystems and Human Well-Being: A Framework for Assessment.* Island Press, Washington, D.C.
O'Shea, John M., and Guy A. Meadows
2009   Evidence for Early Hunters Beneath the Great Lakes. *Proceedings of the National Academy of Sciences of the United States of America* 106:10120–10123.

Pauketat, Timothy R.
  2001   A New Tradition in Archaeology. In *The Archaeology of Tradition: Agency and History Before and After Columbus*, edited by Timothy R. Pauketat, pp. 1–16. University Press of Florida, Gainesville.
Reid, Walter V., Fikret Berkes, Thomas J. Wilbanks, and Doris Capistrano (editors)
  2006   *Bridging Scales and Knowledge Systems: Concepts and Applications in Ecosystem Assessment*. Island Press, Washington, D.C.
Sassaman, Kenneth E.
  2010   *The Eastern Archaic, Historicized*. AltaMira, Lanham, Maryland.
Vayda, Andrew P.
  1995a   Failures of Explanation in Darwinian Ecological Anthropology: Part I. *Philosophy of the Social Sciences* 25:219–249.
  1995b   Failures of Explanation in Darwinian Ecological Anthropology: Part II. *Philosophy of the Social Sciences* 25:360–375.
Wilbanks, Thomas J.
  2006   How Scale Matters: Some Concepts and Findings. In *Bridging Scales and Knowledge Systems: Concepts and Applications in Ecosystem Assessment*, edited by Walter V. Reid, Fikret Berkes, Thomas Wilbanks, and Doris Capistrano, pp. 21–35. Island Press, Washington, D.C.

Contributors

David G. Anderson is a professor of anthropology at the University of Tennessee, Knoxville, Tennessee.

Benjamin M. Auerbach is an assistant professor in the Department of Anthropology, the University of Tennessee, Knoxville, Tennessee.

Anne T. Bader is principal investigator at Corn Island Archaeology, LLC, in Louisville, Kentucky.

James C. Chatters is a senior associate at AMEC Earth & Environmental, Inc., in Bothell, Washington.

Jerome S. Cybulski is curator of physical anthropology in the Archaeology and History Division at the Canadian Museum of Civilization, Gatineau, Quebec, Canada.

Kathy Roler Durand is chair of and associate professor in the Department of Anthropology and Applied Archaeology, Eastern New Mexico University, Portales, New Mexico.

Stephen R. Durand (1952–2009) was a professor of anthropology in the Department of Anthropology and Applied Archaeology, Eastern New Mexico University, Portales, New Mexico, where he taught for eighteen years. His most significant publication was on sourcing construction timbers from Great Houses in Chaco Canyon using stable isotope analysis. He loved fieldwork and spent summers on projects in New Mexico, Nevada, Alaska, and Cyrus. But he was most proud of training and mentoring graduate students.

Sonia E. Guillén is the director of Centro Mallqui in Lima, Peru.

Claire Jantz is assistant professor in the Department of Geography and Earth Science at Shippensburg University, Shippensburg, Pennsylvania.

Richard L. Jantz is professor emeritus and director of the Forensic Anthropology Center at the University of Tennessee, Knoxville, Tennessee.

Brian Kemp is assistant professor in the Department of Anthropology and the School of Biological Sciences at Washington State University, Pullman, Washington.

Kathryn A. King is an instructor in the Department of Sociology and Anthropology at the University of Arkansas, Little Rock, Arkansas.

Paul Marr is a professor in the Department of Geography and Earth Science at Shippensburg University, Shippensburg, Pennsylvania.

Christopher L. Newman is an M.D. candidate at Indiana University School of Medicine, Indianapolis, Indiana.

Jeffrey A. Plunkett is the president of and principal investigator at Accidental Discoveries, LLC, in Noblesville, Indiana.

Kenneth E. Sassaman is Hyatt and Cici Brown Professor of Florida Archaeology in the Department of Anthropology, University of Florida, Gainesville, Florida.

Christopher W. Schmidt is associate professor of anthropology and director of the Indiana Prehistory Laboratory at the University of Indianapolis, Indianapolis, Indiana.

Theodore G. Schurr is associate professor in the Department of Anthropology and consulting curator for the American and Physical Anthropology sections of the Penn Museum of Archaeology and Anthropology at the University of Pennsylvania, Philadelphia, Pennsylvania.

Anna Serrano is pursuing a master of science degree in human biology through the Department of Biology at the University of Indianapolis, Indianapolis, Indiana.

Rachel Lockhart Sharkey is a graduate student in the Department of Anthropology at the University of Indianapolis, Indianapolis, Indiana.

Izumi Shimada is distinguished professor in the Department of Anthropology at Southern Illinois University, Carbondale, Illinois.

Ken-ichi Shinoda is head of the Department of Anthropology, Division of Human Evolution, at the National Museum of Nature and Science, Tokyo, Japan.

David Glenn Smith is a professor in the Department of Anthropology at the University of California, Davis, California.

Meradeth Snow is a doctoral candidate in the Department of Anthropology at the University of California, Davis, California.

James T. Watson is assistant curator of bioarchaeology and assistant professor of anthropology at the Arizona State Museum and School of Anthropology at the University of Arizona, Tucson, Arizona.

Melissa S. Zolnierz is a Ph.D. candidate in the Department of Anthropology at the University of Arkansas, Fayetteville, Arkansas.

Index